D1563880

Plato and Pythagoreanism

Plato and Pythagoreanism

Phillip Sidney Horky

OXFORD
UNIVERSITY PRESS

OXFORD
UNIVERSITY PRESS

Oxford University Press is a department of the University of Oxford.
It furthers the University's objective of excellence in research,
scholarship, and education by publishing worldwide.

Oxford New York
Auckland Cape Town Dar es Salaam Hong Kong Karachi
Kuala Lumpur Madrid Melbourne Mexico City Nairobi
New Delhi Shanghai Taipei Toronto

With offices in
Argentina Austria Brazil Chile Czech Republic France Greece
Guatemala Hungary Italy Japan Poland Portugal Singapore
South Korea Switzerland Thailand Turkey Ukraine Vietnam

Published in the United States of America by
Oxford University Press
198 Madison Avenue, New York, NY 10016

Library of Congress Cataloging-in-Publication Data
Horky, Phillip Sidney.
Plato and Pythagoreanism / Phillip Sidney Horky.
pages cm
Includes bibliographical references (pages) and indexes.
ISBN 978-0-19-989822-0 (alk. paper)
1. Mathematics, Ancient. 2. Pythagorean theorem. 3. Pythagoras and Pythagorean school. 4. Plato. I. Title.
QA22.H67 2013
184—dc23 2012042672

I, too, when I recollect my thoughts, feel a great deal of anxiety as to how at my age I am to make my way across such a vast and formidable sea of words.

—PLATO, *Parmenides* 137a4–5

Nay, by the man who bestowed upon our head the *Tetraktys*,
Possessing the fount and roots of nature ever-flowing.

—the "Pythagorean Oath," AËTIUS 1.3.8

CONTENTS

PREFACE

Ceci n'est pas un livre sur Pythagore. Perhaps enough books have been written on Pythagoras, both in antiquity, and in our modern era, to satisfy even the most voracious Sybaritic appetite. What this book shares in common with those books is the dedicated interest in occluded or fragmented authors, authorities, and ideas. Like Pythagoras, this book's two main protagonists are intellectuals whose life stories are the stuff of lore—even despite the fact that they lived in the fifth and fourth centuries BCE, when history had become well developed as a genre in ancient Greece—and scholars have spent a great deal of time trying to extract gold from the melded histories concerning their lives, activities, and philosophical doctrines. These two figures are Hippasus of Metapontum, the shadowy experimental Pythagorean philosopher and political revolutionary for whom we sometimes have a name, and little more; and Plato of Athens, who always advanced his own ideas in the voices of his characters, such as Socrates, the Eleatic Stranger, and Timaeus, and about whom we possess a plethora of anecdotes, tall tales, and explanations of his philosophical "doctrines." In the history of philosophy, few scholars have connected these two figures, in part because we know so little about the former and so little reliably about the latter. Although this book is not about one great human who, according to ancient traditions, was not supposed to be named, it is about two great humans whose names solicited both praise and scorn from various parts of their contemporary philosophical world. It is about a tradition of philosophical innovation that spans a century and a half in the Greek world (ca. 500–350 BCE), from Sicily and Southern Italy in the west to Athens, Thrace, and Anatolia in the east, and the important figures whose surviving texts allow us to trace a thread of intellectual inquiry that might explain how Plato responded to Pythagorean philosophy. I refer to these figures, who flourished in the three generations subsequent to the death of Pythagoras, as the mathematical Pythagoreans, but they had their own names: Epicharmus of Syracuse, Empedocles of Agrigentum, Philolaus of Croton, Eurytus of Metapontum, and Archytas of

Tarentum. What do these figures share in common? They were all considered in antiquity to have been apostate or pretender Pythagoreans who made public the doctrines of their master illegitimately.

Why, the reader might reasonably ask, should a book about so-called illegitimate Pythagoreanism and Plato be written now? It has been exactly fifty years since Walter Burkert published his groundbreaking *Weisheit und Wissenschaft: Studien zu Pythagoras, Philolaos und Platon* (1962, revised and translated into English in 1972 under the title *Lore and Science in Ancient Pythagoreanism*), and a number of major studies have appeared since Burkert revised our approach to the problem of Plato and Pythagoreanism. In the past two decades alone, the field has shifted dramatically, owing to the influential scholarship of three historians of philosophy and ideas: Carl A. Huffman, whose critical editions and commentaries of the fragments of *Philolaus of Croton* (1993) and *Archytas of Tarentum* (2005) have provided a much-needed platform for discussion; Andrew Barker, whose many articles and books on ancient musicology (especially *The Science of Harmonics in Classical Greece* [2007]) have fundamentally altered the ways scholars understand the scientific methodologies employed by ancient philosophers; and Leonid Zhmud, whose scholarship on the history of Pythagoreanism (*Pythagoras and Early Pythagoreanism* [2012], an update and translation into English of his earlier *Wissenschaft, Philosophie und Religion im frühen Pythagoreismus* [1997]), and the Peripatetic historiography of science (*The Origin of the History of Science in Classical Antiquity* [2006]) have significantly altered the ways we read the ancient historiography of science. Book-length studies of Pythagoras and the history of Pythagoreanism by Charles Kahn (*Pythagoras and the Pythagoreans: A Brief History* [2001]) and Christoph Riedweg (*Pythagoras: His Life, Teaching, and Influence* [2005], originally published in German as *Pythagoras: Leben, Lehre, Nachwirkung* in 2002) have provided much-needed sober interpretation and clear exposition concerning the history of the reception of Pythagoreanism beyond the confines of the ancient world. This book owes a great deal to the work of these scholars, who have continued to keep the discussions flowing and have provided a high-water mark indeed to attain.

In each of these books, one finds dedicated analyses of many things Pythagorean, and many explanations for how Pythagoreanism developed. And it is possible to obtain at least a limited sense of how Pythagoreanism may have influenced Plato's thought, although the precise nature of this influence remains difficult to determine. What worried me most, however—and where I saw a lacuna that needed to be filled—was in these scholars' accounts of what, precisely, "mathematical" Pythagoreanism might be. I wondered this because I wasn't quite sure, especially in the light of Burkert's skepticism, what, if anything, might have been "mathematical" about Pythagoreanism. Was it an interest in numbers? Or perhaps it was early experimentation in what would become deductive mathematics? Possibly it was their focus on harmonics? Or was it a more universal interest in all those parts of knowledge that we would call "scientific"? This puzzle sent me on a quest to try to figure out what anybody in the fourth century BCE, the century in which Pythagoreanism received its many first historical treatments, might have found "mathematical" about the philosophical ideas of the mathematical Pythagoreans.

What I discovered—very much in the light of the scholarship on Pythagoreanism, but with the great help of scholars outside the subdiscipline as well—is that a fuller understanding of what might be "mathematical" about a certain strand of Pythagoreanism would require one to tackle the problem of the Aristotelian division of types of knowledge. One could not simply speculate about the role of Pythagoreanism in Platonic thought without first evaluating Aristotle's classification of the philosophical activities of the Pythagoreans. Somewhat counterintuitively, then, a book entitled *Plato and Pythagoreanism* needed to start with a chapter on Aristotle. The fruits of that endeavor constitute chapter 1 of this book, where I argue that Aristotle in his lost works on the Pythagoreans and in the *Metaphysics* sought to classify the Pythagoreans according to two models of knowledge, knowledge of "the fact" and knowledge of "the reason why." This division becomes crucially important for the analysis of mathematical Pythagoreanism (as distinguished from acousmatic Pythagoreanism), because it actually reflects quite well the evidence that survives from early Pythagoreanism, especially (but not only) the extant fragments of Philolaus of Croton. Once I felt that I had something of a grasp on what Aristotle would have thought "mathematical" about Pythagorean philosophy, I began to wonder whether Aristotle's position on the subject could have been reflected in the treatments of Pythagoreanism by other contemporary intellectual historians and philosophers.

So I pursued a traditional line of inquiry, in which I investigated the early treatment of Pythagoreanism in the Lyceum (by Aristoxenus of Tarentum, Theophrastus of Eresus, and Dicaearchus of Messana) as well as the Early Academy (Speusippus of Athens and Xenocrates of Chalcedon), at least insofar as we might plausibly infer from the later testimonies. Now a great deal of this work had already been covered by Burkert and, in a slightly different light, Zhmud, but I still found that much remained to be said about the treatment of Hippasus of Metapontum, in particular, whose "doctrinal" *testimonia* had been dismissed by all scholars as simply spurious. To my surprise, and as I discuss extensively in chapter 2, the treatments of Hippasus by the Peripatetics and the Academics are starkly differentiated: while the Peripatetics associated Hippasus with the material monism of Heraclitus of Ephesus, the figures in the Early Academy seem to have manufactured a doctrine for Hippasus that assimilated his ideas to Plato's, effectively rendering him as a forerunner in the theory of the Forms. While this portrait cannot be relied on for its historical value in reconstructing Hippasus's own ideas, it at least presents evidence for how figures in the Early Academy associated Hippasus with Platonist doctrines.

If indeed the historical discourse concerning Pythagoreanism was already fully implicated in philosophical antagonism by the third quarter of the fourth century BCE, might there be some alternative historical account—not obviously derived from the accounts of the Peripatetics and Academics—that could provide a test case against the doxographic reports advanced by Aristotle, Speusippus, and others? Taking a cue from the scholarship of the first half of the twentieth century, especially the work of Kurt von Fritz, Augusto Rostagni, and Armand Delatte, I investigated the account of Pythagorean political activities given by Timaeus of Tauromenium, a Western Greek historian of the end of the fourth century BCE who was definitively prodemocracy and somewhat hostile to Aristotle's historiographical procedures. In chapter 3, I evaluate the fragmentary evidence derived from Timaeus's history of the city-states of Sicily and Southern Italy, with special attention to the account of a certain Apollonius preserved by Iamblichus in his work *On the Pythagorean Way of Life* 254–264. The result of this analysis is the introduction of another term, "exoteric" Pythagoreans, which appears to correspond with the mathematical Pythagoreans, on the grounds that both were considered heretical for having published/demonstrated the doctrines of Pythagoras, an act that corresponded with the advent of a "democratic" type of Pythagoreanism. Thanks to Timaeus, it is now possible to see the Peripatetic historical account in relief and to evaluate the evidence accordingly. I have extracted from this comparative analysis of Timaean and Peripatetic histories of Pythagoreanism an account of how the "publication" of the doctrines of Pythagoras corresponds with the "democratization" of philosophical knowledge, an activity that serves as a model for the public use of reason in order to resolve disputes. Thus, I draw the first half of the monograph to a close by developing a new account of mathematical Pythagoreanism that emphasizes the political ideology of making public, by means of scientific demonstrations, the basic tenets of Pythagorean thought.

Armed with a historical framework, in the second half of this study I evaluate the philosophical content of the extant fragments of the mathematical Pythagoreans who were said to have published/demonstrated the doctrines of Pythagoras: Epicharmus of Syracuse, Empedocles of Agrigentum, Philolaus of Croton, Eurytus of Metapontum, and Archytas of Tarentum. I do this in the light of Plato's treatment of the shared ideas each of these figures sought to investigate, especially the concept "number." When I examined the fragments of the earliest of these figures, Epicharmus and Empedocles, I discovered that even by the beginning of the fifth century BCE, mathematical Pythagoreans were posing philosophical problems concerning the "number" of a human being. In particular, they were concerned to evaluate what came to be known as the "Growing Argument," a puzzle that asks one to evaluate one's personal identity in the light of one's quantitative and qualitative growth throughout one's life. In chapter 4, I trace Plato's philosophical responses to the puzzle of Epicharmus's "Growing Argument" in the earlier and middle dialogues of Plato, especially *Euthyphro* and *Cratylus*. Plato's approach to this problem takes a unique turn, since it is rooted in his metaphysical propositions, including the correlative assumptions of participation of sensibles in Forms and imitation as a vehicle for names to obtain the properties of their governing Forms. By attacking a Sophistical version of the "Growing Argument" given by Cratylus, Plato simultaneously appropriates certain principles of ontological predication given by Philolaus of Croton, thus pitting, in effect, the ideas of one mathematical Pythagorean against those of another.

In chapter 5, I assess Plato's recurrent response to the "Growing Argument" in the dialogue that exhibits Plato's most extensive evaluation of the concept of "number," *Phaedo*. There, we see Plato illustrate mathematical Pythagorean argumentative techniques in the figures of Socrates's interlocutors Simmias and Cebes and critique them according to whether or not they exhibit a proper methodological rigor. Once again, the proposition of Forms and teleological causation generates new ways of thinking about number, and when Socrates finally develops the most complete analysis of number that can be found in Plato's oeuvre, in the final argument, he does so chiefly in order to lay the groundwork for a proof of the soul's immortality. The *Phaedo* thus illustrates Plato's most circumspect appropriation of mathematical Pythagorean (especially Philolaic) concepts, and if we consider the theme of the work to reflect a traditional Pythagorean concern with the reincarnation of the soul, we discover that Plato's philosophical methodology departs quite significantly from what can be inferred from the fragments of Empedocles and Philolaus.

While the *Phaedo* presents a fascinating example of how Plato appropriated and superseded his Pythagorean antecedents—essentially beating them at their own game—it is surprisingly indirect in its presentation of the philosophical concepts of the mathematical Pythagoreans. In the sixth and final chapter of this book, I elucidate the ways Plato camouflages his critical responses to Pythagoreanism by using mythological figures to refer to Pythagorean philosophical invention in the middle and later dialogues, especially *Republic*, *Timaeus*, and *Philebus*. The theme of "discovery" takes on Pythagorean overtones, and my exploration of the various culture heroes ("first-discoverers") in these dialogues shows that Plato ensconced his responses to the mathematical Pythagoreans by narrating the stories of mythological philanthropists who suffered punishment for their transgressions, such as Prometheus and Palamedes.

In this final chapter, then, I bring the accounts of mathematical Pythagoreanism given by Aristotle and Timaeus of Tauromenium to bear on Plato's literary treatment of the heretical "first-discoverers," especially the "certain Prometheus" of the *Philebus*, who is credited with passing down the fundamental philosophical method to human beings. Speculation about whether this might refer to Hippasus of Metapontum, the progenitor of mathematical Pythagoreanism, is corroborated by appeal to internal references in Plato's own work (especially the *Timaeus*) and external references in the fragments of Archytas of Tarentum.

This book does not aim to provide a comprehensive account of all the ways Pythagoreanism, broadly conceived, might have influenced Plato's philosophy. Its project is to open up new ways of understanding Plato's intervention in a series of philosophical ideas that we can associate, with some plausibility, with a particular strand of early Pythagorean thought. It is my hope that others will find inspiration to fill in the many gaps that surely remain. This is a book about alleged philosophical pretenders, thieves, and apostates who evade detection just at the moment when we seek to capture them with our minds. Always shifting, these ones are, and frustrating our best attempts to identify them. As the poet said, "these things never cease from continually alternating."

ACKNOWLEDGMENTS

"Whatever naturally is in a process of exchange and never remains the same should be always different from what it had been changed from" (DK 23 B 2 = D.L. 3.9). So a speaker, apparently a debtor, in an unknown comedy by Epicharmus of Syracuse presents the rationale for what came to be known as the "Growing Argument." Epicharmus's character seeks to worm his way out of a contract by saying it was someone else—surely not him—who is under obligation to deliver the goods. But the "Growing Argument" later came to be understood as problematizing the persistent personal identity of a human being as he grows older, gains and loses stuff, takes on new attributes and abandons others. Given the quantitative and qualitative change of parts throughout time, one might ask: am I the same man I was just one moment ago? On reflection, I can't help but think how well Epicharmus's argument epitomizes the nature of *this book*. It was originally presented as a doctoral thesis at the University of Southern California in August 2007, but then it was wholly reformulated at Stanford University, reinvigorated at Harvard University's Center for Hellenic Studies, and brought into the light once again at Durham University. Not a single sentence, not even a single idea, remains unaltered. So the question presents itself: is it the same book?

I will leave it to those who have witnessed its development to decide this question. Individual chapters benefited substantially from the careful reading of many colleagues: Andrew Barker (chapter 6), Rachel Barney (chapter 2), Mauro Bonazzi (chapter 3), George Boys-Stones (entire book), Luca Castagnoli (chapters 1–3), Doug Hutchinson (chapter 1), Don Lavigne (chapter 6), Mariska Leunissen (chapters 1–2), Tony Long (chapter 1), Constantinos Macris (chapter 3), Andrea Nightingale (chapter 6), and Malcolm Schofield (chapter 4). Monte Ransome Johnson graciously read the entire manuscript for Oxford University Press and provided invaluable criticisms as well as suggestions for improvement throughout. The words of the Philosopher could not apply better to anyone else: *οἱ δὲ βουλόμενοι τἀγαθὰ τοῖς φίλοις ἐκείνων ἕνεκα μάλιστα* φίλοι· δι᾽ αὐτοὺς γὰρ οὕτως ἔχουσι καὶ οὐ κατὰ συμβεβηκός (Arist. *EN* 8.3, 1156b9–11). Further generosity was exhibited by the other anonymous reader for Oxford University Press, whose comments at an early stage helped mold the book into its final form. These scholars' attention to detail, as well as broader circumspection, prevented a great number of (though surely not all) erroneous interpretations, inferences, and arguments. Special gratitude is also owed to Carl Huffman, who shared early drafts of the fragments of Aristoxenus from his eagerly awaited edition; Rachel Barney and Malcolm Schofield, who provided versions of what were then unpublished articles on *Metaphysics* A; Christopher Baron, who sent me the preproof draft of his authoritative *Timaeus of Tauromenium and Hellenistic Historiography*; and Leonid Zhmud, who allowed me to see the proofs of *Pythagoras and the Early Pythagoreans* before the book graced the shelves. A debt of gratitude is owed to those who have offered information, encouragement, advice, criticism, or even a sympathetic ear on the journey: Ben Acosta-Hughes, Ahmed Alwishah, Chloe Balla, Jonathan Barnes, Kevin van Bladel, Chris Bobonich, Luc Brisson, Paola Ceccarelli, Del Chrol, Zenon Culverhouse, John Dillon, Jamie Dow, Jackie Feke, Doug Frame, Francesco Fronterotta, Robert Germany, Barbara Graziosi, Sepp Gumbrecht, Tom Habinek, Antony Hatzistavrou, Johannes Haubold, Josh Hayes, Ron Hock, Andrew Hui, Brad Inwood, Richard Janko, Ted Kaizer, John Kirby, Melissa Lane, Valentina di Lascio, Arnaud Macé, Richard Martin, Henry Mendell, Allen Miller, Kathryn Morgan, Greg Nagy, Grant Nelsestuen, Josh Ober, Grant Parker, Chris Pelling, Alexis Pinchard, Susan Prince, Olivier Renaut, Christopher Rowe, Will Shearin, Peter Struck, Chiara Sulprizio, Candace Weddle, David Wible, and David Wray. Special thanks go to Greg Thalmann, my dissertation supervisor, who, like all generous fathers, has always loved, in spite of the growing pains.

Portions of chapters were also presented throughout North America, the United Kingdom, and the Continent at various institutions: the Centre for the Study of the Ancient Mediterranean and Near East and the conference "Ancient Fallacies" at Durham University; the Séminaire de la Société d'Études Platoniciennes, hosted by the Centre National de la Recherche Scientifique, with Université Paris Ouest Nanterre-La Défense and Université de Franche-Comté; the Sunoikisis Research Symposium at Harvard University's Center for Hellenic Studies; the Yorkshire Ancient Philosophy Network, hosted by the Department of Philosophy at the University of Hull; the Department of Classics at the University of Pennsylvania; the Department of Languages, Literatures, and Cultures at the University of South Carolina; and the Collaborative Programme in Ancient and Medieval Philosophy at the University of Toronto. For the opportunities given to put to the test the nascent ideas now captured in print, an immense gratitude is owed to these institutions and to the audiences in attendance.

This book also owes its existence to crucial support from the library staff of Green Library at Stanford University, the Center for Hellenic Studies Library, the Bill Bryson Library at Durham University, and the Bodleian and Sackler Libraries at Oxford University. Its claims would surely have been stillborn if not for the priceless collections of dusty codices preserved by the walls of these institutions and by the philanthropy of their respective staffs, administrators, and donors. If the reader should think that anything written in these pages is worth its weight in paper and ink, may she or he support the preservation of its kin by advocating for the retention of physical books in libraries.

There is also reason to mention those who have helped me bring this book to completion at Oxford University Press, including my editor, Stefan Vranka, his assistant, Sarah Pirovitz, the production editors, Karen Kwak and Amy Whitmer, and the copy editor, Martha Ramsey, in America, as well as Hilary O'Shea, who helped facilitate things in England. I would also like to express my gratitude to Antony Gormley, whose sublime sculpture graces the cover of this book and instantiates wonder about the identity, stability, and measure of man, as well as his assistant at the Studio, Alice O'Reilly.

This book is dedicated to my mother, Lucinda, the healer, my father, Stanley, the scientist, and Eliana, who listens, and calculates—and knows.

ABBREVIATIONS

In general, I have cited Greek authors according to the abbreviations in Liddell and Scott's *Greek-English Lexicon* (revised by Jones with supplement, Oxford, 1968), hereafter referred to as *LSJ*.

For Plato, I have used the various editions of the Oxford Classical Texts and quoted according to the Stephanus pages listed there.

I have tended to use Ross's Oxford editions of the works of Aristotle, which retain Bekker's numbering (*Aristoteles Graece ex recensione Immanuelis Bekkeri*, vols. 1–2, Berlin, 1831), with the notable exception of the fragments of Aristotle, for which I refer to Rose's third edition, published by Teubner (1886).

References to the commentators on Aristotle refer to the editions of the *Commentaria in Aristotelem Graeca* (Berlin, 1892–1909).

For all abbreviations of ancient works, see the *Index Locorum*.

Common abbreviations or short forms of modern textual editions and collections in this book are as follows.

Bastianini and Sedley = *Commentarium in Platonis Theaetetum* (P.Berol. inv. 9782), a cura di G. Bastianini e D. N. Sedley, in *Corpus dei papiri filosofici* III (Florence, 1995), 227–562.

Collard and Cropp = C. Collard and C. Cropp (eds.), *Euripides: Fragments*, Loeb Classical Library, Euripides VIII (Cambridge, Mass., 2009).

Dillon = J. Dillon (ed.), *Iamblichi Chacidensis in Platonis Dialogos Commentariorum Fragmenta* (Leiden, 1973).

Dilts = M. Dilts (ed.), *Heraclides Lembi: Excerpta Politarum* (Durham, N.C., 1971).

DK = H. Diels and W. Kranz (eds.), *Die Fragmente der Vorsokratiker*, 6th ed., vols. 1–3 (Berlin, 1951).

DSH-MRJ = D. S. Hutchinson and M. R. Johnson (eds.), *Aristotle: Protrepticus or an Exhortation to Philosophy*, working paper, www.protrepticus.info/2012v4.pdf.

Düring = I. Düring, *Aristotle's Protrepticus* (Göteburg, 1960).
FGrHist = F. Jacoby (ed.), *Die Fragmente der griechischen Historiker* (Berlin, 1923–58).
FHS&G = W. W. Fortenbaugh, P. Huby, R. W. Sharples, and D. Gutas, eds., *Theophrastus of Eresus: Sources for his Life, Writings, Thought, and Influence*, vols. 1–2 (Leiden, 1992).
Fortenbaugh and Schütrumpf = W. W. Fortenbaugh and E. Schütrumpf (eds.), *Dicaearchus of Messana: Text, Translation, and Discussion* (New Brunswick, N.J., 2001).
Giannantoni = G. Giannantoni (ed.), *Socratis et socraticorum reliquiae*, vols. 1–4 (Naples, 1990).
Graf and Johnston = F. Graf and S. I. Johnston (eds.), *Ritual Texts for the Afterlife: Orpheus and the Bacchic Gold Tablets* (New York, 2007).
IP = M. Isnardi Parente (ed.), *Senocrate—Ermodoro: Frammenti* (Naples, 1981).
K.-A. = R. Kassel and C. Austin (eds.), *Poetae Comici Graeci*, vols. 1–8 (New York, 1998–).
KRS = G. Kirk, J. Raven, and M. Schofield (eds.), *The Presocratic Philosophers*, 2nd ed. (Cambridge, 1982).
Mazzarelli = C. Mazzarelli (ed.), "Raccolta i interpretazione delle testimonianze e dei frammenti del medioplatonica Eudoro di Alessandria," *Rivista di Filosofia neo-scolastica* 22 (1985), 197–209, 535–555.
Müller = C. Müller (ed.), "Heraclides Lembus," in *Fragmente historicorum graecorum*, vol. 3, pp. 167–171 (Paris, 1849).
Nauck = A. Nauck (ed.), *Tragicorum Graecorum Fragmenta*, 2nd ed. (Leipzig, 1889).
Pendrick = G. J. Pendrick (ed.), *Antiphon the Sophist: The Fragments* (Cambridge, 2002).
Radt = S. Radt (ed.), *Tragicorum Graecorum Fragmenta*, vol. 4 (Gottingen, 1977).
Smith = A. Smith (ed.), *Porphyrii Philosophi Fragmenta* (Stuttgart, 1993).
SVF = H. von Arnim (ed.), *Stoicorum veterum fragmenta* (Leipzig, 1903–21).
Tarán = L. Tarán (ed.), *Speusippus of Athens* (Leiden, 1981).
Thesleff = H. Thesleff (ed.), *The Pythagorean Texts of the Hellenistic Period* (Åbo, 1965).
Wehrli = F. Wehrli (ed.), *Die Schule des Aristoteles*, vols. 1–10 (Basel, 1949–59).

For the majority of the Presocratics and Sophists, fragments and testimonia have been marked with the classification used by Diels-Kranz, which is designated in the text by DK, then the author number, and finally by the classification of item (A for Testimonium, B for Fragment, and C for Imitation); e.g., DK 68 B 4: Diels-Kranz (DK), Anaxagoras (68), Fragment 4 (B 4).

I have used the editions of Carl Huffman for the fragments of Philolaus (*Philolaus of Croton: Pythagorean and Socratic*, Cambridge, 1993) and Archytas (*Archytas of Tarentum: Pythagorean, Philosopher, and Mathematician King*, Cambridge, 2005).

For Iamblichus, I have used the Teubner editions of Deubner (*On the Pythagorean Life*) and Festa (*On the General Mathematical Science*), both revised by Klein, as well as Pistelli (*Protrepticus, Introduction to the Arithmetic of Nicomachus*). I have cited Iamblichus's works according to section number, Teubner page, and line number.

Throughout this book F stands for "Fragment" or "Fragments," and T stands for "Testimonium" or "Testimonia"; these designations are only used where DK is not being employed or, in some rare cases, when DK does not classify the item according to the A/B/C scheme.

For ancient texts, book, chapter, and section citations are generally written in Arabic numerals and separated by periods (e.g. Stob. *Ecl.* 1.3.2, rather than I. iii.2), except in the cases of major authors such as Plato and Aristotle, where I cite book and/or section followed by a comma and then Stephanus or Bekker page and line numbers (e.g. Pl. *R.* 7, 533b1–3 or Arist. *Metaph.* 1.5, 986b1–4). For the sake of convenience, I have latinized Greek names (e.g. Empedocles of Agrigentum, rather than Empedokles of Akragas) throughout this book.

Translations of Plato are often based on those given in J. M. Cooper and D. S. Hutchinson (eds.), *Plato: Complete Works* (Cambridge, 1997).

Plato and Pythagoreanism

1

Aristotle on Mathematical Pythagoreanism in the Fourth Century BCE

In this first chapter, I will attempt to describe the kinds of Pythagoreans who may have existed from the sixth through fourth centuries BCE and the philosophical activities in which they seem to have engaged by appeal to the evidence preserved by Aristotle. The goal is to identify the characteristics that distinguished the mathematical Pythagorean *pragmateia*—where we may tentatively describe a *pragmateia* here (for Aristotle) as both the object of a philosophical inquiry and the treatment of the same object—from the *pragmateia* of the rival acousmatic Pythagorean brotherhood in Magna Graecia. This goal is part and parcel of the larger project that will occupy the entirety of this book: to trace the history of mathematical Pythagoreanism from a variety of informed ancient perspectives. My claim in this chapter is that Aristotle, especially in *Metaphysics* A and the lost writings on the Pythagoreans (preserved in a fragmentary state without significant modifications in Iamblichus's work *On the General Mathematical Science*),[1] establishes this distinction by appeal to the divergent philosophical *methodologies* of each group: the mathematical Pythagoreans, who are the same as the "so-called Pythagoreans" in *Metaphysics* A, employ superordinate[2] mathematical sciences in establishing something that approximates demonstrations that explain the "reason why" (τὸ διότι) they hold their philosophical positions, whereas the acousmatic Pythagoreans, who are distinguished from the "so-called" Pythagoreans in *Metaphysics* A, appeal to basic, empirically derived "fact" (τὸ ὅτι) in defense of their doctrines. Furthermore, I suggest, Aristotle criticizes the *pragmateia* of the mathematical Pythagoreans for improper methodological procedure: while the

1. I refer to Aristotle's "works on the Pythagoreans" on the grounds that we cannot know for sure what work Iamblichus was using to extract his descriptions of the Pythagoreans. Titles are attested for *On the Pythagoreans* (one book), *Against the Pythagoreans* (one book), *On the Philosophy of Archytas* (three books), and *Summary of the Timaeus and of the Works of Archytas*.

2. I employ the term "superordinate" to refer to those sciences Aristotle considered superior to the "subordinate" or "one beneath the other" (θάτερον ὑπὸ θάτερον) sciences (*APo*. 1.7, 75b15–16), following Johnson 2009.

demonstrations offered by the mathematical Pythagoreans represent a significant philosophical innovation over the uncritical reflection on the so-called "facts" by the acousmatic Pythagoreans, the mathematical Pythagoreans' activity of hasty assimilation across categories leads to confusions in logic and metaphysics. Analysis of the extant fragments of Philolaus of Croton (among others) gives evidence for the kind of approach to understanding the universe that Aristotle associates with the mathematical Pythagoreans, and it becomes likely that the targets of Aristotle's disapproval are those Pythagoreans who undertook to perform basic demonstrations of the Pythagorean definitions of things as preserved in the *acusmata* attributed to Pythagoras. It becomes possible, then, to inquire further as to whether Aristotle's classification might have any value for a reconstruction of the philosophical methodologies of earlier Pythagoreans in the first half of the fifth century BCE.

The Pythagoreans of the fifth century BCE probably did not see themselves as a community unified by philosophical and political doctrines. Rather, insofar as we can reconstruct their history, there arose an internal conflict among the Pythagoreans who were living in the southern part of Italy, which appears to have effected a split between the ascetic Pythagoreans who lived in the western part of Italy (and fled to Asia Minor) and the intellectualist Pythagoreans who occupied the eastern part of the Italian peninsula, near Tarentum. Differences in approach to the philosophical "life" and its activities can already be detected in the comic fragments that survive from the early part of the fourth century BCE, as Christoph Riedweg has shown.[3] With Aristotle, I suggest, we find a rather elaborate account of the division of the early Pythagoreans into two groups—traditionalist acousmatics (*οἱ ἀκουσματικοί*) and progressive mathematicians (*οἱ μαθηματικοί*). What the terms "acousmatic" and "mathematical" mean will require a careful examination of Aristotle's descriptions, a project that will occupy chapters 1 and 2.[4] While most modern scholars have

3. Riedweg 2005: 108–109.

4. It is extremely difficult to correlate the bifurcation into "acousmatic" and "mathematical" Pythagorean with the tripartite subsections that developed in the Hellenistic world (*σεβαστικοί, πολιτικοί, μαθηματικοί*). Delatte (1922: 22–28) took seriously the possibility of the tripartite organization, to which earlier and later traditions as well as the so-called Hellenistic pseudo-Pythagorean writings adhere closely. Burkert (1972: 193 n. 6) suggests that the triad is a chronological grouping that aligned with the terms *Πυθαγορικοί, Πυθαγόρειοι, Πυθαγορισταί* and corresponded with the "pupils, pupils of pupils, external advocates (*ἔξωθεν ζηλωταί*)" (Anon. Phot. 438b = Thesleff 237.7–12), and whose philosophical interests in Aristotelian terms are associated respectively with theology, human affairs (i.e. politics), and mathematical sciences, including geometry and astronomy. I suspect that these chronological associations are all developed, at least in some way, out of the historical writings of Timaeus of Tauromenium, whose treatment of Pythagoreanism I will discuss in chapter 3. Zhmud (2012: 183–185) considers all these distinctions to be dated much later, probably from the first century CE.

been willing to accept the classifications of acousmatic and mathematical Pythagoreans of the fifth century BCE, they nevertheless assume that certain contradictory elements within their own constructed "Pythagoreanism" might be misinterpretation or confusion on the part of ancient critics like Aristoxenus, Dicaearchus, or Timaeus,[5] all Hellenistic commentators whose accounts at least partially derive from the descriptions of Pythagoreanism in the writings of Aristotle.[6] I would like to present an alternative account. Since Aristotle, our most comprehensive early source for a history of Pythagoreanism, differentiated two groups of Pythagoreans along methodological lines (or so I will argue), we should admit the possibility that these apparently contradictory elements in our own reconstruction reflect *actual* divisions within the community. Indeed, the primary criterion for distinguishing acousmatic from mathematical Pythagoreans, as I will show, is each group's *pragmateia* (πραγματεία), a term that must be further contextualized in order to make sense of precisely how Aristotle draws the line.

ARISTOTLE ON THE *PRAGMATEIAI* OF THE PYTHAGOREANS

It is my contention that Aristotle differentiates two groups of Pythagoreans according to the *pragmateia* of their respective philosophies. What does the term *pragmateia* mean for Aristotle? It will be useful to start with an operating definition, which can then be developed in the course of our argument: in Aristotle's usage, the *pragmateia* of a philosopher or philosophical group is both the *object* of their philosophical inquiry and the unique *treatment* of that object in their philosophy.[7] Some possible meanings for Aristotle listed in *LSJ*: "system" (*Metaph.* 1.6, 987a30 and 1.5, 986a8), "philosophical argument or treatise" (*Top.* 1.1, 100a18 and 1.2, 101a26; *Phys.* 2.3, 194b18; *EN* 2.2, 1103b26), and "subject of such a treatise" (*Phys.* 2.7, 198a30). Similarly, Bonitz (1970) lists several possible meanings, among which we see: *rei alicuius tractatio via ac ratione instituta* (*Pol.* 3.1, 1274b37), *interdum non tam tractationem rei quam rationem rei tractandae* (*Rh.* 1.15, 1376b4), or even *quaestio* (*APo.* 2.13, 96b15). We can assume

5. Here I refer to the historian Timaeus of Tauromenium, who is not to be confused with Timaeus of Epizephyrian Locri, the fictional eponymous authority in Plato's dialogue.

6. Most recent scholars accept the distinction between acousmatic and mathematical Pythagoreans as original with Aristotle, e.g. Burkert (1972: 192–207), Huffman (1993: 11–12 and 2010), Kahn (2001: 15), McKirahan (1994: 89–93), and Riedweg (2005: 106–108); an exception is Zhmud (2012: 169–206), who wonders whether the division is original with Nicomachus, but nevertheless accepts the basic terminology along these lines.

7. Some other scholars' definitions of Aristotelian *pragmateia* are "philosophic activity" (Burkert/Minar 1972: 194) and "enterprise" (Steel 2012: 181). Unfortunately, Aristotle nowhere explicitly defines *pragmateia*.

some semantic overlap, in the sense that for Aristotle, there was a fluid relationship between these meanings. *Pragmateia* is apparently first used in a technical manner by Archytas of Tarentum, who posits it as the "treatment" or "investigation into" an object of mathematics:

> Logistic [ἁ λογιστικά] seems to be far superior indeed to the other arts in regard to wisdom, and in particular [it seems] to deal with [πραγματεύεσθαι] what it wishes more clearly [ἐναργεστέρω] than geometry. Again in those respects in which geometry is deficient, logistic puts demonstrations into effect [ἀποδείξιας ἐπιτελεῖ] and equally, if there is any *pragmateia* of shapes [εἰ μὲν εἰδέων τεὰ πραγματεία], [logistic puts demonstrations into effect] with respect to what concerns shapes as well.
>
> (ARCHYTAS F 4 HUFFMAN = Stobaeus *Proem*; translation after Huffman)

Archytas seems to use the abstract term *pragmateia* as well as the verb *pragmateuesthai* to refer to both the object of philosophical investigation and the treatment suitable to that object. This usage is in contrast to that of Plato, where *pragmateia* more generally means "the business of" (e.g. *Grg.* 453a2–3, *Theaet.* 161e4)[8] without any technical philosophical usage. It thus becomes possible that Aristotle inherited this special use of *pragmateia* and terms related to it from Archytas himself.[9] The idea that Aristotle might have adopted the technical terminology for the categorization of objects of philosophy and particular treatment of those objects from a Pythagorean is significant, since, as I will argue, Aristotle himself uses the term *pragmateia* as a marker that establishes characteristic distinctions between acousmatic and mathematical Pythagoreans according to the treatment of the objects of their philosophical inquiry.[10] The larger implications of the difference between the *pragmateiai* of the mathematical and acousmatic Pythagoreans have a direct significance for this study,

8. Noted by Huffman (2005: 251).

9. Still, there is one place (*R.* 7, 528d1–3) where Plato uses the term *pragmateia* in relation to mathematics. Glaucon asks Socrates if the "geometry" is to be considered the "study of the plane" (τοῦ ἐπιπέδου πραγματεία). In the context of Plato's criticisms of Pythagoreanism, especially of Archytas, it is probable that Glaucon is using a term inherited from Pythagorean mathematics here.

10. That is, if we should consider Archytas to have been a Pythagorean. I count him as one, at least in a conditional sense, for reasons I lay out in chapter 3. Huffman has inferred from the fact that Aristotle wrote three books on Archytas and two books on the Pythagoreans, and from the fact that Aristotle never calls Archytas a "Pythagorean," that Archytas's "importance was not limited to the Pythagorean tradition" (2005: 128).

since, as I will show, the figure credited with establishing the distinctive *pragmateia* of the mathematical Pythagoreans, Hippasus of Metapontum (ca. 520?–440 BCE?), may have also played a central role in the political factionalization that occurred in the Pythagorean community in the second quarter of the fifth century BCE.[11]

Who was this "Hippasus of Metapontum"? A substantial portion of this book will deal with this elusive and enigmatic figure, and I will begin by contextualizing him with the broader classification of the mathematical and acousmatic Pythagoreans advanced by Aristotle. The consensus view, which follows Walter Burkert in his extremely influential study *Lore and Science in Ancient Pythagoreanism* (1972), is that Hippasus of Metapontum was a mathematical Pythagorean (μαθηματικός). What is more troubling, though, is that neither Burkert nor those who follow him are sure how to define a Pythagorean μαθηματικός or his philosophical activities.[12] This provides an opportunity for us to pursue a more complete understanding of the Pythagorean μαθηματικός, especially in the light of Aristotle's classification of two types of Pythagoreans.[13] The relevant evidence for this comes in a tricky passage from Iamblichus's work *On the General Mathematical Science*, in which Iamblichus is summarizing[14] portions of Aristotle's lost works on the Pythagoreans:

> There are two types of the Italian, also called the Pythagorean, philosophy. For there were also two kinds of people who treated it, namely the acousmatics and the mathematicians. Of these two, the acousmatics were recognized to be Pythagoreans by the others [the mathematicians], but they did not recognize the mathematicians [as Pythagoreans], nor did they think that the *pragmateia* [of the mathematicians] derived from Pythagoras, but rather that it derived from Hippasus.

11. Iambl. *DCM* 25, 76.16–77.24 and *VP* 257–258, 138.14–139.9. I will discuss these specific passages more extensively in chapters 2 and 3, respectively.

12. Burkert 1972: 192–201. Similarly followed by Huffman (2005), Riedweg (2005), and Kahn (2001). Zhmud (2012: 255–258) emphasizes the role that ἀποδείξεις play in Pythagoras's teaching of the μαθηματικοί in Iamblichus's account (also see Zhmud 2006: 132, where he refers to Hippocrates of Chios, Archytas, and Eudoxus as "typical" μαθηματικοί).

13. Riedweg's account (2005: 106–108) is probably the best synthetic account outside of Burkert (1972), although we should recognize the care with which Burnyeat (2005a) examined the philosophical context in Aristotle (without analysis of the political aspects of the reported schism). Burnyeat thus leads the way for my study.

14. As I will suggest below, Iamblichus goes on to quote the work directly.

> Δύο δ' ἐστὶ τῆς Ἰταλικῆς φιλοσοφίας εἴδη, καλουμένης δὲ Πυθαγορικῆς. δύο γὰρ ἦν γένη καὶ τῶν μεταχειριζομένων αὐτήν, οἱ μὲν ἀκουσματικοί, οἱ δὲ μαθηματικοί. τούτων δὲ οἱ μὲν ἀκουσματικοὶ ὡμολογοῦντο Πυθαγόρειοι εἶναι ὑπὸ τῶν ἑτέρων, τοὺς δὲ μαθηματικοὺς οὗτοι οὐχ ὡμολόγουν, οὔτε τὴν πραγματείαν αὐτῶν εἶναι Πυθαγόρου, ἀλλὰ Ἱππάσου.
>
> (IAMBLICHUS, *On the General Mathematical Science* 25, 76.16–22)

Now Burkert synthesizes the material derived from Aristotle's works on the Pythagoreans and preserved by Iamblichus[15] in order to demonstrate two significant points: first, that *all* followers of Pythagoras were adherents of the *acusmata*, also called *symbola*,[16] a set of orally transmitted sayings passed down from Pythagorean teacher to student in the period of silence that apparently attended the first five years of their educational curriculum, and second, that what distinguished the ascetic acousmatic Pythagoreans (ἀκουσματικοί) from the progressive mathematical Pythagoreans (μαθηματικοί) was each group's unique philosophical and political *pragmateia*:

> Aristotle recognizes among the Pythagoreans a twofold πραγματεία: on the one hand, the Πυθαγορικοὶ μῦθοι, metempsychosis, the Pythagoras legend, and the *acusmata*, and on the other a philosophy of number connected with mathematics, astronomy, and music, which he never tries to trace back to Pythagoras himself and whose chronology he leaves in abeyance.[17]

Furthermore, Burkert argues that Aristotle categorized the Pythagorean *acusmata* according to whether or not they answered these three questions: τί ἔστι (what is?), τί μάλιστα (what is to the greatest degree?), and τί πρακτέον (what is to be done?).[18] While the implications of this fascinating tripartite categorization both for Aristotelian philosophy and for Pythagoreanism could extend far

15. Zhmud's arguments (2012: 174) that suggest Clement of Alexandria as the *source* for the division into ἀκουσματικοί and μαθηματικοί are not decisive. For one thing, it remains for Zhmud to explain the philosophical language of the passage quoted by Iamblichus. See below in chapters 1 and 2 for my alternative treatment of the evidence.

16. See Zhmud 2012: 173 with n. 16.

17. Burkert 1972: 197.

18. See Burkert 1972: 167–169, with Iambl. *VP* 82, 47.11–13, and Delatte 1915: 274–307. Burkert rightly reminds us that these "orally transmitted maxims and sayings" were also called *symbola*. Recently, Struck (2004: 96–110) has attempted a comprehensive study on symbolic or enigmatic communication in antiquity, although his book also does not treat the third kind of *acusma*.

beyond this study, throughout this chapter I focus chiefly on the third classification, namely on those things that fall under the category τί πρακτέον.

Burkert explicates those *acusmata* that fall under the category "what is to be done" by focusing, almost entirely, on ethical imperatives and ritual activity.[19] He demonstrates their significance for the establishment of a Pythagorean way of life as an "amazing, inextricable tangle of religious and rational ethics."[20] This is a valuable approach to understanding one important aspect of the philosophical lifestyle ascribed to the Pythagoreans, because it reveals the religious semantics of *pragmateia*. Burkert's study also reflects its own Aristotelian intellectual lineage since, as Iamblichus argues (in the Aristotelian analysis of the "what is to be done" injunctions that follows on their listing), what is divine (τὸ θεῖον) is the first principle and origin (ἀρχή).[21] But I suspect that there is more to Aristotle's classification of the two Pythagorean *pragmateiai* than Burkert discusses. For Aristotle, as for Archytas, the term *pragmateia* was chiefly associated with philosophical methodology, and not only with theology, although the latter is implicated in the former. Can we gain some traction on the philosophical activities of the Pythagoreans by examining more closely this implication of theology in philosophical activity? One passage from Iamblichus's work *On the Pythagorean Life*, probably derived from a Peripatetic account of Pythagoreanism, helps to show the way:

> All such *acusmata*, however, that define what is to be done or what is not to be done [περὶ τοῦ πράττειν ἢ μὴ πράττειν], are directed toward the divine [ἐστόχασται πρὸς τὸ θεῖον], and this is a first principle [ἀρχή], and their whole way of life is arranged with a view to following God [ὁ βίος ἅπας συντέτακται πρὸς τὸ ἀκολουθεῖν τῷ θεῷ], and this is the rationale [λόγος] of their philosophy.
>
> (IAMBLICHUS, *On the Pythagorean Way of Life* 86–87, 50.18–21; translation after Dillon and Hershbell 1991)

One of the great challenges of this passage is to extract what, if anything, traces back to the fourth century BCE. We may never be absolutely certain.[22] The reference

19. Burkert 1972: 174–192. Similarly followed by Kahn (2001: 9–10) and Riedweg (2005: 63–67).

20. Burkert 1972: 185.

21. Iambl. *VP* 86, 50.18–19. Note that Aristotle makes a similar claim at *Metaph*. 1.2, 983a6–11, on which see Nightingale 2004: 236–237.

22. Zhmud (2012: 189) thinks this passage derives from Nicomachus, but I think it is an overstatement to describe the differences between the various passages of *VP* 81, 87–89, and 82–86 as "self-evident," especially since, as Zhmud himself admits (p. 191), there are "clear signs of editorial emendations by Iamblichus." A related problem here is the grammar

to the *acusmata*—especially those that deal with "what one is to do" injunctions—*sounds* Aristotelian, as does the ascription of divinity to the first principle. A likely source for this part of the text, as I argue in chapter 2, is Aristoxenus of Tarentum, who speculated about the Pythagorean first principle in related ways in his *Pythagorean Precepts*.[23] In the passage that immediately precedes this one, however, the attempt to define a "first principle" (ἀρχή) and a "reason" or "rationale" (λόγος) for the Pythagorean philosophy as related to the first principle is characteristic of Aristotle's method of describing and critiquing earlier philosophical systems. We might, for example, recall the beginning of the *Nicomachean Ethics* (1.4, 1095a30–b14), where Aristotle questions whether it is better to employ arguments (λόγοι) that *derive from* first principles (ἀπὸ τῶν ἀρχῶν) or those that *lead to* first principles (ἐπὶ τὰς ἀρχάς). In this digression, Aristotle appears to distinguish his own philosophical method from Plato's by arguing that we should begin from what is already known and familiar to us, namely, the "what is" or "fact" (τὸ ὅτι), which he also calls a "first principle" (ἀρχή).[24] With regard to first principles, Aristotle's approach here stands in contrast to the approach attributed to the Pythagoreans in Iamblichus's work *On the Pythagorean Way of Life* 86–87, which attributes to Pythagoreans the sorts of λόγοι that reduce to the first principle, namely the divine.[25]

But which Pythagoreans, acousmatics or mathematicians, was Iamblichus describing in this passage? Or was he talking about the *pragmateia* of all the Pythagoreans? There is no standard scholarly position on this question, in part because scholars have been unclear about which sections derive from the Peripatetic source, or how much Iamblichus has doctored the text.[26] It is likely,

---

of "aiming" (ἐστόχασται). While it is the case that Aristotle speaks of "aiming at" objects such as a "good" (*Pol.* 1.1, 1252a4), "pleasure" (*Metaph.* 6.2, 1027a3) or "the mean" (*EN* 2.9, 1109a30–2), the object at which one aims is always in the genitive case, whereas in Iamblichus it is in the πρὸς + accusative phrase.

23. See chapter 2, section entitled "Pythagoreanism and the Axiology of What Is 'Honorable.'"

24. ἀρχὴ γὰρ τὸ ὅτι· καὶ εἰ τοῦτο φαίνοιτο ἀρχούντως, οὐδὲν προσδεήσει τοῦ διότι. On the relationship between the "fact" and the "why," see Burnyeat 1981: 118 and, more recently, Zhmud 2006: 136.

25. In this way, the Aristotelian passage preserved in *VP* 86–87 may have formed the basis for (or referred to the same system described by) Aristoxenus's account of the *Pythagorean Precepts*, especially F 33 Wehrli (= Iambl. *VP* 174–176, 97.23–98.24) and F 34 Wehrli (= Stob. *Ecl.* 4.25.45), which describe the ontological stratification of being for the "Pythagoreans." See Huffman 2006: 112 and 2008: 107–108. Theophrastus (*Metaph.* 11a26–b12) also speaks of Plato and the "Pythagoreans" as reducing to the first principles, on which see Horky: forthcoming.

26. See Burkert 1972: 196 n. 17.

I suggest, that the reference is to the Pythagoreans in general, and not to a particular group, in this passage. While it is true that the distinction between acousmatic and mathematical Pythagoreans immediately precedes this passage in Iamblichus's text, there are three reasons for interpreting this passage as referring to Pythagoreans more generally. First, Iamblichus separates a long passage where he discusses the distinctions between two groups of Pythagoreans (*On the Pythagorean Way of Life* 81–86, 46.23–50.17) by a poignant "however" (*μέντοι*), suggesting that he has completed discussion of the split between two groups of Pythagoreans.[27] Second, there is nothing specific to suggest that we should identify the system of religious order described as acousmatic or mathematical: this is unsurprising, since it is generally agreed that the mathematical Pythagoreans accepted the religious and ethical precepts of the acousmatic Pythagoreans.[28] Finally, when Iamblichus returns to discussing the *acusmata* later in the treatise (*On the Pythagorean Way of Life* 137, 77.13–19), he repeats this passage almost verbatim and describes it as illustrating the principles of religious worship of the gods as attributed to Pythagoras and to his followers (*Πυθαγόρας τε καὶ οἱ ἀπ'αὐτοῦ ἄνδρες*). Thus, the broader description of the *pragmateia* of the Pythagoreans (as formulated by Iamblichus in his work *On the Pythagorean Way of Life* 86–87) focuses on two important aspects that I will continue to discuss in this study: the hierarchy of the cosmos, which one honors by understanding that the divine is the first principle that must be pursued in order to attain the good; and the hierarchy of a political organization, which is analogous to the cosmic hierarchy. In this way, when Iamblichus's Peripatetic source characterized the universal Pythagorean *pragmateia*, he seems to have exploited both the religious and political senses of the term *ἀρχή*.[29]

Close attention to *philosophical methodologies*, however, might give us a better insight into the rationales that distinguished the *pragmateiai* of the different Pythagoreans. When he describes the rationale (λόγος) for the maxims that

27. Iamblichus synthesizes the descriptions of the two groups: the first group, the *ἀκουσματικοί*, are said initially (*VP* 82, 47.4–6) to practice a philosophy "without demonstration and without argument" (*ἀναπόδεικτα καὶ ἄνευ λόγου*) and are later (*VP* 86, 50.9–12) associated with those who undertake philosophical activity that is properly "Pythagorean" (*Πυθαγορικαί*); and the *μαθηματικοί*, who offer up "probable reasons" (*ἐικοτολογίαι*), are called "some [others] from outside" (*ἔνιοι ἔξωθεν*). As I will show, I believe the distinction given earlier between *ἀκουσματικοί* and *μαθηματικοί* to be Aristotelian, whereas I suggest that the later differentiation between those "inside" and "outside" the school may derive from Timaeus of Tauromenium. See chapter 3.

28. See Huffman 2010, Riedweg 2005: 106–107, and Kahn 2001: 15.

29. Aristoxenus is explicit in exploiting both meanings by reference to the Pythagoreans and is the likely source here. See chapter 2, section entitled "Pythagoreanism and the Axiology of What Is 'Honorable.'"

answer the question *τί πρακτέον*, Iamblichus (*VP* 86, 50.6–13) distinguishes the use of rationales by the more conservative Pythagoreans from the use by those people whose philosophical activities he claims are "non-Pythagorean" (*οὐκ εἰσὶ Πυθαγορικαί*) and who are also called "outsiders" (*ἔξωθεν*). Are those figures designated "outsiders" the same as the mathematical Pythagoreans?

The evidence concerning the "esoteric" and "exoteric" Pythagoreans in Iamblichus's work is ambivalent, but it is not likely, I suggest, that Aristotle understood the division along insider and outsider lines.[30] Rather, as I will show in chapter 3, the source for the passages that distinguish "exoteric" from "esoteric" Pythagoreans in Iamblichus's work *On the Pythagorean Way of Life* appears to be the late fourth-/early third-century BCE Western Greek historian Timaeus of Tauromenium, who posited a division between those Pythagoreans who were more advanced in their learning (inside) and those who did not advance beyond a certain level (outside).[31] Even so, the source for this part of *On the Pythagorean Way of Life* 86 still evinces a division along philosophical grounds. These "exoteric" Pythagoreans differ from the "esoteric" Pythagoreans specifically because they "attempt to attach a likely rationale/account" (*πειρωμένων προσάπτειν εἰκότα λόγον*) to the ethical injunctions that constitute the Pythagorean *acusmata*.[32] The "likely account" (*εἰκοτολογία*) that Iamblichus's source attributes to those people who are "non-Pythagoreans" or "exoteric" in this passage represents a more sophisticated approach to wisdom traditions such as those of Pythagoras or the Seven Sages, but it is not "mathematical" in the strong sense, at least if we are to judge by the examples given. The sorts of "likely account" given by the "exoteric" Pythagoreans are focused on practical—indeed, even

30. Of course, Aristotle himself referred to some of his writings as "exoteric" (*ἐξωτερικοὶ λόγοι*), which, at *EE* 1.8, 1217b22 he sets in contrast to those writings that he calls "philosophical" (*οἱ κατὰ φιλοσοφίαν λόγοι*). Much has been said about this distinction, and little is agreed on (for two divergent recent accounts, see Gerson 2005: 47–76 and Zanatta 2008: 26–35). What is of value for this study is that the version of the "exoteric"/"esoteric" division found in Iamblichus's works is *never explicitly* drawn by Aristotle and, therefore, probably owes its origins to someone else. For a useful study of the relationship between the terms "exoteric/esoteric" and "acousmatic/mathematical" Pythagorean, see von Fritz 1960: 8–10.

31. In chapter 3, I explore at much greater length Timaeus of Tauromenium's criticisms of Aristotle's history of the Pythagoreans.

32. The term *εἰκὼς λόγος*, which is technical, receives a great number of conflicting treatments in antiquity. In Plato's *Timaeus* (30b8), it refers to the "likely story" that cannot, on Morgan's reading (2000: 275), be verifiable by appeal to empirical knowledge. It is interesting to note that Ps.-Archytas's *On Intelligence and Perception* (F 1 Thesleff = Stob. 1.41.5) refers to *εἰκοτολογίαι* in reference to political treatises, namely things that deal with "affairs" (*πράξιας*).

political—reasoning in a way not unlike the εἰκὼς λόγος given by Timaeus of Epizephyrian Locri in Plato's *Timaeus* and developed in some of Aristotle's works, including the *Politics*.[33] According to Iamblichus's source, those who were described as "exoteric" Pythagoreans exhibited different types of *logos*, including cultural-historical explanation ("one should not break bread" *because*, in the past, people used to come together in order to eat a single loaf of bread, as foreigners do) and normative-religious ("one should not break bread" *because* one ought not to establish the sort of omen that occurs at the beginning of the meal by means of breaking and crushing bread).[34] Such examples suggest that the "exoteric" Pythagoreans whose *pragmateia* involved cultural-historical or normative-religious types of *logos* appealed to fifth century BCE sorts of explanation, such as those we find in the writings of Herodotus or the writers of the Hippocratic Corpus.[35] They appear, in this account, to resemble more the Pythagorists of Middle Comedy who know how to make clever arguments by using various fallacious devices (ἐπισοφιζομένων), or even the sorts of Presocratics whose speculations formed the basis for the character of Socrates in Aristophanes's *Clouds*, than highly regarded practitioners of wisdom. Still, accusations of illegitimate claims to wisdom are as old as Pythagoras himself, and they were of interest to Timaeus of Tauromenium: our source for Heraclitus's slander of Pythagoras, in which he refers to Pythagoras as a "prince of lies" (κοπίδων ἀρχηγός), is Timaeus himself.[36]

33. See Burnyeat 2005b, who emphasizes the reasonableness or appropriateness (the "ought": δεῖ) that constitutes the goal to which the practitioner of the εἰκὼς λόγος aims. I consider Plato's *Timaeus* to be an "exoteric" Platonic dialogue, in the sense that it makes public *and explains* what might otherwise be considered "unspeakable" ideas in a fourth-century BCE context to an indistinguished audience.

34. It is worth noting that the information preserved here is almost exactly the same as that attributed by Diogenes Laertius to Aristotle's work *On the Pythagoreans* (F 195 Rose = D.L. 8.33–35). It is possible, then, that Iamblichus was looking at Aristotle's text while recording this information or, for that matter, that the historian Timaeus of Tauromenium had access to Aristotle's text while drawing up his own list of the *acusmata* (on which see chapter 3).

35. For Herodotean ἱστορία and its contexts, see Lateiner 1989: 15–17 and Thomas 2000: 21–27; for Presocratic and Hippocratic ἱστορία see Schiefsky 2005: 19–35; more generally, for philosophically related uses of ἱστορία before Plato, see Riedweg 2005: 94–95 and Darbo-Peschanski 2007.

36. FGrHist 566 F 132 (see DK 22 B 81). The term ἐπισοφίζομαι occurs in Iamblichus and in post-Iamblichean texts, but it is also attested in the Hippocratic corpus (*Art.* 14) with reference to clever doctors who demonstrate their cleverness by attaching a piece of lead to a fractured bone in order to stabilize it. See Burkert 1972: 174 with n. 64 and 200. I would add, however, that such "cleverness" is attached to the Tarentine Pythagoreans whose rhetorical *logoi* are satirized in two plays, both entitled *The Tarentines*, written by the

So *On the Pythagorean Way of Life* 86–87 presents us with a paradigmatic case of the problems involved in sorting out the sources of Iamblichus's information concerning the classification of the Pythagoreans: not only must we deal with the terminology of *at least* two different historians (Aristotle and possibly Timaeus, not to speak of Aristoxenus or Nicomachus), we have to be sensitive to how Iamblichus might have confused the accounts. Despite this hindrance, we can gain some traction on the question of the philosophical activities of the various Pythagoreans as Aristotle figured them by appeal to a passage, preserved by Iamblichus fortunately with some direct quotation:

> (A) There are two types of the Italian, also called the Pythagorean, philosophy. For there were also two kinds of people who treated it: the acousmatics and the mathematicians. Of these two, the acousmatics were recognized to be Pythagoreans by the others [the mathematicians], but they did not recognize the mathematicians [as Pythagoreans],[37] nor did they think that the *pragmateia* [of the mathematicians] derived from Pythagoras, but rather that it derived from Hippasus. Some say that Hippasus was from Croton, while others say from Metapontum.[38] And, of the Pythagoreans, those who concern themselves with the sciences [*οἱ περὶ*

fourth-century BCE comedians Alexis of Thurii (F 223 K.-A.: *Πυθαγορισμοὶ καὶ λόγοι / λεπτοὶ διεσμιλευμέναι τε φροντίδες / τρέφουσ' ἐκείνους*) and Cratinus the Younger (F 7 K.-A.: *ἔθος ἐστὶν αὐτοῖς . . . διαπειρώμενον / τῆς τῶν λόγων ῥωμῆς ταράττειν καὶ κυκᾶν / τοῖς ἀντιθέτοις, τοῖς πέρασι, τοῖς παρισλώμασιν, / τοῖς ἀποπλάνοις, τοῖς μεγέθεσιν, νουβιστικῶς*). We can thus posit a popular tradition, not necessarily derived from Aristotle, that attributes sophisms of a rhetorical sort to the Tarentine Pythagoreans. Note, too, that Cratinus employs terms both rhetorical and mathematical, such as *πέρας* and *μέγεθος*, translated by Edmonds as "end" and "sublimity." The former is attested in a rhetorical sense in the Aristotelian *Rhetoric to Alexander* (32, 1439a38), where it is described as the conclusion that rounds off an exhortation. The latter appears in Aristotle's *Rhetoric* (3.9, 1409a36), with reference to periodic sentences that can be measured, as well as in Dionysius of Halicarnassus (*Comp.* 17), as "sublimity." It is difficult to know precisely what Cratinus the Younger intended their meaning to be.

37. Iamblichus elsewhere (*VP* 87, 51.7–12), in a passage that is attached to the same one given in *DCM*, attributes to a certain acousmatic Pythagorean "Hippomedon" the claim that Pythagoras originally gave demonstrations of the precepts, but that, due to the laziness of those who passed them down, ultimately only the precepts remained. Unfortunately, it is difficult to confirm this information, since (1) there are textual problems here (see Deubner's text); (2) we know almost nothing else about this Hippomedon; (3) it is possible that Iamblichus has confused "acousmatic" with "mathematical" Pythagorean here, as he did earlier at *VP* 81, 46.26–47.3 (see Burkert 1972: 193 n. 8).

38. It is not clear to me whether this sentence is Iamblichus's insertion or original with his source, who is probably Aristotle.

τὰ μαθήματα][39] recognize that the others [i.e. the acousmatics] are Pythagoreans, and they declare that they themselves are even more [Pythagorean], and that the things they say [ἃ λέγουσιν] are true. And they[40] say that the reason [αἰτία] for such a disagreement is this:

(B) "Pythagoras came from Ionia, more precisely from Samos, at the time of the tyranny of Polycrates, when Italy was at its height, and the first men of the city-states became his associates. The older of these [men] he addressed in a simple style, since they, who had little leisure on account of their being occupied in political affairs, had trouble when he conversed with them in terms of sciences [μαθήματα] and demonstrations [ἀποδείξεις]. He thought that they would fare no worse if they knew *what* they ought to do [εἰδότας τί δεῖ πράττειν], even if they lacked the explanation [ἄνευ τῆς αἰτίας] for it, just as people under medical care fare no worse when they do not additionally hear *the reason why* they ought to do [διὰ τὶ πρακτέον] each thing in their treatment. The younger of these [men],

39. The term μαθήματα is extremely difficult to translate, and no single translation will do justice. Alternatives include "learning" or "mathematics," but I think Burkert (1972: 195 and 207 n. 80) is correct in defining this term as the branches of learning the Greeks called arithmetic, geometry, astronomy, and music. We should note that Archytas specifically refers to his predecessors as τοὶ περὶ τὰ μαθήματα (Archytas F 1 Huffman) and attributes to them innovations in scientific method, especially concerning numbers, geometry, music, and the motions of the stars. In chapter 6, I argue that Archytas is referring chiefly to Hippasus when he speaks of his predecessors.

40. Who is the subject of this φασίν? Stylistically, there is a minor change of tune from the previous section, which had focused on whether or not the acousmatics or mathematicians "recognized" one another (various forms of ὁμολογεῖν), where a distinction is drawn between the acousmatics who "did not recognize" (in the imperfect tense) the mathematicians as Pythagoreans, and the mathematicians who "recognize" (in the present tense) the acousmatics as Pythagoreans. The appearance of the phrase οἵ περὶ τὰ μαθήματα in that earlier section suggests the possibility, indeed, that the information might derive ultimately from Archytas (see the previous note). And, as I argue in chapter 3, Archytas and other mathematical Pythagoreans wrote about their predecessors. But the appearance of concern with "reason" or "cause" (αἰτία), which is followed up in the portion that seems to be quoted directly (B), which focuses on causation, suggests that someone who formulated a philosophical engagement with causation is responsible for the information that follows. From what remains of Archytas's fragments, there is no obvious interest in causation as such; but Eudemus's account of Archytas's physics (A 23 Huffman) suggests that he did believe that inequality and unevenness were causes of motion. And he was concerned with demonstration as well (F 4 Huffman). Still, we cannot be sure that Eudemus has not mapped Peripatetic terminology onto Archytas's ideas about physics. The most obvious candidate for the subject of this φασίν, then, remains Aristotle, as Burkert originally argued (1972: 457), and as Burnyeat has confirmed (2005a: 40–43). Possibly this material derives from one of Aristotle's works on Archytas. Thanks to Monte Johnson for pressing me to clarify my position on this issue.

however, who had the ability to endure the education, he conversed with in terms of demonstrations and sciences. So, then, these men [i.e. the mathematicians] are descended from the latter group, as are the others [i.e. the acousmatics] from the former group."

(C) And concerning Hippasus, they say that while he was one of the Pythagoreans, he was drowned at sea for committing heresy, on account of being the first to publish, in written form [διὰ τὸ ἐξενεγκεῖν καὶ γράψασθαι],[41] the sphere, which was constructed from twelve pentagons. He acquired fame for making his discovery, but all discoveries were really from "that man" [as they called Pythagoras; they do not call him by name][42] . . . well, then, such are basically the characteristic differences between each philosophical system and its particular science.[43]

(Iamblichus, *On the General Mathematical Science* 25, 76.16–78.8)

This passage of Iamblichus, which is the central evidence for Aristotle's version of the factionalization of the Pythagorean brotherhood,[44] further supports my claim that what primarily distinguished the acousmatic and mathematical Pythagoreans was the object of their philosophical inquiry and treatment of that object (*pragmateia*). The passage can be divided into three sections: (A), which, while not obviously direct quotation, is nonetheless derived, in great part (if not wholly), from Aristotle's lost writings on the Pythagoreans; (B), which is apparently direct quotation from Aristotle; and (C), which is also likely to be derived from Aristotle.[45] In the section apparently quoted directly from one of Aristotle's lost works on the Pythagoreans (B), what distinguishes the acousmatic from the mathematical Pythagoreans is *type of knowledge*: the

41. For a more precise analysis of what this phrase means in context, see chapter 2, section entitled "Aristotle on Hippasus of Metapontum."

42. There is likely to be an interpolation here, which originally came from the *History of Arithmetic* of Eudemus of Rhodes. See Zhmud 2006: 187.

43. A very similar version found at Iambl. *VP* 87–89, 51.12–52.14, but—notwithstanding the confusion of acousmatic and mathematical, discussed in note 37, and the interpolation probably from Eudemus—there Iamblichus substitutes the "followers of Pythagoras" (τῶν ἀνδρῶν τῶν ἀκροωμένων) for "sciences" (τῶν μαθημάτων) and "we have ascertained" (παρειλήφαμεν) for "such are the characteristic differences" (τοιαῦτά ἐστι τὰ συμβεβηκότα). The presence of the Aristotelian term συμβεβηκότα in *DCM* probably indicates the more original text.

44. We can compare this account with that given by Iamblichus at *VP* 247, 132.18–21, whose provenance is unclear (possibly Nicomachus).

45. These divisions accord with the switch to indirect discourse and return to direct discourse. I will discuss (C) more extensively in chapters 2 and 3.

acousmatic Pythagoreans only have knowledge of the fact of "what one is to do" (τί δεῖ πράττειν), but the mathematical Pythagoreans, whose understanding is more advanced, have knowledge of the "reason why they are to do" (διὰ τὶ πρακτέον) what they should do.[46] This methodological distinction between "fact" (ὅτι) and "reason why" (διὰ τί) is originally Aristotelian, and it thus corroborates my suggestion that passage (B), and possibly the contingent passages (A) and (C), derive from Aristotle.[47] Indeed, the distinction between the "fact" (ὅτι) and the "reason why" (διότι) is central in Aristotle's controversial description of the knowledge of mathematicians in the *Posterior Analytics*:

> The reason why [τὸ διότι] is superior to[48] the fact [διαφέρει τοῦ ὅτι] in another way, in that each is studied by means of a different science. Such is the case with things that are related to one another in such a way that one is subordinate to the other, e.g. optics to geometry, mechanics to stereometry, harmonics to arithmetic, and star-gazing to astronomy. Some of these sciences bear almost the same name, e.g. mathematical and nautical astronomy are called "astronomy," and mathematical and acoustical harmonics are called "harmonics." In these cases it is for those who concern themselves with perception to have knowledge of the facts [τὸ ὅτι εἰδέναι], whereas it is for the mathematicians to have knowledge of the reason why [τὸ διότι εἰδέναι]. For the latter grasp[49] demonstrations of the causes [τῶν αἰτίων τὰς ἀποδείξεις], and they often do not know the facts [τὸ ὅτι], just as people who study the universal often do not know some of the particular instances *for lack of observing them*.[50] The objects of their study are the sort that, although they are something different in substance, make use of forms [κέχρηται τοῖς εἴδησιν]. For mathematics is concerned with forms; its objects are not said of a particular substrate.
>
> (ARISTOTLE, *Posterior Analytics* 1.13, 78b34–79a8)

46. The distinction is also identified by Iamblichus at *VP* 82, 47.4–10.

47. Some scholars (e.g. Zhmud 2012: 186 and, following him, Afonasin 2012: 31 n. 75) have speculated that this whole passage is chiefly derived from Nicomachus; but there is simply no evidence of Nicomachus adopting the Aristotelian differentiation between subordinate and superordinate sciences, which, I argue, underlies the differentiation of types of Pythagorean philosophical activity in this passage. Nor is there any extant evidence adduced by Zhmud to show that Nicomachus himself was concerned with the epistemic status of demonstration or proof.

48. Or, possibly, "differs from." But the language of subordination here suggests that Aristotle was using the common Greek idiom διαφέρει + genitive to mean "is superior to" or "excels." See *LSJ* s.v. διαφέρω 3.4.

49. Translating ἔχουσι literally, but the sense might be something like "able [to make]."

50. My italics.

This description of the so-called subordinate sciences develops a useful analogue for how acousmatic Pythagoreans differ from the mathematicians. The philosophy of the acousmatics, which is described by Iamblichus (*On the Pythagorean Way of Life* 82, 47.4–6) as consisting of "*acusmata* undemonstrated [ἀκούσματα ἀναπόδεικτα] and without argument [ἄνευ λόγου]," focuses on knowledge of "what" to do (ὅτι πρακτέον), not the reasons "why" to do it. By contrast, the mathematical Pythagoreans obtain the same characteristics as the mathematicians described in the *Posterior Analytics*, who have knowledge of the "reason why" and are able to grasp and produce "demonstrations" of the causes of the objects of their study. Aristotle's characterization of mathematicians as people who make use of demonstrations in their philosophical *pragmateia* parallels that of the mathematical Pythagoreans in the Aristotelian passage (B) quoted in *On the General Mathematical Science* 25, 77.4–18, although, importantly, there is no reference to Aristotle's peculiar understanding of mathematical "forms" or "substance" in Iamblichus's text. If the work quoted from was composed very early in Aristotle's career, before he undertook new approaches to ontology in the *Categories*, it would not be surprising that we do not hear about such problems. Be that as it may, my analysis of the passages that preserve some material from Aristotle's lost works on the Pythagoreans in Iamblichus's work *On the Pythagorean Way of Life* and *On the General Mathematical Science* reveals strong links to the differentiation of the two types of science in the *Posterior Analytics*, which leads to the supposition that Aristotle saw the main differentiating factor between the acousmatic and the mathematical Pythagoreans as demonstration.

ON THE "SO-CALLED" AND MATHEMATICAL PYTHAGOREANS

The establishment of sections (A) and (B) from Iamblichus's work *On the General Mathematical Science* 25, 76.16–77.18 as derived generally from Aristotle's writings on the Pythagoreans is very important for our understanding of mathematical Pythagoreanism, as Aristotle constructed it, because it corroborates a claim that has often been suggested but never explicitly argued for by scholars:[51] that the "so-called" Pythagoreans (οἱ καλούμενοι Πυθαγόρειοι) to whom Aristotle refers in *Metaphysics* A (1.5, 985b23 and 1.8, 989b29), *On the Heavens*

51. See Burkert 1972: 30 with nn. 8–9 and 51–52, who is followed by Huffman (1993: 31–35). Huffman's suggestion that others who might be "so-called" Pythagoreans would include Hippasus, Lysus, and Eurytus is plausible, although I doubt that those who proposed the theory of *sustoicheia* would be included. The most extensive analysis of this problem was undertaken by Timpanaro Cardini (1964: 6–19), but she concludes erroneously, I would argue, that there is no distinction between the various types of Pythagoreans named in Aristotle's *Metaphysics*.

(2.1, 284b7 and 2.13, 293a20–21), and *Meteorology* (1.6, 342b30 and 1.8, 345a14) are, indeed, one and the same with the mathematical Pythagoreans described in the lost works on the Pythagoreans.[52] Given my new approach to thinking about the *pragmateia* of the mathematical Pythagoreans, that is, the object of their philosophical investigations and their particular treatment of that object, it is worth considering whether there might be parallels to draw with the "so-called" Pythagoreans in those texts.

Let us examine a famous passage from the first book of Aristotle's *Metaphysics*, which one might assume (with Jaeger, Ross, and Owens)[53] to have been written rather early in Aristotle's career, when he was still under Academic influence:

> The "so-called" Pythagoreans employ first principles and elements [*ταῖς ἀρχαῖς καὶ στοιχείοις χρῶνται*] more abstrusely[54] than some of the physicists. The reason is that they took their first principles from non-perceptible objects: for the objects of mathematics [*τὰ μαθηματικὰ τῶν ὄντων*],[55] apart from those that concern astronomy, belong to the class of things lacking in motion. And yet they discuss and wholly make the object of their philosophical inquiry [*πραγματεύονται*] nature. For they generate heaven, and they observe what happens concerning its parts, attributes, and functions, and they lavish these things with first principles and causes, and as such they are in agreement with the other natural scientists that what actually exists is what is perceived and that "so-called" heaven contains it. But, as we mentioned, the causes and the first principles, which they say are sufficient to rise up above the horizon [*ἐπαναβῆναι*][56] to the higher parts

52. I will deal primarily with the passages in *Metaphysics* A, for the sake of their strong connections with the fragments of Aristotle's lost works on the Pythagoreans. It should be noted that the term "so-called" is not particularly innovative for Aristotle, given that skepticism concerning people who called themselves after Pythagoras can be detected in the writings of Isocrates and Antisthenes. See chapter 2.

53. See Owens 1951: 85–89; Jaeger 1948: 171–176; Ross 1924, vol. 1: xv.

54. *ἐκτοπωτέρως*, following Asclepius's commentary (*in Metaph.* p. 65.29–35 Kroll) and the most recent edition of Primavesi 2012.

55. As I translate this very tricky phrase. Literally, it means something closer to "the mathematicals among the things," which coordinates in potentially interesting ways with Philolaus's (F 6 Huffman) phrase "the being of things" (*ἁ ἐστὼ τῶν πραγμάτων*). See below in the section entitled "Mathematical Pythagoreanism and the 'Objects of Mathematics.'"

56. This translation is preferable to Tredennick's "capable of application to the remoter class of realities" or Ross's "sufficient to act as steps even up to the higher realms of reality," neither of which accounts for the technical language of astronomy reported here. In a passage of the

of reality, are better suited even for these than for arguments concerning nature. Nevertheless, they say nothing about how there will be motion, if the only things premised are limit and limitless, and odd and even, nor about how there can be generation and destruction, nor the activities of objects that move through the heavens, without motion and change.

Further, if someone were to grant to them that spatial magnitude derives from these things, or if this were to have been demonstrated by them [δειχθείη τοῦτο], still how will some bodies be light and others heavy? For, given what they assume and maintain, they are speaking no more about mathematical bodies than about perceptible bodies. Hence they have said nothing whatsoever about fire or earth or any other bodies of this sort, since, in my opinion, nothing they say is peculiar to perceptible bodies.

Moreover, how is one to understand that both the attributes of number and number itself are the causes of things that exist and come to be throughout the heavens—both from the beginning and now—and that there is no other number than this number out of which the cosmos is composed? For, whenever they place opinion and opportunity in such and such a region, and injustice and separation or mixture a bit higher or lower, and they make a demonstration on the grounds that [ἀπόδειξιν λέγωσιν ὅτι] each of these is a number—but there already happens to be a plurality of magnitudes composed [of numbers] in that place, because the attributes correspond to each of these places—is, then, the number in heaven, which one is supposed to understand as each [of these abstractions], the same [as the one in the lower region], or is it a different number?

(ARISTOTLE, *Metaphysics*, 1.8, 989b29–990a29)

Obviously, there is a great deal to unpack in this extended discussion of the "so-called" Pythagoreans and their relationship to the mathematical Pythagoreans. I would like to highlight just a few aspects of Aristotle's argument that are relevant to this analysis.

Aristotle seeks a technical language in order to respond to what he takes as the fundamental aspects of the "so-called" Pythagorean philosophical system. He points out category confusion in "so-called" Pythagorean philosophy: while their first principles are all derived from the mathematical (i.e. non-perceptible)

Meteorology (1.6, 342b30–35), Aristotle describes how the "some" of "so-called" Pythagoreans believe that Mercury is, like comets, one of the Planets that "does not rise far above the horizon" (τὸ μικρὸν ἐπαναβαίνειν), and therefore its appearances are invisible, as it is seen in long intervals.

sciences, the objects of their philosophical inquiry (i.e. their *pragmateia*) are things that have been generated and possess motion, namely phenomena. This would be unsurprising, especially if the "so-called" Pythagoreans made the object of their investigations the motions of the heavens, the superior science of which would be, according to the *Posterior Analytics*, stereometry. But the example he gives involves number: how can number, which is a non-perceptible entity, both (1) be superordinate (i.e. a "first principle") and reside in the highest part of reality and (2) be identical with something in a lower substrate, like opinion or opportunity? In accordance with Aristotle's establishment of the proper objects to the various sciences in the *Posterior Analytics*, this "so-called" Pythagorean approach represents a confounding of the sciences that deal with the "reason why" (τὸ διότι) and the sciences that deal with the "fact" (τὸ ὅτι). That is to say, it leads to confusion about what kind of science the "so-called" Pythagoreans practice, since they employ the principles of theoretical mathematics in order to explain natural phenomena.

Regardless of Aristotle's criticisms of the "so-called" Pythagoreans, we can see that they were thought to have undertaken demonstrations of some sort, which suggests to us that they are the same as the mathematical Pythagoreans Aristotle described in his lost works on the Pythagoreans. If we are to trust Aristotle's evidence here, then the mathematical Pythagoreans described in *Metaphysics* A may have provided at least two types of demonstrations: (1) that all entities that are derived from number are themselves numbers, on the grounds that all entities possess the attributes of number, and, possibly, (2) that spatial magnitude is derived from their first principles, namely the objects of mathematics.[57] Other demonstrations ascribed to the "so-called" Pythagoreans in Aristotle's works are suggestive, if incomplete, evidence for the explanatory methods employed by these philosophers.[58] In this way, they are distinguished from the acousmatic Pythagoreans, whose philosophical *pragmateia* is said to have been focused uniquely on the "fact" (τὸ ὅτι), that is, that which is particular, mutable, perceptible, and known through experience alone.[59] However,

57. That Aristotle mentions the proof concerning magnitude suggests it is possible that someone could have (or did) try to demonstrate this.

58. Among those that I will not discuss further: at *Cael.* 2.2, 284b6–8, they have a *logos*—it is unclear how it is demonstrated—that argues that there is a right and left side in heaven; at *Mete.* 1.8, 345a14–19, we hear of two kinds of arguments attributed to "so-called" Pythagoreans: the first is mythological (the Milky Way is a path on the grounds that it is the path of one of the stars that fell at the time of the fall of Phaethon), and the second is based on stereometric speculation and natural science (the sun, which was once borne through the circle that is the Milky Way, created a path when it moved out of this orbit by scorching the region).

59. See McKirahan 1992: 242 and Johnson 2009: 336; τὰ φαινόμενα, however, includes not only perceptible objects such as the heavenly bodies but also λεγόμενα and ἔνδοξα. See below.

Aristotle's focus on the role of mathematics in the *pragmateia* of the "so-called" Pythagoreans raises an important question: if the *pragmateia* of the mathematical Pythagoreans involves application of mathematical principles to the objects of nature, how is this system distinguished from the *pragmateia* of the acousmatic Pythagoreans (if, indeed, that system is to be found in the *Metaphysics* at all)?

We can approach this problem by investigating Aristotle's descriptions of the first principles ascribed to the "so-called" Pythagoreans. In an earlier passage of *Metaphysics* A, where the "so-called" Pythagoreans appear for the first time in the text, we get a more precise description of what Aristotle took their principles to be:

> In the time of these men [i.e. Leucippus and Democritus] and before them[60] the "so-called" Pythagoreans were the first to latch onto mathematics. They advanced mathematics and, by being brought up in it, they began to believe that the principles of mathematics [ἀρχὰς τούτων [τῶν μαθημάτων]] were the principles of all things in existence [ἀρχὰς τῶν ὄντων πάντων]. And since numbers are first among these [i.e. beings][61] by nature, they seemed to see many resemblances [ὁμοιώματα] in numbers to things that are and things that come into being, rather than in fire and earth and water. For example: *this* attribute of numbers was justice, *that* was soul and mind, and another opportunity, and all the rest, so to speak, in the same way. Moreover, because they saw that the attributes and ratios of musical scales consisted in numbers—well, since other things seemed to be modeled [ἀφωμοιῶσθαι] on numbers in their nature in its entirety, and numbers seemed to be primary of all nature, they began to assume that the elements of numbers were the elements of things that are and that the whole of heaven was musical scale [ἁρμονία] and number [ἀριθμός]. Whatever resemblances to the attributes and regions of heaven and the entire order of the cosmos they were able to show [δεικνύναι] to be in numbers and musical scales, they collected [συνάγοντες] and fitted

60. Or, as Schofield (2012: 144) translates: "Among these thinkers." It is true that ἐν τούτοις could mean "among them," but it is difficult to square this with the temporal sense of πρὸ τούτων that follows. Alexander of Aphrodisias (*in Metaph*. p. 37.6–16 Hayduck) felt the need to explain this phrase as well, and he glosses: "Concerning the Pythagoreans, he says that some were born before Democritus and Leucippus, and some lived about the time of them (κατὰ τούτους)." He further explains that Pythagoras himself lived "a bit" (ὀλίγον) before Democritus and Leucippus, but that "many of the Pythagorean students (ἀκουσάντων) flourished at the same time as them."

61. See Schofield 2012: 144, with n. 8.

> them together [ἐφήρμοττον]. And if something were to be missing anywhere, they hastened to supplement it [προσεγλίχοντο][62] in order that their entire *pragmateia* might hang together. For example, since the number 10 is thought to be perfect and to encompass the entirety of numbers in their nature, they assert that there are ten things in heavenly orbit; but since there are only nine that are actually visible throughout the heavens, they invent a tenth, the Counter-Earth.
>
> We have treated this subject in greater detail elsewhere.[63] But the object of our discussion is to learn from them, too, what principles they posit, and how these correspond to the causes we have discussed. Well, then, evidently these men too believe that number is a principle, both in terms of matter for things that are and in terms of their attributes and states. And they take the elements of number to be the odd and the even, and, of these, the former to be limited [πεπερασμένον], and the latter unlimited [ἄπειρον]; the one is constituted of both of these (since it is both odd and even); number is derived from the one; and, as we've already said, the whole heaven is numbers.
>
> (ARISTOTLE, *Metaphysics*, 1.5, 985b23–986a21; translation after Schofield 2012)

This passage gives us a sense of what Aristotle thought the philosophical method and the first principles of the "so-called" Pythagoreans to be. In particular, it identifies what Aristotle understood to be a problem in their attempts to provide demonstrations.[64] According to Aristotle, the "so-called" Pythagoreans contaminate their understanding of the sensible facts by "hastening to supplement" whatever might be lacking in the empirical data with theoretical knowledge assumed by the "reason why"; within context of the classification of types of knowledge discussed earlier in *Metaphysics* A, it is not surprising that this

62. Schofield (2012: 144) has "they bent their efforts." This uncommon word appears at *Metaph.* 14.3, 1090b31, where Aristotle likewise complains that some Platonists (Xenocrates or Speusippus? See Annas 1976: 209–210) hasten to apply mathematics to the Forms (προσγλιχόμενοι ταῖς ἰδέαις τὰ μαθηματικά). At *Cael.* 2.13, 293a27, Aristotle also accuses some of the "so-called" Pythagoreans of "attracting the data to certain rationales and opinions of their own [πρός τινας λόγους καὶ δόξας αὑτων τὰ φαινόμενα προσέλκοντες]."

63. It has been discussed at *Cael.* 2.13, 293a18–b15, an extremely challenging passage that has presented many difficulties for scholars; it also appeared in Aristotle's lost writings on the Pythagoreans, especially in F 203 Ross, preserved by Alexander. For my analysis of this passage from *De Caelo*, see chapter 2, section entitled "Pythagoreanism and the Axiology of What Is 'Honorable.'"

64. See Schofield (2012: 153), who also notes that Aristotle does seem to ascribe to the Pythagoreans a logic.

characterization of "so-called" Pythagorean philosophical method obtains.[65] To put it simply, according to Aristotle, the "so-called" Pythagoreans adapt the immediate facts (τὸ ὅτι) to fit the explanation (τὸ διότι) in an ad hoc manner. The example given involves the bodies of the heavens: nine bodies can be perceived by the senses, but, since the Pythagoreans assume the number 10 as the perfect number, and since all things are number, there must be ten heavenly bodies. This example also reveals Aristotle's second substantial criticism of the philosophical system of the "so-called" Pythagoreans: they hastily and carelessly compare things in order to secure relationships between their first principles and observed phenomena. Such an activity leads the Pythagoreans, in Aristotle's estimation, to leave out the efficient and final causes. The "so-called" Pythagoreans' a priori philosophical methodology, which flies in the face of observation, is further described as a "fitting together" (ἐφήρμοττον), a word whose semantics are related both to investigation of the heavens elsewhere in Aristotle and to the concept of "musical scale" (ἁρμονία) more generally in Greek culture.[66] One might therefore hear an echo of a Pythagorean concept. Indeed, in Philolaus's work *On Nature* (F 7 Huffman = Stob. *Ecl.* 1.21.8), he claims that "the first thing fitted together [τὸ πρᾶτον ἁρμοσθέν], the one in the center of the sphere, is called the hearth."[67] And as Carl Huffman has shown, Philolaus is to be included among the "so-called" Pythagoreans, and, moreover, that Aristotle had Philolaus chiefly in mind when criticizing their methodology.[68]

In pursuit of the metaphysics of Philolaus and the other "so-called" Pythagoreans, Aristotle returns to the interrogation of what is primary in their *pragmateia*. He claims that the "so-called" Pythagoreans posit the objects of mathematics as the first principles and elements of everything. The chief example he gives is number, which is apparently primary in two senses: as formal and material

65. Compare Arist. *Metaph.* 1.1, 981a11–32, on which see McKirahan 1992: 242.

66. See Arist. *MA* 1, 698a10–14, where ἐφαρμόττειν is translated by Nussbaum as "be in harmony." Lennox (2001a: 10 n. 23) would translate in a more decidedly methodological way as "apply" and notes that "Aristotle likely has in mind the application of universal accounts *via* proof to the particulars, since it was in order to understand them that the search for the universal began."

67. Luca Castagnoli points out to me that the language is the same, but that Aristotle's criticism—which emphasizes how a priori principles and phenomena are "fitted together" harmonically—diverges from the actual meaning of Philolaus's fragment.

68. Huffman 1993: 225, with reference to this passage, but without a sufficient discussion of Aristotle's description of "so-called" Pythagorean methodology. For a very good general analysis of the relationship between Philolaus's fragments and this passage, see Huffman 1993: 177–193.

cause.[69] Aristotle's concern with form and matter is reinforced by the description of those things that are primary as either first principles (ἀρχαί) or elements (στοιχεία). They are first principles with regard to their function as formal causes, but they are elements with regard to their function as the "that out of which" things in the universe are constructed.[70] These Aristotle defines according to a hierarchy based on priority: the odd and even are apparently prior to the one, since the one is constructed out of them, and the one is prior to number, since numbers are constituted from the one. And since all perceptibles have number as a property, they somehow derive from number itself.

This passage, however, stimulates us to consider whether another set of principles are ontologically prior to the odd and the even: limit and unlimited. After all, Aristotle thinks that the attributes of limit are present in the odd (i.e. it "has been limited" [πεπερασμένον]), whereas the even itself is unlimited (ἄπειρον).[71] A hierarchy of entities, even among first principles, is implicit in this passage, and the "so-called" Pythagoreans might be thought to attempt to provide explanations by means of demonstration that limit and unlimited are prior to odd and even. Apparently, the hierarchy is given its order on the assumption that the prior principles must act on those things given definition by them. In the case of limit and unlimited, we have good evidence from the genuine fragments of Philolaus of Croton that Aristotle, even if he was distorting Philolaus's thought, was essentially presenting a verifiable account of how some of the more sophisticated Pythagoreans undertook demonstration by employing mathematical objects as principles in demonstrations that involved perceptible objects.

One example of this approach to demonstration is a particularly difficult fragment of Philolaus preserved by Stobaeus:

69. See Zhmud 2012: 436–437. For an excellent exposition on number as material cause here, see Schofield 2012: 145–147.

70. The subject of the relationship between principles and elements is well-trodden ground, and I don't wish to pursue this question too far. It does not appear to me that, in *Metaphysics* A, Aristotle distinguishes explicitly between these, or that he has discovered a clear means of distinguishing them, but that does not mean that they are simply synonymous either. When he composed book Λ (12.6, 1071b22–26), he distinguished between first principles as external and elements as inherent. It is suggestive in book Δ (5.1, 1013a7–10) that he defines ἀρχή in several ways, among them (1) the thing as a result of whose immanent presence something first comes into being, and (2) that from which something comes into being, although it is not present in it. When defining στοιχεῖον in the same book (5.3, 1014a26–30), he clarifies that it is an immanent, indivisible entity out of which other things are composed and draws reference to the "elements of sound." Was Aristotle referring to Pythagorean mechanical attempts to obtain the basic elements of the concords here?

71. This is confirmed at *Metaph.* 1.8, 990a8–12.

> It is necessary [ἀνάγκα] that the things that are be all either limiting [περαίνοντα], or unlimited [ἄπειρα], or both limiting and unlimited, but not in every case unlimited alone. Well then, since [ἐπεὶ] it is apparent [φαίνεται] that they are neither from limiting things alone, nor from unlimited things alone, it is clear, then [τἆρα], that the cosmos and the things in it were fitted together [συναρμόχθη][72] from both limiting and unlimited things. Things in their activities also make this clear [δηλοῖ δὲ καὶ τὰ ἐν τοῖς ἔργοις]. For, some of them from limiting [constituents] limit, others from both limiting and unlimited [constituents] both limit and do not limit, others from unlimited [constituents] will appear to be unlimited.
>
> (F 2 Huffman = Stobaeus, *Eclogae* 1.21.7a; translation after Huffman 1993)

Careful examination of this fragment demonstrates that Aristotle's criticism is not off the mark: Philolaus undertakes some sort of demonstration by reducing perceptibles to the objects of mathematics. This is already suggested in the first few lines and is quite explicit in the statement that one can see that limiters and unlimited things constitute the cosmos when one detects them in "activities" (τὰ ἔργα), a word that seems to refer to the attributes that we can perceive.[73] It is also implicit in the language used to discuss how things could be the way they are, as Philolaus uses particles (e.g. ἐπεί and τἆρα) that suggest modal relations and appeals to philosophical concepts used commonly in Aristotelian demonstrations, such as necessity (ἀνάγκα).

If I am correct in thinking that the "so-called" Pythagoreans as described in Aristotle's *Metaphysics* A are one and the same as the mathematical Pythagoreans in his lost writings on the Pythagoreans, then we should expect to find a description of the *pragmateia* of the acousmatic Pythagoreans, who did not engage in demonstrations of some sort. Indeed, in a passage that immediately follows on the long passage (*Metaph.* 1.5, 985b24–986a21), our hypotheses are corroborated: we get a very concise description of what appears to be the *pragmateia* of the acousmatic Pythagoreans:

72. The appearance of this term, along with other terms related to ἁρμόζω, in other fragments of Philolaus (e.g. F 1 Huffman = D.L. 8.85: ἁ φύσις δ' ἐν τῷ κόσμῳ ἁρμόχθη ἐξ ἀπείρων τε καὶ περαινόντων; F 7 Huffman = Stob. *Ecl.* 1.15.7: τὸ πρᾶτον ἁρμοσθέν ἑστία καλεῖται) is suggestive evidence for the correlation between Aristotle's description of the demonstration of the "so-called" Pythagoreans, which involves "fitting together in addition" (ἐφήρμοττον) the "resemblances to the attributes and regions of heaven and the entire order of the cosmos."

73. On translating this difficult word, see also the account of Huffman 1993: 111–112.

People other than these very people [i.e. the "so-called" Pythagoreans][74] say [λέγουσιν] that there are ten principles, which they name in two elementary columns of cognates [τὰς κατὰ συστοιχίαν λεγομένας]:

Limit	Unlimited
Odd	Even
One	Plurality
Right	Left
Male	Female
Rest	Motion
Straight	Curved
Light	Darkness
Good	Evil
Square	Oblong . . .[75]

The Pythagoreans declared how many and what sorts [πόσαι καὶ τίνες] of contraries there were. Thus, from both of these authorities [i.e. Alcmaeon of Croton and the Pythagoreans] we can gather this much, that the contraries are the first principles of things in existence; but how many and what sorts these are [we can gather] from [only] one of these authorities [i.e. from the Pythagoreans]. Nevertheless, *how* [πῶς] these principles can be brought together [συνάγειν] and referred to our aforementioned list of causes has not been clearly articulated [σαφῶς οὐ διήρθρωται] by them, but they seem to arrange [ἐοίκασι τάττειν] the elements under the grouping of matter; for they say that substance is composed and fashioned out of these underlying elements.

(ARISTOTLE, *Metaphysics* 1.5, 986a22–b8)

Aristotle, I suggest, seems to distinguish these "Pythagoreans" (as well as Alcmaeon of Croton) from the "so-called" Pythagoreans by appeal to their respective treatments of the first principles. The scientific pursuit of *these* "Pythagoreans" only goes so far as to (1) postulate the number and types of contraries, and (2) put them in an order. They put their principles in an order based on contrariness, and with no further attention to definition, nor any

74. Schofield (2012: 155–157) identifies this group as the "*sustoichia* theorists" and sees them as differentiated from the previously described group by the fact that "he ends up finding it difficult to ascribe any significant contribution from Alcmaeon and the *sustoichia* theorists to his current project." I suggest that the main reason for this is the fact that they do not obviously contribute to a science of demonstration.

75. Following Huffman (1993: 10–11), I have excised anything that deals explicitly and solely with Alcmaeon of Croton, whose status as a "Pythagorean" is questionable.

attempt to provide a demonstration for this organization. The activity of "arrangement" (τάξις), of course, occupies a significant role in Aristotle's philosophy, and, as he says at *Topics* 8.1, 155b9–10 (of the arrangement of questions), it is the activity peculiar to the dialectician, whose practice as such is contrasted against that of the philosopher, who engages in "demonstration" (ἀπόδειξις).[76] Such an arrangement cannot be considered a type of demonstration, nor is there any evidence of these "Pythagoreans" offering an explanatory "reason why" (τὸ διότι) the elements of their so-called Table of Contraries are arranged the way they are.[77] In the case of these "Pythagoreans," there is no attempt to show, for instance, *how* or *why* the limiter limits things in existence, or to provide an explanation for the systematization that is given.[78] To put it another way, the *pragmateia* associated with these "Pythagoreans" does not help a student to "grasp the demonstration as a demonstration, coming to see its premises *as* the causes and explanations of its conclusion."[79] From Aristotle's point of view, the "Table of Contraries" constitutes the sort of "perceptible" that falls under the umbrella term τὰ λεγόμενα: the "Table of Contraries" appears to function (for Aristotle's purposes) as data derived from observation (φαινόμενα).[80] It does not seem that Aristotle believes that the information given in the "Table of Contraries" listed here could be used as premises to generate demonstrations, even if it still has some residual value for Aristotle's own inquiry—otherwise it simply wouldn't be included. With regard to Aristotle's project in *Metaphysics* A, the Table itself functions as a sort of φαινόμενον in two ways: first, to the Pythagoreans who espouse it, it functions as a type of λεγόμενον, namely, what is passed down orally from Pythagorean teacher to student, an ipse dixit injunction like the *acusmata*. In this sense, the "Table of Contraries" does not represent anything other than the empirically derived "facts" that are immediate and familiar, at least for these "Pythagoreans." Second, for Aristotle himself, the Table and its

76. Arist. *Top*. 1.1, 100a25–31. This subject is, of course, a contentious point among scholars. But for my purposes, it serves only to exhibit Aristotle's attempt to distinguish two types of reasoning: that which proceeds by appeal to demonstration and that which proceeds by appeal to ordering.

77. For a useful treatment of "nondemonstrative" science as that which allows premises to multiply infinitely, see Smith 2009: 54.

78. See Huffman 1993: 47 n. 1.

79. As eloquently put by Robin Smith (1997: xvii).

80. I am adapting the famous argument of G. E. L. Owen (1986: 242–243) to include the opinions of previous philosophers in *Metaphysics* A as the sorts of ἔνδοξα or λεγόμενα that could be construed as φαινόμενα.

contents function as an ἔνδοξον, a reputable opinion that Aristotle is able to employ in the course of his own predemonstrative inquiry (ἱστορία).[81] Given Aristotle's lack of attribution of any sort of reasoning that involves demonstrations to these "Pythagoreans" and the implication that their "Table of Contraries" is to be considered a φαινόμενον, we can speculate with some reason that Aristotle considered these "Pythagoreans" to be the same as the acousmatic Pythagoreans discussed in his lost writings on the Pythagoreans.

There remains a third and final passage in *Metaphysics* A that refers to Pythagoreans of one or the other sort. Initially, it might seem unclear to which group Aristotle is referring. This text, I suggest, is also crucial to our understanding of Pythagoreanism, as reconstructed and appropriated by Aristotle, because it illuminates another way Pythagoreans engaged in their *pragmateia*, that is, through definitions:

> But while the Pythagoreans have claimed in the same way that there are two principles, they made this addition, which is peculiar to them, namely that they thought that the limited and the unlimited were not uniquely different substances[82], such as fire and earth and anything else of this sort, but that the unlimited itself and the one itself were the substance of the things of which they are predicated, and hence [διό] that number was the substance of all things. Concerning these issues, then, they expressed themselves in this way. And concerning essence [περὶ τοῦ τί ἐστιν], they began to make statements and definitions [λέγειν καὶ ὁρίζεσθαι], but their treatment was too simple [λίαν ἁπλῶς ἐπραγματεύθησαν]. For they both defined superficially and thought that the substance of the thing [ἡ οὐσία τοῦ πράγματος] was that to which a stated term would first be predicable, e.g. as if someone were to believe that "double" and "two" were the same because "two" is the first thing of which "double" is predicable. But surely to be "double" and to be "two" are not the same things. If that were to be the case, one thing would be many—a consequence that they actually drew.
>
> (ARISTOTLE, *Metaphysics* 1.5, 987a13–27)

While it is true that Aristotle refers to this group as "the Pythagoreans," and not the "so-called" Pythagoreans, it is nevertheless probable that this is a description

81. On the role of the "predemonstrative inquiry" in Aristotle's scientific works, see Lennox 2001b: 40–46.

82. Taking φύσεις in the sense later defined by Aristotle in book Δ (5.6, 1014b35–37) and only because it makes sense of Aristotle's use of the term "substance" (οὐσία) in the next sentence.

of a mathematical Pythagorean *pragmateia*.[83] This group of Pythagoreans is not simply listing first principles as contraries and assuming them as elemental to all things in existence. According to Aristotle, this group of Pythagoreans "began to make statements and definitions" and engaged in a primitive analysis concerning the essence (lit. the "what is" [περὶ τοῦ τί ἐστιν]), although their "treatment" (ἐπραγματεύθησαν, i.e., the application of their methods to the *pragmata* of their inquiry) was too simple. There also appears to be some preservation of an argumentative technique: these Pythagoreans thought that, since the "unlimited" and the "one" are the substance (οὐσία) of the things of which they are predicated, *therefore* number is the substance of all things. These are quite important innovations in philosophy for Aristotle, by contrast with the monists and pluralists, whose philosophy sought to describe the world without providing definitions by appeal to metaphysics and logic.[84]

The accumulation of evidence concerning the *pragmateia* of the mathematical Pythagoreans from *Metaphysics* A corroborates and further expands my two hypotheses, namely (1) that Iamblichus in his work *On the General Mathematical Science* 25 has excerpted a section from Aristotle's lost works on the Pythagoreans that accounts for the different *pragmateiai* of the mathematical and the acousmatic Pythagoreans, and (2) that those mathematical Pythagoreans described by Aristotle in his lost works on the Pythagoreans are the same as the "so-called" Pythagoreans of *Metaphysics* A and elsewhere in his texts.

MATHEMATICAL PYTHAGOREANISM AND THE "OBJECTS OF MATHEMATICS"

Given the detailed account above of the ways Aristotle distinguishes the *pragmateia* of the "so-called" Pythagoreans in *Metaphysics* A, we can now come back to Aristotle's account as preserved by Iamblichus in his work *On the*

83. There could be a very good reason for this. As Cherniss (1944: 192, with n. 112) suggests, this passage appears to have been inserted later by Aristotle. If, as I think, Aristotle only distinguished between the "so-called" (i.e. mathematical) Pythagoreans and the "Pythagoreans" (i.e. acousmatic) in his earlier treatments of the history of philosophy, which would include the crucial passage (1.5, 985b24–986b8) that demonstrates the differences, and if later on he only concerned himself with the philosophy of the mathematical Pythagoreans, then it would be unsurprising for him to refer to the mathematical Pythagoreans here as "Pythagoreans" *simpliciter*.

84. For a comprehensive analysis of this passage, now see Schofield 2012: 161–165. I follow Schofield in believing that Aristotle probably has Philolaus's F 6 directly in mind, but I also note the significance (again) of the term ἡ οὐσία τοῦ πράγματος, which may have a resemblance to Philolaus's ἁ ἐστὼ τῶν πραγμάτων. I will discuss the mathematical Pythagorean responses to predication further in chapters 4 and 5.

General Mathematical Science 25. Another revealing passage, also thought by Walter Burkert, Carl Huffman, Myles Burnyeat, and Oliver Primavesi to have been derived from Aristotle,[85] continues from the excerpt I have discussed:[86]

> (D) The Pythagoreans devoted themselves to mathematics. They both admired the accuracy of its arguments, because it alone among things that humans practice contains demonstrations [*εἶχεν ἀποδείξεις ὧν μετεχειρίζοντο*], and they saw that general agreement is given in equal measure to theorems concerning attunement, because they are [established] through numbers, and to mathematical studies that deal with vision, because they are [established] through diagrams. This led them to think that these things and their principles are quite generally the causes of existing things. Consequently, these are the sorts of things to which anyone who wishes to comprehend things in existence—how they are—should turn their attention, namely numbers and geometrical forms of existing things and proportions, because everything is made clear [*δηλοῦσθαι*] through them. So, then, by attaching the powers of each thing to the causes and primaries—only things that were less opportune or less honorable than them—they defined other things, too, in nearly the same manner. (E) Therefore, their education in numbers and the objects of mathematics [*τὰ μαθήματα τῶν πραγμάτων*] seemed to come through these subjects and in this general sketch. Such was also the method of demonstrations [*ἡ μέθοδος τῶν ἀποδείξεων*] among them, which both arose out of such principles and thereby attained fidelity and security in their arguments.
>
> (IAMBLICHUS, *On the General Mathematical Science* 25, 78.8–26)

The information preserved by Iamblichus in section (D) suggests that Aristotle, in his lost works on the Pythagoreans, continued to refer to the mathematical

85. See Burkert 1972: 50 n. 112, followed by Primavesi (2012: 251–252). Burkert (447–448), however, claims that Iamblichus or someone else has made spurious insertions in two places: "and to mathematical studies that deal with vision, because they are [established] through diagrams" and "and geometrical forms of existing things." Important correctives have been offered by Burnyeat (2005a: 38–43), who appeals to Arist. *APo*. 1.13, 79a7–8 in arguing that nothing should be excised here. This is in keeping with the stylistic traits of Iamblichus when he quotes from Aristotle, as recently analyzed by D. S. Hutchinson and Monte Ransome Johnson: he tends to preserve large blocks of material without modifying them (2005: 281–282). Zhmud (2007: 84–95) speculates without extensive direct evidence that Nicomachus is the source here; but even he admits that Nicomachus has nothing to say about demonstration, which is a central topic throughout *DCM* 25.

86. See above in the section entitled "Aristotle on the *Pragmateiai* of the Pythagoreans."

Pythagorean *pragmateia*. Indeed, there is good reason to believe, with Burnyeat, that Aristotle is criticizing specifically the activities of the mathematical Pythagorean Archytas of Tarentum here, although we should not assume that Archytas's philosophy is the only object of Aristotle's criticism.[87] It is also striking that, in section (E), Iamblichus refers to the educational curriculum of the mathematical Pythagoreans as dealing with the "objects of mathematics" (*τὰ μαθήματα τῶν πραγμάτων*), a peculiar phrase that is unattested anywhere else in Iamblichus's oeuvre, or, for that matter, in what remains of Greek philosophy or mathematics. We do, however, see something very close to it in Aristotle's description in *Metaphysics* A of the *pragmateia* of Plato,[88] which is considered a successor to the philosophical *pragmateia* of the "Italians,"[89] although with some modifications:

> Therefore, Plato named these other sorts of entities "Ideas," and he [said that] perceptibles are all called after them and in accordance with them. For the many things that bear the same name as the forms exist by virtue of participation [*κατὰ μέθεξιν*] in them.[90] With regard to participation, he changed the name only: for whereas the Pythagoreans claim that objects in existence exist by way of imitation of numbers [*μιμήσει τῶν ἀριθμῶν*], Plato says by way of participation [*μεθέξει*], modifying the name. As to what participation or imitation is, however, they left it to us to seek it out together.
>
> Furthermore, Plato claims that in addition to perceptibles and Forms is a middle type of entity, the objects of mathematics [*τὰ μαθηματικὰ τῶν πραγμάτων*], which differ from perceptibles in being eternal and immutable, and from Forms in that many [objects of mathematics] are similar, whereas each Form itself is unique.
>
> (Aristotle, *Metaphysics* 1.6, 987b7–18)

87. Burnyeat 2005a. Contra Huffman (2005: 568), who thinks that Aristotle could not have referred to Archytas as a "Pythagorean." Given the explicit reference to the use of visual diagrams, we should also consider admitting figures like Eurytus of Tarentum, whose approach to definition of objects by means of pebble arithmetic was known to Aristotle. On Eurytus, see chapter 4, section entitled "Growing and Being: Mathematical Pythagorean Philosophy before Plato."

88. Identified explicitly as such at *Metaph.* 1.6, 987a30.

89. Arist. *Metaph.* 1.6, 987a29–31. Aristotle there draws comparisons with Plato and the Italians, although he more generally states that Plato succeeded the "aforementioned philosophies" (*μετὰ τὰς εἰρημένας φιλοσοφίας*). He is somewhat unclear here, but in regard to the inheritance of modes of definition given at *Metaph.* 13.4.3, 1078b17–23, Aristotle explicitly lists those who influenced Plato's inquiry into essence as Socrates, Democritus, and the "earlier" (*πρότερον*) Pythagoreans.

90. This is a notoriously difficult passage. I have adopted the text of Ross.

This is one of the more problematic passages in the history of ancient philosophy, and the task to identify with precision the objects of mathematics, as intermediaries between Plato's Forms and perceptibles, is not made easier by Aristotle's admitted confusion.[91] Part of the problem here is that the term *τὰ μαθηματικὰ τῶν πραγμάτων*, like the explicit ascription of a theory of imitation (*μίμησις*) *in numbers* to the Pythagoreans, is an Aristotelian construction that cannot be found anywhere in ancient philosophy outside Aristotle and his immediate associates.[92] It is not clear from this passage whether Aristotle would consider *τὰ μαθηματικὰ τῶν πραγμάτων* to be distinguished from other terms he uses to describe the objects of mathematics, especially the relatively common simple formulation *τὰ μαθηματικά*, which he uses often in reference to the ontological theories of Plato, Speusippus, and Xenocrates.[93] We have seen, of course, that Aristotle mentions the "objects of the mathematics" (*τὰ μαθηματικὰ τῶν ὄντων*) by reference to the first principles of the "so-called" Pythagoreans (*Metaph.* 1.8, 989b32). Generally, Aristotle does not seem to distinguish between *τὰ πράγματα* and *τὰ ὄντα* in referring to the "things"

91. See Ross's useful discussion of the problems that arise from this passage and for a history of their treatment from antiquity to the early twentieth century (1924: 161–168). Jaeger concerned himself with the principle of intermediary, without focusing on the objects of mathematics (1948: 91, with n. 2). Cherniss (1944: 75–78) denied that the objects of mathematics as intermediates existed for Plato and considered the ascription of this by Aristotle possibly to have been a misunderstanding of a passage from Plato's *Republic* (551a–e). Tarán (1981: 23 n. 120) followed Cherniss but saw the ascription of intermediary objects of mathematics to Plato as a point of contrast to Speusippus's postulation of separate and unchangeable mathematical numbers/ideas and magnitudes. Burkert (1972: 43–45) plausibly connects "imitation" to Aristotle's descriptions of "resemblances" (*ὁμοιώματα*) at *Metaph.* 1.5, 985b27 and concludes that "again and again it becomes clear that the Pythagorean doctrine cannot be expressed in Aristotle's terminology." Denyer (2007: 302–304) has argued in favor of the presence of intermediate mathematicals in Plato's epistemology but without reference to Pythagoreanism. Most recently, Steel (2012: 183) has aptly noted: "if some (as Cherniss) may complain about an excessive Pythagorising of Plato, one can as well point to a Platonisation of the Pythagorean doctrine of numbers."

92. *τὰ μαθηματικὰ τῶν πραγμάτων* occurs nowhere among the Peripatetic fragments. Aristoxenus (F 23 Wehrli) speaks of Pythagoras "likening all things to numbers" (*πάντα τὰ πράγματα ἀπεικάζων τοῖς ἀριθμοῖς*), on which see chapter 2. Theophrastus ascribes to Plato and the "Pythagoreans" a theory of *μίμησις*, in which sensibles within the universe are understood to imitate the first principles (*Metaph.* 11a26–11b7). But, as I've shown elsewhere (Horky: forthcoming), this theory should be ascribed to Xenocrates or, at most, to the "Pythagoreans" as seen through Xenocrates's point of view.

93. E.g. *Metaph.* 8.1, 1042a11–12; 12.1, 1069a35; 13.1, 1076a33; 13.2, 1077a16; 13.3, 1077b33, etc. Aristotle will speak of mathematicals that are "separate from" + genitive (e.g. *Metaph.* 13.2, 1076a33-34: *κεχωρισμένα τῶν αἰσθητῶν*) or "intermediate of" + genitive (e.g. *Metaph.* 11.1, 1059b6: *μεταξύ τε τῶν εἰδῶν καὶ τῶν αἰσθητῶν*).

that exist, but the unusual complication of *τὰ μαθηματικά* with either *τὰ πράγματα* or *τὰ ὄντα* in the genitive plural is a peculiarly Aristotelian formulation, and, moreover, is localized to discussions of Plato or the mathematical Pythagoreans (and, importantly, not Speusippus or Xenocrates) in *Metaphysics* A. As it turns out, in fact, the relatively unusual formulation *τὰ μαθηματικὰ τῶν πραγμάτων* most closely resembles the language of the mathematical Pythagorean Philolaus of Croton (F 6 Huffman = Stob. *Ecl.* 1.21.7d), who, when he spoke of the "being of things" (*ἁ ἐστὼ τῶν πραγμάτων*), was referring to the entity by virtue of which limiters and unlimiteds, the mathematical principles of his philosophy, could be thought to exist.[94] It is therefore probable that Iamblichus, in mentioning *τὰ μαθήματα τῶν πραγμάτων* in passage (B) from *On the General Mathematical Science*, was still looking at Aristotle's treatises on the Pythagoreans, perhaps written contemporaneously with *Metaphysics* A and, importantly, earlier than the treatments of the Pythagoreans in M or N. This is significant, because it suggests that it was Aristotle who celebrated the mathematical Pythagoreans for having achieved some credibility in their method of demonstration, even if they were overzealous in their pursuit of a unified philosophical system.

CONCLUSIONS

We have seen that the fundamental difference between acousmatic and mathematical Pythagoreanism as formulated by Aristotle lies in the latter group's attempts to make use of some sorts of demonstrative argumentation in order to provide explanations for their ideas. While acousmatic Pythagoreans apparently made no attempts to engage in demonstrations, mathematical Pythagoreans engaged in investigations that employed the principles of mathematics in order to make sense of the world they experienced. Their demonstrations tended to be derived from the principles of mathematics, including limiter and unlimited, as attested in the genuine fragments of Philolaus of Croton. It is also possible that their demonstrations were axiomatic and took the form of diagrams, as in the case of the speculative optical theories of Archytas of Tarentum.[95] Doubtless other types of Pythagorean demonstration have been lost to

94. The term *ἁ ἐστὼ τῶν πραγμάτων* has been used as grounds for dismissal of this fragment as authentic, especially since the term itself is replicated in the spurious Περὶ ἄρχων of Ps.-Archytas. But the authenticity of Philolaus's fragment has also been defended in various ways by Nussbaum (1979: 101) and Huffman (1993: 131–132). The term *ἡ τῶν πραγμάτων οὐσία*, which appears in Plato's *Cratylus*, also occurs by reference to Philolaus, as I argue in chapter 4, section entitled "Plato and Mathematical Pythagorean 'Being' before the *Phaedo*."

95. See Burnyeat 2005a: 45–51.

us. The plurality of the objects of their study made it difficult for Aristotle to characterize their philosophical system and to locate it squarely within the inquiry portion of his history of philosophy. The mathematical Pythagoreans were apparently also prone to establish relationships of similarity between numbers and perceptibles. What is more, as I will show, they posited an ontological order that was based on attributes that were strongly related to social organization within the *polis*, such as the notion of "what is more honorable," thus suggesting an organic relationship between the terms of political order and of ontological hierarchy. This important aspect of Aristotle's description of the *pragmateia* of the mathematical Pythagoreans is the subject of the first portion of chapter 2.

2

Hippasus of Metapontum and Mathematical Pythagoreanism

In the last chapter, I argued that Aristotle distinguished between and identified the peculiar characteristics of the philosophical systems (*pragmateiai*) of two groups of Pythagoreans whom he called "acousmatic" and "mathematical." Aristotle's description of the *pragmateia* of the mathematical or "so-called" Pythagoreans cannot be considered apart from his own classification of scientific knowledge: the mathematical Pythagoreans employed the superordinate (or superior) science of geometry in order to provide the "reason why" (τὸ διότι) things in the universe are the way they are. The acousmatic Pythagoreans, whose philosophical systems were based on the fact (τὸ ὅτι), did not engage in such sophisticated explanatory strategies in order to account for their observations about things in the universe. Moreover, as I argued, Aristotle in the lost writings on the Pythagoreans was invested in distinguishing the object of the philosophical activities of the mathematical and acousmatic Pythagoreans according to the threefold classification of "what is" (τί ἔστι), "what is to the greatest degree" (τί μάλιστα), and "what is to be done" (τί πρακτέον). The first two classes are important to Aristotle, to the extent that they deal with defining or describing things; and, as I will discuss in chapters 4–5 of this book, they formed the background for Plato's dialectical response to Pythagoreanism.[1] This chapter will be smaller in scope, as it is devoted to pursuance of a more comprehensive understanding of the philosophical doctrines assigned to the so-called progenitor of the mathematical Pythagoreans, Hippasus of Metapontum (ca. 520–440 BCE?) by the later doxographical tradition. The value of these doctrines for any actual understanding of Hippasus's philosophy varies considerably, in large part because certain aspects of the philosophy attributed to him appear to be imaginative reconstructions of Middle Platonists. This means that several of my arguments will conclude negatively, only indicating

1. The same might also be argued more extensively for Aristotle, for example, in the *Topics* and *Categories*; but that subject is beyond the scope of this book.

what Hippasus's doctrines could not have been. Still, some aspects of the philosophical concepts associated with Hippasus's philosophy reveal attempts to situate his doctrines within the history of philosophy being pursued in the Lyceum by Aristotle, Theophrastus, and Aristoxenus, as well as in the Early Academy by figures such as Speusippus. This chapter will aim to present and assess the relevant doxographical material in order to get a grip on the various ways figures in the Lyceum and in the Academy may have presented the philosophy of Hippasus of Metapontum, with attention to how such presentations of his doctrines are appropriated to fulfill the peculiar interests of each of the figures who discusses his philosophical ideas. We will conclude by suggesting that the some figures in the Early Academy appear not only to have assimilated the philosophy they attributed to Hippasus to their own philosophical doctrines but also to have filled out the picture that makes Hippasus a Pythagorean.

PYTHAGOREANISM AND THE AXIOLOGY OF WHAT IS "HONORABLE"

Aristotle's primary criticism of the mathematical Pythagoreans, as I argued in chapter 1, is that in their pursuit of an axiological hierarchy of things in the universe, their employment of the objects of mathematics led them to assume the existence of things that cannot be accounted for in the phenomena, such as the infamous Counter-Earth deployed by Philolaus of Croton. This reduction of perceptibles into mathematical objects brought the mathematical Pythagoreans to correct reality in a way that flies in the face of experience. Such categorical confusion is difficult to understand, both for us and (apparently) for Aristotle. Aristotle's examples of the Counter-Earth (*Metaph.* 1.5, 985b24–986a21) and the confusion of "two" and "double" (*Metaph.* 1.5, 987a13–27) in particular demonstrate how Pythagorean metaphysics might have been improperly brought to bear on physics, what Aristotle calls "resemblances" (*τὰ ὁμοιώματα*) and characterizes as types of "imitation" (*μίμησις*). As I hope to demonstrate in this section, the testimonies from Aristotle, his associates Aristoxenus of Tarentum (ca. 375–300 BCE) and Theophrastus of Eresus (ca. 371/0–287/6 BCE), and the Platonist Speusippus of Athens (ca. 410–339/8 BCE) that describe an axiological hierarchy based on comparative grades of what is "honorable" provide us with an avenue for understanding how metaphysics might have been brought to bear on religion and politics in the mathematical Pythagorean *pragmateia*.

We will recall Iamblichus's description of the "rationale" of the Pythagoreans' philosophy, derived, as I suggested, from a Peripatetic account:

> All such *acusmata*, however, that define what is to be done or what is not to be done [*περὶ τοῦ πράττειν ἢ μὴ πράττειν*], are directed toward the

divine [ἐστόχασται πρὸς τὸ θεῖον], and this is a first principle [ἀρχή], and their whole way of life is arranged with a view to following God [ὁ βίος ἅπας συντέτακται πρὸς τὸ ἀκολουθεῖν τῷ θεῷ], and this is the rationale of their philosophy.

(Iamblichus, *On the Pythagorean Way of Life* 86, 50.18–21; translation after Dillon and Hershbell 1991)

The testimony here is significant, because it provides a description of how the Pythagoreans may have established a "rationale" (λόγος) that integrated ethical and religious spheres.[2] The Pythagoreans' approach to philosophy (*pragmateia*), moreover, is characterized by an "aiming at the divine" (ἐστόχασται πρὸς τὸ θεῖον), which is understood to be the source from which all things come, and as that which must be pursued.[3] What is meant here by "aiming at the divine" in one's practice of daily life? Can we see "aiming at the divine" in a dialectical relationship with other approaches to the first principles?

We are aided to some extent by the testimony of Aristotle on the "so-called" Pythagoreans' approach to cosmology in Aristotle's work *On the Heavens*. There, we find that Aristotle describes the mathematical Pythagorean philosophical activity as organizing the phenomena according to the principle of what is "honorable":

Concerning [the earth's] position there is some divergence of opinion. Most of those who hold that the whole universe is limited say that it lies at the center, but this is contradicted by those from around Italy who are "so-called" Pythagoreans. These affirm that the center is occupied by fire, and that the earth is one of the stars, and it creates night and day as it travels in a circle about the center. In addition they invent another earth, lying opposite our own, which they call by the name "Counter-Earth," not in pursuit of accounts or explanations that conform with the appearances [οὐ πρὸς τὰ φαινόμενα τοὺς λόγους καὶ τὰς αἰτίας ζητοῦντες], but attempting to align the phenomena to certain reasons and opinions of

2. Interestingly, in the passage that follows, Iamblichus goes on to illustrate this claim: "For human beings act ridiculously when they seek the goods anywhere besides the gods, and similarly just as if some citizen in a land ruled by a king were to worship a subordinate governor [ὕπαρχον], overlooking him who is the lord of all [ἀμηλήσας αὐτοῦ τοῦ πάντων ἄρχοντος]." As Dillon and Hershbell note (1991: 111 n. 7), Philo (*Decal.* 61) uses the same phrase to speak about the Persian king and his satraps. As Bonazzi argues (2008: 249), Philo elsewhere (*Migr.* 128) appropriates the "Pythagorean" dictum of "following god" to the concept of living in accordance with nature. It is difficult to know whether Philo and Iamblichus are looking at the same Peripatetic authority.

3. On this passage, see chapter 1, section entitled "Aristotle on the *Pragmateiai* of the Pythagoreans."

> their own through attraction [ἀλλὰ πρός τινας λόγους καὶ δόξας αὑτῶν τὰ φαινόμενα προέλκοντες καὶ πειρώμενοι συγκοσμεῖν]. There are many others too who might agree that it is wrong to assign the central position to the earth, men who see proof not in the appearances but rather in abstract theory. *These reason that the most honorable body ought to occupy the most honorable place, that fire is more honorable than earth, that a limit is a more honorable place than what lies between its limits, and that a center and outer boundary are the limits.*[4] Drawing resemblances from these things [ἐκ τούτων ἀναλογιζόμενοι], they say it must be not the earth, but rather fire, that is situated at the center of the sphere.
>
> (ARISTOTLE, *On the Heavens* 2.13, 293a17–b1; translation after Guthrie 1939)

Generally, this passage affirms what I have said before about the *pragmateia* of the mathematical, or "so-called" Pythagoreans, in chapter 1, but with a significant addendum ascribed to a subset of these called the "many others": their scientific inquiry consists of primitive types of demonstrations that involve what we might describe as "saving the honorable" (by contrast to Aristotelian "saving the phenomena").[5] The criterion for this group of mathematical Pythagoreans' conceptualization of the universe is how "honorable" the object is in relation to how "honorable" its location in the universe might be; in essence, what they offer is a type of demonstration for the placement of the fire at the center of the universe based on a priori assumptions about what is "honorable."[6] Who are these "many others" whose philosophy (1) adheres to the basic tenets of Philolaus's cosmology (especially that a fire is placed at the center of the universe), and (2) whose "demonstration" adopts a priori assumptions about the status of each object's "honor" in the universe?[7]

The answer to this question requires bringing to bear on this study a group who, I would argue, played an important role in the transmission of the history of Pythagoreanism: the Platonists of the Early Academy. When I write about these Platonists, I am referring to those immediate students and/or associates of Plato who, in the decades following his death in 347 BCE, undertook to

4. τῷ γὰρ τιμιωτάτῳ οἴονται προσήκειν τὴν τιμιωτάτην ὑπάρχειν χώραν, εἶναι δὲ πῦρ μὲν γῆς τιμιώτερον, τὸ δὲ πέρας τῶν μεταξύ, τὸ δ' ἔσχατον καὶ τὸ μέσον πέρας. Italics mine.

5. Simplicius (*in Cael.* p. 513.13–32 Heiberg) thinks that the passage dealing with "what is honorable" refers to Pythagoreans.

6. See Lloyd 1966: 52–53.

7. This question has, to my knowledge, only been asked explicitly by Huffman (1993: 244–245), who does not advance an answer to the problem, and Burkert, on whom see below.

develop philosophical positions that were at least significantly derived from Plato's teachings and/or writings. Many of these figures played a role in the articulation of Plato's ideas for posterity, and several of them actually wrote works on Plato's philosophy in an attempt to attribute "doctrines" to him after his death. Diogenes Laertius (3.46) gives us a comprehensive list of these figures that includes Speusippus of Athens (ca. 410–339/8 BCE), Xenocrates of Chalcedon (ca. 396/5–314/3 BCE), Heraclides of Pontus (ca. 390–after 322 BCE), and Philip of Opus (fourth century BCE); and to that list we might add the rather shadowy figure Hermodorus of Syracuse (fourth century BCE), who is mentioned by Philodemus in the fragmentary *History of the Academy* (Col. VI).[8] These figures, each in his own way, participated in the transmission and appropriation of Platonic thought, as well as in the characterization of Pythagoreanism in the second half of the fourth century BCE. Thus, we will need to take serious account of them in our examination of the history of Pythagoreanism, in no small part because their works on the Pythagoreans have the potential to offer an alternative to the accounts of Aristotle that have occupied much of this study up to this point. Another group of sources, contemporary with but slightly younger than the Platonists and Aristotle, which offers a great deal of evidence for our investigation into the *pragmateia* of the mathematical Pythagoreans is Aristotle's associates in the Lyceum, Theophrastus of Eresus (ca. 371/0–287/6 BCE) and Aristoxenus of Tarentum (ca. 375–300 BCE). Theophrastus, who followed Aristotle as the head of the Lyceum, is especially important for at least two reasons: first, his fragments and extant treatises reveal a knowledge of Pythagoreanism that, as I've argued elsewhere,[9] is informed by the Platonists' (probably Xenocrates's) works on the Pythagoreans, and thus he presents a view of Pythagoreanism that is not chiefly derived from his teacher Aristotle's influential but ultimately self-serving construction of Pythagorean philosophy; and second, his so-called "doxographical" works, which survive in abbreviated and somewhat modified forms in the collection often associated with the figure known as Aëtius, also reveal important differences from

8. The most recent text of Philodemus's *History of the Academy* is Dorandi 1991. I leave Polemo, head of the Academy from 314–276 BCE, off this list. It is difficult to know, based on the testimony of Clement (*Strom.* 7.32.9), why Polemo adopted an ethics based in vegetarianism; certainly Xenocrates before him might have been the direct influence (rather than Pythagoreanism itself). Polemo did criticize those who practiced theoretical philosophy without application, attacking those who write handbooks on "harmony" without practical knowledge (D.L. 4.18). This evidence alone, however, is not enough to show that he wrote about Pythagoreanism. For a more comprehensive discussion of the associates of the Platonic Academy, see Dillon 2003: 13–16. On Hermodorus, in particular, see especially Horky 2009: 79–92 and Dillon 2003: 198–204.

9. Horky: forthcoming.

Aristotle's treatments of the Pythagoreans.[10] Aristoxenus, for his part, is a figure equally important for the history of Pythagoreanism, because he not only (1) was influenced by Pythagoreanism (having learned about it from his "father" Spintharus of Tarentum, a younger contemporary of Archytas),[11] but also (2) was an Aristotelian philosopher, who, I suggest, synthesized Pythagorean ideas with Peripatetic concepts and terminology. Moreover, Aristoxenus's works on Pythagoreanism (a *Life of Pythagoras* as well as a *Life of Archytas*, a work called the *Pythagorean Precepts*, and various accounts of the history of Southern Italy) and reveal a kind of deep engagement with Pythagoreanism that was, significantly, likely to have been informed by local knowledge and experiences in Tarentum.

Aristoxenus's *Pythagorean Precepts* (Πυθαγορικαὶ ἀποφάσεις), which Carl Huffman has recently argued must be considered genuine, develop the metaphysical theory of "aiming at the divine" attributed by Iamblichus to the mathematical Pythagoreans.[12] In doing so, Aristoxenus extends Aristotle's ideas further by associating this tendency with not only theories of metaphysics, as Aristotle had emphasized, but also politics and individual ethics:

> To have the attitude toward the divine that it exists and is so disposed to the human race that it looks attentively on it, and does not neglect it: this the Pythagoreans learned from [Pythagoras], and deemed to be useful. For we need supervision, of such a sort that we shall not at all dare to rebel against it; and such is the supervision that arises from the divinity, if indeed the divine is such as to be worthy of rule over everything [εἴπερ ἐστὶ τὸ θεῖον τοιοῦτον <οἷον> ἄξιον εἶναι τῆς τοῦ σῦμπαντος ἀρχῆς]. For they declared, correctly, that the living being is by nature prone to insolence, and unstable in its impulses, desires, and the rest of its emotions. It needs, thus, such supremacy and threatening from which there derives some self-control and order [σωφρονισμός τις καὶ τάξις]. They thought, then, that every one, being conscious of the complexity of his own nature [συνειδότα τὴν τῆς φῦσεως ποικιλίαν], must never be forgetful of piety toward and worship of the divine [μηδέποτε λήθην ἔχειν τῆς πρὸς τὸ θεῖον ὁσιότητός τε καὶ θεραπείας], but always put before his mind the

10. For a useful summary of the doxographical traditions that stem from Theophrastus, see Mansfeld 2012 (with bibliography).

11. On the grounds that the Suda (F 1 Wehrli) identifies both Spintharus and Mnesius as possible names of fathers of Aristoxenus, Wehrli thought that Spintharus was the teacher (but not biological father) of Aristoxenus. Visconti (1999: 36–63) suggests that the Suda might have confused the real name of the father with an epithet (i.e. Spintharus "the rememberer").

12. In two publications: Huffman 2008 and 2006. But also see Zhmud 2012: 65 n. 17.

fact that the divinity looks attentively at and watches over the conduct of human beings.

(Aristoxenus F 33 Wehrli = Iamblichus, *On the Pythagorean Way of Life* 174–175, 97.23–98.14; translation after Dillon and Hershbell 1991)

Aristoxenus emphasizes that, for the Pythagoreans, "some self-control and order" derive from a sort of comprehensive self-knowledge (συνειδότα) of human nature, which is designated as "complex" (ποικιλία), probably in a pejorative sense. The term ποικιλία requires a bit of contextualization: Aristotle only tangentially refers to a person being ποικίλος once, in the *Nicomachean Ethics* (1.10, 1101a8–14), where he describes how a person can be "variable" and "quick to change" (εὐμετάβολος) as a consequence of bad luck, just like Priam. Prior to Aristotle, Plato (*Republic* 8, 561e4) refers to a "democratic" person as ποικίλος, with reference to the complex constitution of his character; in this context—the description of the various types of human and political constitutions that are possible—Plato associates the quality of being "variegated" with the anarchic aspect of the "democratic" type (e.g. *Republic* 8, 557c4–9), and this quality cannot therefore be considered simply praiseworthy for Plato (see *Republic* 8, 559d8–e2, 568d5). Aristoxenus's usage, to be sure, in no way suggests the specific political context of Plato's argument, nor can we assume that Plato would subscribe to the idea that the natural constitution of every human being is ποικίλος.[13] Self-knowledge as described by Aristoxenus involves understanding of the nature of human beings and compels the person to worship the divine properly, on the grounds that humans are incapable of correcting their nature without help from the divine. Aristoxenus's theory of Pythagorean "self-knowledge," as it is presented in F 33 Wehrli, is far less detailed than Plato's and involves ideas about the nature of "self-control" that are more familiar in the ethics of Archytas of Tarentum, from the *Life of Archytas* also written by Aristoxenus.[14]

13. See Rowe 1998: 212–213, with reference to *Phaedrus* 277c2–3. Rowe astutely connects the quality of being ποικίλος with the beast of appetite from the *Republic* 9 (588c7). In this light, it is worth adducing a similar passage from the *Phaedrus* (230a3–7), where Socrates amusingly speculates about how to interpret the Delphic injunction: "I inquire . . . into myself, to see whether I am actually a beast more complex [πολυπλοκώτερον] and more violent than Typhon, or both a tamer and simpler creature, sharing some divine and un-Typhonic portion by nature."

14. See Archytas's attacks on bodily pleasure and praise of self-control (σωφροσύνη/ *temperantia* or the quality of being σώφρων) in T A 9 (Ath. 12, 545a = Aristoxenus F 50 Wehrli), A 9a (Cic. *Sen.* 12.39–41), and A 11 (Ael. *VH* 14.19) Huffman. It should be noted, however, that Archytas does not assume that self-control resulted from self-knowledge coupled with proper worship of the gods, although T A 9 Huffman suggests that the "lawgivers" who might have been Pythagorean celebrated as deified Justice, Temperance, and Self-Control.

As in the philosophy of Archytas, the application of intellectual activity (self-examination) to ethical behavior and political activity is explicit, but the emphasis in Aristoxenus's *Pythagorean Precepts* is not on this application itself but on how this application leads one to recognize the supremacy—one is tempted to say "priority," as I will show later—of the divine. Aristoxenus continues to develop this idea by establishing a hierarchy of beings based on the principle of what is more divine in the passage that follows.[15]

> After the divine and the daemonic [they thought it was necessary] to give greatest regard to parents and laws [*ποιεῖσθαι λόγον γονέων τε καὶ νόμων*], not feignedly, but by conforming oneself to these things out of conviction. They approved of abiding by the traditions and laws of their ancestors, even if they should be somewhat worse than those of others.
>
> (Aristoxenus F 34 Wehrli = Stobaeus, *Eclogae* 4.25.45; translation after Huffman 2008)

Huffman's analysis of this passage has confirmed that Aristoxenus attributes to the Pythagoreans a fairly common Greek practice of honoring the gods first, followed by parents, and then laws; following Wehrli, he cites both Plato's *Laws* (717a–b, 884a–885a, 930e) and Xenophon's *Memorabilia* (4.19–20) as comparative evidence.[16] But his interpretation misses the mark slightly in two ways: first, it does not sufficiently highlight the extent to which Plato's account in *Laws* 717a–b, which is the closest *comparandum*, goes far beyond Aristoxenus's in establishing a hierarchy of worship,[17] and second, it does not locate the hierarchy within a broader project of "aiming at the divine,"[18] which is implicit in Aristoxenus's account and betrays a Peripatetic conceptual superstructure. For Aristoxenus, the cognitive activity of self-examination is coextensive with the conforming of the self to the parental traditions and laws of the city-state; it is

15. In Iambl. *VP* 175, 98.15–18, but I have chosen to employ the same passage as cited by Stobaeus on the grounds that there are likely to be fewer modifications to the original text, at least in this circumstance.

16. Huffman 2008: 108.

17. Plato's hierarchy of worship is constructed according to a hierarchy of what is more worthy of piety, consisting of (1) Olympian gods, (2) Chthonic gods, (3) *daemons*, (4) heroes, (5) private rites of the ancestral gods, and finally (6) honors for parents.

18. Still, Plato (*Leg.* 717a3–b2)does speak in a self-consciously metaphorical way about discovering the "target" (*σκοπός*) at which to "aim" (*στοχάζεσθαι*) their weapons, where the "target" is "piety" (*εὐσεβεία*) and the primary "missile" (*βέλη*) is "honors" (*τιμαί*) rendered in proper order. There is little overt appeal to a complex metaphysics in Plato's employment of what is apparently a poetic trope.

in this way that Aristoxenus brings ethics to bear on political activity, and all within the larger goal of "aiming at the divine."

But what is the relationship between Aristoxenian "aiming at the divine" and Aristotelian ordering of the universe according to what is more "honorable"? Among those fragments printed in Wehrli's edition, we have one reference to what is *τίμιος* being a primary factor in the determination of the inherent values of entities within the larger philosophical system of the Pythagoreans. This occurs in a work of Aristoxenus perhaps entitled *On Arithmetic* (Περὶ ἀριθμητικῆς), cited by Stobaeus:[19]

> Pythagoras seems to have honored, most of all, the *pragmateia* that concerns numbers, and to have advanced it by withdrawing it from the use of merchants and likening all things to numbers.
>
> *τὴν δὲ περὶ τοὺς ἀριθμοὺς πραγματείαν μάλιστα πάντων τιμῆσαι δοκεῖ Πυθαγόρας καὶ προσαγαγεῖν εἰς τὸ πρόσθεν, ἀπαγαγὼν ἀπὸ τῆς τῶν ἐμπόρων χρείας, πάντα τὰ πράγματα ἀπεικάζων τοῖς ἀριθμοῖς.*
>
> (Aristoxenus F 23 Wehrli = Stobaeus, *Eclogae* 1, proem 6)

I am not particularly interested, as are others,[20] in evaluating this fragment for the evidence it presents for our knowledge of Pythagoras; on the contrary, my immediate goal here is to understand what it tells us about Pythagoreanism as that might have been understood by a Peripatetic who inherited Aristotelian baggage regarding the philosophical study (*pragmateia*) of the Pythagoreans, and who apparently had local knowledge of Tarentine Pythagoreanism.[21] Now Aristoxenus's treatment of Pythagoras emphasizes the "honoring" (*τιμῆσαι*) of numbers over all other things (*μάλιστα πάντων*), and the shifting of "numbers" from the particular provenance of tradesman to philosophy is thought to involve the philosophical activity of "likening all things" (*πάντα τὰ πράγματα ἀπεικάζων*) to them. This specific activity explains what Aristoxenus had previously said, namely, that Pythagoras brought the study of numbers

19. Burkert (1972: 414, with n. 77) assesses the language of this statement and concludes that "the introductory sequence, with its meticulous formulation, looks like an exact quotation." This claim is strengthened by the appearance of an explanatory particle (*γάρ*) in the following sentence, followed by an explanation and definition that rings Platonist: "for number possesses the other things and is the relationship of all numbers to one another."

20. E.g. Riedweg 2005: 90.

21. See Zhmud 2006: 218–221.

forward by means of assimilation to other objects.[22] Implicit here is the association of the Pythagorean philosophical system (πραγματεία) with the activity of dealing with things (τὰ πράγματα) in certain ways. This formulation of Pythagoras's *pragmateia* is familiar from Aristotle's terse and ultimately aporetic description of the mathematical[23] Pythagorean *pragmateia* in *Metaphysics* A, an intertext for Aristoxenus that has been recognized since Frank.[24] What is distinctive here, however, as Burkert notes,[25] is that Aristoxenus has shifted the attribution of this "advancement" in mathematics back from the Pythagoreans to Pythagoras himself. I shall have something more to say about the history of the "advancement" of mathematics with regard to the mathematical Pythagoreans a little later in this chapter. For the present inquiry, it is enough to acknowledge that this fragment evidences Aristoxenus's interest to explain a Pythagorean *axiology of the "honorable"* by appeal to strategies of assimilation between numbers and things.[26]

Commentators[27] on this fragment have not recognized another important *testimonium* concerning the principle of "honoring" numbers that relates it to the idea of the "first principle" (ἀρχή) of the Pythagoreans, which is to be found in a section of the *Pythagorean Precepts* quoted by Iamblichus in his work *On the Pythagorean Way of Life*:[28]

22. Compare Arist. *Metaph.* 1.5, 985b23–986a8. On this passage, see chapter 1, section entitled "On the 'So-called' and Mathematical Pythagoreans."

23. Literally, there, the "so-called" Pythagorean *pragmateia*. But see chapter 1, where I argue that "so-called" Pythagoreans in *Metaphysics* A and the mathematical Pythagoreans referred to in Aristotle's fragmentary writings on the Pythagoreans are one and the same.

24. Frank 1923: 260 n.1. The text is Arist. *Metaph.* 1.6, 987b7–20. The important bit is where Aristotle claims, "With regard to 'participation,' Plato only changed the name: for whereas the Pythagoreans claim that entities exist by means of imitation of numbers [μιμήσει τῶν ἀριθμῶν], Plato says by means of participation [μεθέξει], modifying the name. As to what participation or imitation is, however, they left it to us to seek it out together." On this *testimonium*, see chapter 5, section entitled "What Is Wisest? Three." It is also worth comparing Thphr. *Metaph.* 33, 11a27–b10, on which see Horky: forthcoming and Dillon 2002: 185–186.

25. Burkert 1972: 414.

26. An axiology of "the honorable" would constitute a systematic science of values based in a priori assumptions about specifically what is more "honorable." The best discussion of Aristotle's implication of teleology in axiology is Johnson 2005 (see especially pp. 90–93 and 289–295).

27. E.g. Frank (1923: 260 n. 1) and Burkert (1972: 414, with n. 77).

28. That this section derived from the *Pythagorean Precepts* was suggested to me by Carl Huffman.

> They asserted that the first principle in everything is one of the most honorable things in knowledge, experience, and in generation likewise; and again [likewise] in the household, city, and army; but that the nature of the first principle is difficult to discern and comprehend. For, in the sciences, when looking at the parts of the system, it is a task of no ordinary intellect to comprehend and to discern correctly what sort [might be] the principle of these things. . . . The same goes for "principle" in the other sense [i.e. "ruling" principle]: for neither a household nor a city-state is well managed when not subject to the rule and authority of a genuine commander and master. For authority to arise it is necessary for both, the ruler and the ruled, to be equally willing. Just so, they declared that teaching is correctly imparted when it takes place voluntarily, and both the teacher and the student are willing. For if either of the two resists in any way, the proposed work can never be duly completed.
>
> *ἀρχὴν δὲ ἀπεφαίνοντο ἐν παντὶ ἕν τι τῶν τιμιωτάτων εἶναι ὁμοίως ἐν ἐπιστήμῃ τε καὶ ἐμπειρίᾳ καὶ ἐν γενέσι, καὶ πάλιν αὖ ἐν οἰκίᾳ τε καὶ πόλει καὶ στρατοπέδῳ. δυσθεώρητον δ' εἶναι καὶ δυσσύνοπτον τὴν τῆς ἀρχῆς φύσιν· ἔν τε γὰρ ταῖς ἐπιστήμαις οὐ τῆς τυχούσης εἶναι διανοίας τὸ καταμαθεῖν τε καὶ κρῖναι καλῶς βλέψαντας εἰς τὰ μέρη τῆς πραγματείας, ποῖον τούτων ἀρχή. . . . τὸν αὐτὸν δ'εἶναι λόγον καὶ περὶ τῆς ἑτέρας ἀρχῆς· οὔτε γὰρ οἰκίαν οὔτε πόλιν εὖ ποτε ἂν οἰκηθῆναι μὴ ὑπάρξαντος ἀληθινοῦ ἄρχοντος καὶ κυριεύοντος τῆς ἀρχῆς τε καὶ ἐπιστασίας ἑκουσίως. ἀμφοτέρων γὰρ δεῖ βουλομένων τὴν ἐπιστατείαν γίνεσθαι, ὁμοίως τοῦ τε ἄρχοντος καὶ τῶν ἀρχομένων, ὥσπερ καὶ τὰς μαθήσεις τὰς ὀρθῶς γινομένας ἑκουσίως δεῖν ἔφασαν γίνεσθαι, ἀμφοτέρων βουλομένων, τοῦ τε διδάσκαντος τε καὶ τοῦ μανθάνοντος· ἀντιτείνοντος γὰρ ὁποτέρου δήποτε τῶν εἰρημένων οὐκ ἂν ἐπιτελεσθῆναι κατὰ τρόπον τὸ προκείμενον ἔργον.*
>
> (Iamblichus, *On the Pythagorean Way of Life* 182–183, 101.20–102.14; translated after Dillon and Hershbell 1991)

Aristoxenus's *testimonium* is important for two reasons. First, it extends the idea of the classification of objects in the universe according to what is more "honorable" to the first principle itself and provides an argument that the discernment and understanding of this first principle are difficult, but necessary, to achieve. Second, it demonstrates how Aristoxenus took on the double exegesis of the *ἀρχή*, that is, as both (1) ontological and logical first principle and (2) principle of political, military, and household rule. This triad—along with the striking epistemological *comparanda* of "knowledge, experience, and generation"[29]—is understood to form the "parts of the system" (τὰ μέρη τῆς πραγματείας) of the Pythagoreans that must have an underlying ἀρχή, although it is not obvious from this fragment alone whether the ἀρχή is one and the same for each part or whether each part has its own peculiar ἀρχή.

The principle that the ἀρχή is a "most honorable" thing and, thereby, forms a source for and at which other objects aim is, I would argue, a product originally of Platonic thought that has been expanded into a systematic mode of explanation in the writings of Aristotle and, as I will show, the early Platonist Speusippus of Athens. In Plato's corpus (including the spurious dialogues and letters), the adjective τίμιος and comparative/superlative forms of that word occur sixty-one times (not a particularly high number), and of those occurrences, it appears most frequently in the *Laws*, nineteen times (32 percent). Metaphysical and/or logical uses of this term almost never occur there, and the predominant semantics involve evaluation of political status for citizens.[30] In the *Timaeus*, a dialogue that is *named after* the "honorable man" (Τίμαιος), τίμιος-words only occur once (45a3), in a biological context, namely when the gods create the human body and place the head in the "more honorable" (τιμιώτερον) position.[31] This is enough to show that evaluation according to what is "honorable" functions as a mode of explanation in Plato's thought, if not fully realized as such. A better comparison of the use of τίμιος-words in Plato's writings to that in Aristoxenus's *Pythagorean Precepts* comes at the programmatic prelude in book 5 of the *Laws* (726a1–734e2), where the Athenian Stranger establishes a hierarchy based on what is "more honorable," starting with the gods (with their attendants)[32], then proceeding to soul, and finally body. There, the Athenian Stranger goes so far as to define τιμή as "to cleave to what

29. For the importance of triads to Pythagoreanism, see chapter 5, section entitled "What Is Wisest? Three."

30. E.g. Pl. *Leg.* 730d1–4; 808c2–6; 829d1–4.

31. This passage is sometimes cited (e.g. Lennox 2001a: 268) as proof that Aristotle's employment of the term "honorable" in biological contexts may be Platonic. This may be true, but the more general use of the term "honorable" and its cognates in Aristotle's philosophy goes far beyond this singular usage in Plato's writings. See below.

32. To whom does Plato refer when he mentions "those who attend to the gods" (τοὺς τούτοις ἑπομένους)? Earlier (*Laws* 717a–b; see above), the Athenian Stranger had referred to the *daemons* as coming after the Olympian and Chthonic gods. Philip of Opus (*Epinomis* 984d5–e3), if he is the author of that text, expands on Plato's description by calling them *daemons*, who, being "next after" (μετά) the "most honorable" (τιμιωτάτους) gods, are made of air and occupy the middle position between the gods and humans.

is better and, where practicable, to make as perfect as possible what is worse."[33] Thus, when Plato inscribes these objects into a hierarchy of goods, he has something in mind quite different from Aristoxenus's description of the Pythagoreans' ἀρχή.

We can, however, trace more complex and nuanced treatments of the "honorable" back to those associates of Plato who undertook systematically to describe the philosophical system of Plato: the early Platonists. I shall have more to say about Xenocrates of Chalcedon later on in this chapter, but for present purposes, note that he does not obviously ascribe to the Pythagoreans an axiology of the "honorable," nor does he espouse such a thing himself.[34] Speusippus of Athens is a different matter altogether, as we discover when we read Theophrastus's treatment of the "honorable" in his *Metaphysics*:

> If [the aforementioned claim] does not apply,[35] we must still admit certain limits to the (1) "for the sake of" and (2) "toward the best" [claims], and not posit these for all cases without qualification; since even claims of the following kind leave some room for doubt, both when expressed (2a) without qualification and (2b) in individual cases—without qualification, that (2a) "nature strives for the best in all cases and, when possible, makes things share in the eternal and in the orderly"; and likewise the same applies in the case (2b) of animals, for "where the better is possible, there it never fails, e.g. the windpipe is before the gullet (for it is more honorable), and the mixture in the middle ventricle of the heart is best, because the middle is most honorable"; and similarly (1) for all

33. Pl. *Leg.* 728c6–8: τιμὴ δ' ἐστὶν ἡμῖν, ὡς τὸ ὅλον εἰπεῖν, τοῖς μὲν ἀμείνοσιν ἕπεσθαι, τὰ δὲ χείρονα, γενέσθαι δὲ βελτίω δυνατά, τοῦτ' αὐτὸ ὡς ἄριστα ἀποτελεῖν. This is a summary of an earlier passage (726a6–727a7), in which the Athenian Stranger claims: "thus when I say that next after the gods—our masters—and those who attend to them, a man must honor his soul, my suggestion is correct. But hardly a man among us honors it correctly, but he only thinks he does. For honor, I suppose, is a divine good, and none of the evil things is honorable, and if one thinks that he is magnifying his soul by flattery or gifts or indulgence—when he in no way makes it better from worse—he thinks that he is honoring it, but actually he is achieving nothing."

34. There are two pieces of evidence in which Xenocrates speaks about Pythagoras: F 87 IP, in which Xenocrates is said by a certain "Heraclides" to have claimed that Pythagoras discovered that musical intervals have a numerical structure; and F 221, in which Xenocrates is coupled with a certain Epimenides and Eudoxus in claiming that Pythagoras was the son of Apollo and Parthenis.

35. Accepting van Raalte's emendation of the manuscripts' text from τοῦθ' to τοῦ γ'.

> things that are "for the sake of order." For even if the desire [of nature] is such, *this* [claim] (3), at any rate, shows that there is much—actually, more than much—that does not obey or receive the good: for "what is animate is scanty, and what is inanimate is infinite, and the being of these inanimate things is both momentary and better." But, in general, [the claim (3a) that] "the good is something rare and is in few things, but evil is an abundant multiplicity"—and that it does not solely consist in indeterminateness and, so to speak, in the form of matter, as is the case for the things of nature, is [the claim] of a most ignorant person. For quite random [is the claim] of those like Speusippus who (3b), speaking of the entirety of being, make the honorable, which [is found] in the space at the center, something rare, whereas [they do make what is not honorable] what lies at the extremities on either side.[36]
>
> (THEOPHRASTUS, *Metaphysics* 31–32, 11a1–25)

As has been made clear by James Lennox in his studies of Aristotle's biological works,[37] Theophrastus's rebuke of claims (1) and (2) is chiefly directed against Aristotle, although it might refer to some Platonists as well.[38] And as Monte Johnson has argued, Peripatetic criticisms of the methodology of teleological explanation are rooted in a skepticism concerning uncritical appeal to the final cause.[39] But we ourselves should not thereby assume that Theophrastus targeted only his Academic associates: textual evidence suggests otherwise. Indeed, Aristotle employs an argument comparable with (2b), in which the windpipe is said to be before the esophagus on the grounds that "generally, with regard to above and below, before and behind, and right and left, the better and more honorable part [τὸ βέλτιον καὶ τιμιώτερον] is always placed uppermost, in

36. If indeed this is how we are to make sense of the incomplete and frustratingly abrupt close of the sentence that reads: *τὰ δ' ἄκρα καὶ ἑκατέρωθεν*. I agree with Tarán (1981: 446–448) that the δ' connotes an adversative relationship between this phrase and the statement antecedent to it (*εἰκῇ γὰρ οἱ περὶ τῆς ὅλης οὐσίας λέγοντες ὥσπερ Σπεύσιππος σπάνιόν τι τὸ τίμιον ποιεῖ τὸ περὶ τὴν τοῦ μέσου χώραν*). I suggest—and I believe this is a totally novel interpretation, although it is only slightly different from Tarán's—that what Theophrastus is contrasting here is what is "honorable" versus what is understood to be "not honorable," i.e. what is "evil" at the edges. This interpretation has the benefit of preserving the train of thought in Theophrastus's argument, namely, distinguishing what is more "honorable" from what is less so.

37. Lennox 2001b: 266–272.

38. See Dillon 2002: 184–185.

39. Johnson 2005: 35–39.

front, and to the right, unless something greater stands in its way."[40] The combination of "better" with "more honorable" is a recurring *topos* in Aristotle's writing, and it raises the question to what extent should arguments attributed to the Pythagoreans by Aristoxenus that involve the metaphysics of the "honorable" be considered originally Aristotelian?

In order to advance on this question, we will need to take into account the fact that arguments from what is "better" and more "honorable," while somewhat numerous in Aristotle's writings,[41] may derive from his earlier thought, when he was arguably under greater influence of other philosophers such as Plato than in his later writings. Indeed, the earliest attestations of an axiology of the "honorable" in Aristotle's writings are in two texts often considered to have been written under Academic sway, the *Categories* and the *Protrepticus*; these texts provide us with a slightly better, if not fully satisfactory, sense of how he assumed an axiology based on the "honorable" as an a priori assumption in his overall philosophical program. In the *Categories*, Aristotle describes four uses[42] of the term "prior" (πρότερον): when it means (1) "prior in age/time," (2) "prior in a sequence" (i.e. the number 1 is prior to the number 2), (3) "prior in a series" (i.e. the point is prior to the line), and finally—the loosest sense, employed by "the many"—(4) "naturally prior":

> In addition to the aforementioned uses, the better and the more honorable seems to be prior in nature [τὸ βέλτιον καὶ τὸ τιμιώτερον πρότερον εἶναι τῇ φύσει]. And the many are accustomed to say of those whom they consider honorable and whom they love greatly that they are "prior" in their hearts. And this, indeed, is nearly the strangest of all the uses.
>
> (Aristotle, *Categories* 12, 14b3–7)

40. Arist. *PA* 3.3, 665a23–26. Lennox (2001b: 254) cites *IA* 5, 706b11–16 as explanation for this claim, on the grounds that a "location is "valuable" because an "origin" (e.g. of perception or locomotion) is found there. So assertions about the value of locations are parasitic on those about the value and function of organs present there." Indeed, this appears to resemble a circular argument, which in fact it is: "And it is reasonable that the ἀρχαί too come from these parts; for the ἀρχή is honorable, and the above is more honorable than the below, as is the front than the back and the right than the left. It is also true to reverse the proposition concerning these things and say that because the ἀρχαί are in these places, they are therefore more honorable than the opposite parts."

41. E.g. *Top.* 3.3, 118b20–27; *Cael.* 2.5, 288a2–12; *EN* 1.13, 1102a15–26; *Metaph.* 9.9, 1051a4–19 and 11.7, 1064a37–b6.

42. Aristotle goes on (*Cat.* 12, 14b10–24) to document another usage, namely, the use of "prior" in describing types of predication, which he also describes as a type of "natural" priority (πρότερον . . . τῇ φύσει).

What Aristotle means by "naturally prior" here is not obvious, and he speaks obliquely about the way the "many" employ the *topos* "better and more honorable" (τὸ βέλτιον καὶ τὸ τιμιώτερον)—although he himself uses such terms with just as little apparent clarification about their meaning as do "the many." In the *Protrepticus*, however, Aristotle had given a more comprehensive description of what it means to be *naturally* "best and most honorable" (φύσει . . . τὰ βέλτιστα καὶ τιμιώτατα) by reference to Pythagoras himself:

> Moreover, the animals are surely things that have come to be by nature, either absolutely all of them or the best and most honorable of them; for it makes no difference if someone thinks that most of them have come into being unnaturally because of some corruption or wickedness. But certainly a human is the most honorable of the animals down here; hence it's clear that we have come to be both in nature and according to nature.[43] [And this, among existing things, is that for the sake of which nature and god have brought us into existence.][44] Pythagoras, when asked "what is it?" [τί δὴ τοῦτό ἐστι],[45] responded "to observe the heavens" [τὸ θεάσασθαι τὸν οὐρανόν], and he used to claim that he himself was an observer of nature [θεωρὸν τῆς φύσεως], and that it was for the sake of this [i.e. observation of nature] that he had passed into life [παρεληλυθέναι εἰς τὸν βίον].
>
> (IAMBLICHUS, *Protrepticus* 9, 50.27–51.10 = B 16 and B 18 Düring; translated after DSH-MRJ)

Aristotle here understands a kind of comprehensive natural teleology that is united under the umbrella of what is "better" and "more honorable." In particular, this axiology is associated strongly with the final cause, speculation about which Aristotle seems to have associated with Pythagoras, the "observer of nature" (θεωρὸς τῆς φύσεως), and with his particular way of life (βίος).[46] We

43. There is possibly a break in the fragment of Aristotle here, as noted by Düring (1961: 53–55) and DSH-MRJ (2012: 46–47).

44. Quite possibly, this is an insertion by Iamblichus that marks where he skips over a portion of Aristotle's text.

45. Due to the fact that the "what is it?" *acusma* was a standard question for the Pythagoreans, as I argued in chapter 1, I include this portion in the original text of Aristotle. We cannot know for sure what specifically Pythagoras was defining in Aristotle's original, though.

46. On Aristotle's treatment of *theoria* in this passage, see Nightingale 2004: 193–194. For a defense of this reference to Pythagoras as original to Aristotle's text, see Zhmud 2012: 56, with n. 108.

can see a more fully elaborated axiology of this sort in Aristotle's *Topics*, where he extends the argument by suggesting that what belongs to a god is "better" and "more honorable" than what belongs to a human.[47] The appeal to what is "honorable" functions in various ways, not the least (for our purposes) in constructing a *scala naturae*, which represents an adaptation of (at least) Plato's hierarchy of beings as expressed in the *Laws*. The employment of these concepts remains problematic in Aristotle's later writings as well, as he attempted to shape these holdovers from his earlier thought into something that could still play a determined role in his philosophy.[48] This has significant implications for Aristotle's general approach to teleology, as Geoffrey Lloyd recognized almost a half century ago, and as Monte Johnson and Mariska Leunissen have emphasized in various ways more recently.[49] I do not have space to go further into this topic, but suffice it to say that the impetus for Aristotle's assumption of an axiology of the "honorable" might be a holdover from his time in the Academy, either in consideration of Platonic or of Pythagorean ideals.[50]

There is, of course, another source worth considering for Aristoxenus's ascription of what I'm calling an axiology of the "honorable" to the Pythagoreans in the *Pythagorean Precepts*: Speusippus of Athens. Recall in Theophrastus's critique of the arguments "for the sake of" and that "aim toward the best" that he shifts the criticism away from Aristotle in claims (1) and (2) and toward Speusippus, whom Theophrastus characterizes as having believed (3b) that within the entirety of being, the honorable, which is at the center, is something rare, whereas what lies at the extremities on either side is, apparently, not honorable.[51] Scholars have taken passage (3b) to refer in its entirety to Speusippus's philosophy,[52] and,

47. *Top*. 3.1, 116b12–15. It is unclear whether Aristotle, in *Metaphysics* A (1.3, 983b32–984a5), accepts the mode of argumentation from what is "most honorable," i.e. the theologians.

48. See, for example, *PA* 2.1, 646a24–646b2, where Aristotle distinguishes two types of priority that are strongly related to his teleology: what is prior in generation and what is prior in nature, which belongs to final causes. Mariska Leunissen (2010: 21) understands the latter to be a type of "primary teleology," which explains features that "can be exhibited to be necessary prerequisites for natural beings such as animals to perform the functions specified in their form." Assumptions about what is "better" and more "honorable" (see *Metaph*. 11.7, 1064a36–b6) affect the subdivisions of theoretical philosophy and present problems for conceiving of Aristotle as anti-Platonic, as Merlan has argued (1968: 59–87).

49. Lloyd 1966: 52–61; Johnson 2005: 138–140; Leunissen 2010: 59 and 62.

50. Recall the association of this theory with a group of the "so-called" Pythagoreans in *On the Heavens* 2.13, 293a15–b1, cited and discussed above.

51. Such is how I have reconstructed the argument. See above.

52. See Dillon 2002: 185; Tarán 1981: 445–449; Cherniss 1944: 559.

with good reason, to see passage (3a) as the inference that Theophrastus draws from it in order to connect the Aristotelian-Academic arguments in (1) and (2) to Speusippus's thought. When we combine Theophrastus's criticism of Speusippus's doctrine of the "honorable" as what is "rare" and at the center—in contrast to what is not "honorable" and lies at the extremities—with what Aristotle, in his work *On the Heavens* 2.13, 293a15–b1 quoted above, postulated about the "many others" who espouse Philolaic cosmology, it becomes likely that the group described by Aristotle as "those who reason that the most honorable body ought to occupy the most honorable place, that fire is more honorable than earth, that a limit is a more honorable place than what lies between its limits, and that a center and outer boundary are the limits," most likely refers to Speusippus of Athens. Such a conclusion is corroborated by the total absence of axiological uses of "honorable" in the genuine fragments of the central mathematical Pythagoreans whose writings survive: Philolaus of Croton and Archytas of Tarentum. If some members of the Early Academy advocated for the cosmological system of Philolaus, then this would suggest that they had access to either (1) genuine texts of (at least) Philolaus, or (2) the teachings of the mathematical Pythagoreans via oral sources. Thus arises a new wrinkle in my argument, one that scholars of Pythagoreanism have been forced to acknowledge since Burkert and Philip published their formative studies on Pythagoreanism in the 1960s:[53] in the above account of *On the Heavens*, Aristotle appears to include Platonists in the group/s that he nominates the "so-called" Pythagoreans.

This new wrinkle, as I've designated it, presents problems for all scholars of the history of Pythagoreanism, since it forces us to consider the possibility that (1) Aristotle's account of Pythagoreanism is at least partially mediated by the writings on the Pythagoreans by Speusippus (*On Pythagorean Numbers*) and Xenocrates (*Pythagorean Things*; perhaps also *On Numbers*, *On Geometry*, *On Dimensions*, *Speculation of Numbers*, and *On Mathematics*), and (2) the information found in accounts of Aristotle's students Theophrastus of Eresus and Aristoxenus of Tarentum might also be indebted to Early Academic accounts, either directly, or as filtered through Aristotle's writings themselves.[54] Consequently, we will have to be extra careful in our pursuit of knowledge about Pythagoreanism by attending to the modalities of construction—and the attendant philosophical assumptions—of each of these early historians of

53. Burkert 1972 and Philip 1966.

54. It was Frank's opinion (1923: 258) that Speusippus was the main source of Aristotle's information concerning the Pythagoreans. I've recently argued that Theophrastus's account of Platonic and Pythagorean metaphysics in his *Metaphysics* is chiefly derived from Xenocrates's dogmatic writings (Horky: forthcoming).

Pythagoreanism. A further consequence of this that will be treated in chapter 3 is that we will want to seek out, if possible, historical sources not obviously derived from Aristotle and/or the Platonists, in order to establish a dialectic between the Academic/Peripatetic sources and these other sources. For our purposes here, a test subject for the various modalities that the construction of mathematical Pythagoreanism took on would be with Hippasus of Metapontum (ca. 520–440 BCE?), an enigmatic figure who was considered the progenitor of mathematical Pythagoreanism. I will turn in the next section to the construction of Hippasus by early historians of philosophy.

ARISTOTLE ON HIPPASUS OF METAPONTUM

We can assume with some certainty that Hippasus first appears as a figure in the history of philosophy in Aristotle's writings.[55] Aristotle refers to Hippasus of Metapontum (ca. 520–440 BCE?) twice in his extant works: first (*Metaphysics* 1.3, 984a7–8), in an uncontroversial way, as one of the natural philosophers who believed that fire is the first principle of and prior to all other corporeal elements. His doctrine is coupled with that of Heraclitus there and placed in a general dialectic with other natural philosophers, those who held that water is the first principle (Thales and Hippon of Samos), air (Anaximenes and Diogenes of Apollonia), and these three along with earth (Empedocles). Important for my purposes, Aristotle bases his claim that these figures postulate that their respective corporeal elements are most prior on the grounds that "what is most ancient is most honorable, and what is most honorable is what we swear by" (*τιμιώτατον μὲν . . . τὸ πρεσβύτατον, ὅρκος δὲ τὸ τιμιώτατόν ἐστιν*).[56] This overall appeal to an element's priority by what is more "honorable," which is central to Aristoxenus's characterization of Pythagorean first principles, is influential as the earliest example we have of any description of Hippasus's philosophy. In Hippasus's thought, Aristotle seems to be saying, fire is the first principle and the most prior of the elements because it is the "most honorable" and "most ancient."[57]

55. It is, of course, possible that Hippias of Elis referred to Hippasus in his doxographical compendium or in his inquiry into (history of?) discoveries in geometry, on which see Kerferd 1981: 48. Glaucus of Rhegium included him in his history of music (F 90 Wehrli = DK 18 F 12).

56. Arist. *Metaph.* 1.3, 983b33–984a3.

57. It should be noted that Aristotle himself casts doubt on the historical validity of the application of this methodological principle to all the natural philosophers and theologians he lists. Rachel Barney (2012: 90) speculates that this information might be derived from Hippias's catalogue of related ideas and argues that "Homer may well have had opinions on some of the same questions as Thales, and that they were allegorically expressed is not an insuperable barrier to interpretation. What Homer did not have was an *argument*" (italics original).

The second reference to Hippasus in Aristotle's works, as I argued in chapter 1, is quoted in Iamblichus's work *On the General Mathematical Science*. That passage suggests that Aristotle himself considered Hippasus's role in the establishment of the mathematical or "so-called" Pythagorean *pragmateia*, a philosophical system whose methodology was focused on demonstration of the "reason why" (διότι), to have been foundational:

> (A) There are two types of the Italian, also called the Pythagorean, philosophy. For there were also two kinds of people who treated it: the acousmatics and the mathematicians. Of these two, the acousmatics were recognized to be Pythagoreans by the others [the mathematicians], but they did not recognize the mathematicians [as Pythagoreans],[58] nor did they think that the *pragmateia* [of the mathematicians] derived from Pythagoras, but rather that it derived from Hippasus. Some say that Hippasus was from Croton, while others say from Metapontum.[59] And, of the Pythagoreans, those who concern themselves with the sciences [οἱ περὶ τὰ μαθήματα] recognize that the others [i.e. the acousmatics] are Pythagoreans, and they declare that they themselves are even more [Pythagorean], and that the things they say [ἃ λέγουσιν] are true. And they say that the reason [αἰτία] for such a disagreement is this:
>
> (B) "Pythagoras came from Ionia, more precisely from Samos, at the time of the tyranny of Polycrates, when Italy was at its height, and the first men of the city-states became his associates. The older of these [men] he addressed in a simple style, since they, who had little leisure on account of their being occupied in political affairs, had trouble when he conversed with them in terms of sciences [μαθήματα] and demonstrations [ἀποδείξεις]. He thought that they would fare no worse if they knew *what* they ought to do [εἰδότας τί δεῖ πράττειν], even if they lacked the explanation [ἄνευ τῆς αἰτίας] for it, just as people under medical care fare no worse when they do not additionally hear *the reason why* they ought to [διὰ τὶ

58. Iamblichus elsewhere (*VP* 87, 51.7–12), in a passage that is attached to the same one given in *DCM*, attributes to a certain acousmatic Pythagorean "Hippomedon" the claim that Pythagoras originally gave demonstrations of the precepts, but that, due to the laziness of those who passed them down, ultimately only the precepts remained. Unfortunately, it is difficult to confirm this information, since (1) there are textual problems here (see Deubner's text); (2) we know almost nothing else about this Hippomedon; (3) it is possible that Iamblichus has confused "acousmatic" with "mathematical" Pythagorean here, as he had done earlier at *VP* 81, 46.26–47.3 (see Burkert 1972: 193 n. 8).

59. It is not clear to me whether this sentence is Iamblichus's insertion or original with his source, who is probably Aristotle.

πρακτέον] each thing in their treatment. The younger of these [men], however, who had the ability to endure the education, he conversed with in terms of demonstrations and sciences. So, then, these men [i.e. the mathematicians] are descended from the latter group, as are the others [i.e. the acousmatics] from the former group."

(C) And concerning Hippasus, they say that while he was one of the Pythagoreans, he was drowned at sea for committing heresy, on account of being the first to publish, in written form [διὰ τὸ ἐξενεγκεῖν καὶ γράψασθαι], the sphere, which was constructed from twelve pentagons. He acquired fame for making his discovery, but all discoveries were really from "that man" (as they called Pythagoras; they do not call him by name) . . . well, then, such are basically the characteristic differences between each philosophical system and its particular science.

(Iamblichus, *On the General Mathematical Science* 25, 76.16–78.8)

I have already discussed much of the information from passages (A) and (B) in chapter 1, but a question was left hanging there: is the information preserved under passage (C) to be considered original with Aristotle, or is Iamblichus deriving it from another source, such as the Middle Platonist Nicomachus of Gerasa?[60] Prima facie from comparative internal evidence here—that is, given the repetition of "they say" in passage (B), which appears to be direct quotation of Aristotle,[61] and passage (C), as well as the lack of a change of subject for the source quoted—it would appear that passage (C) simply refers to the same source, namely Aristotle.[62] Moreover, the appeal to "publishing" (as I've translated the term ἐξενεγκεῖν, from the Greek verb ἐκφέρω) is more fruitfully understood when compared with two passages in Aristotle's work. First, Aristotle complains in *Metaphysics* Z (7.15, 1040b2–4) that the advocates of the Ideas (i.e. Plato and Xenocrates) never "brought forth" (ἐκφέρει), i.e. produced, a definition (ὅρος) of the Idea. Second, in the *Poetics* (1, 1447b17), Aristotle claims that we need to classify writings of natural science that are "published" (ἐκφέρωσιν) in meter, such as those of Empedocles, as poetry.[63] The idea seems

60. As is held by Zhmud (2012: 186–188). But his analysis is far from conclusive, as he himself admits (2012: 191): "Whether Nicomachus was the author of the story of the *mathematici* and *acusmatici* remains open to question." He elsewhere speculates (2012: 275–276) that, if Nicomachus was indeed the source for this information, he might have had access to Eudemus or another fourth-century BCE source.

61. See above in chapter 1, section entitled "Aristotle on the *Pragmateiai* of the Pythagoreans."

62. As is assumed by DSH-MRJ in their forthcoming edition of Aristotle's *Protrepticus*.

63. How Empedocles might be embedded in the broader act of "publishing" Pythagorean doctrines I will discuss in chapter 3.

to be that the activity of "bringing forth" into the public realm is not simply an exposition of something otherwise kept internal to a certain in-group, that is, not *only* a revelation of the secrets of the mysteries (although it might be that as well).[64] Rather, it appears that, for Aristotle, this act of "bringing forth" goes hand in hand with the comprehensive organization of that information that can be valuable for *definition of* the object under investigation.[65]

If I am right about this way of understanding ἐκφέρειν—as well as about the idea that Aristotle believed that Hippasus and other mathematical Pythagoreans practiced certain (albeit simplistic) types of demonstration—then we are prompted to consider the meaning of publication of the sphere, as constructed from twelve pentagons, "in written form" (γράψασθαι). Should we understand that Hippasus was distributing the secrets of Pythagoreans to noninitiates by

64. Profanation of the mysteries, then, might be assumed to date the Hippasus story to which Iamblichus refers here to a later era, when such a concern over "publication" of the mystic secrets (e.g. the "secret" of incommensurability) to noninitiates would have been considered impious (see *VP* 88, 52.5: ὡς ἀσεβήσας), e.g. in the periods circumscribed by Middle Platonism. Note, for example, that Dillon and Hershbell (1991: 111 n. 11) speculate that the information about Hippasus might be derived from Nicomachus of Gerasa, especially given the fact that, for Nicomachus (as Burkert 1972: 461 noted), Pythagorean philosophy was ἄρρητος ἐν τοῖς στήθεσι διαφυλαχθεῖσα (at Porph. *VP* 57 = Iambl. *VP* 252, 135.19–20). But ἀσέβεια *only* appears in Iamblichus's entire corpus with reference to Hippasus, and nowhere in the extant writings of Nicomachus. It must be admitted that Plutarch (*Alc.* 19.1, 22.3) refers to the sort of ἀσέβεια with reference to the profanation of the Eleusinian mysteries allegedly committed by Alcibiades and Andocides, who apparently spoke things that were meant not to be spoken to noninitiates. This might be thought to point to a Middle Platonist invention of the "divulging" of the mysteries. But Plutarch's source here is likely to be from the late fifth century BCE, either Andocides himself (e.g. *On the Mysteries* 11, 19) or the anonymous writer of *Against Andocides*, sometimes considered to be Lysias (51). To that end, we should add late fifth-century BCE evidence from Ps.-Euripides's *Rhesus* (962–973) and Aristophanes's *Frogs* (1030–1035), which ascribe to Orpheus the revelation of the secret mysteries to humankind, as well as the testimony of the story of Aeschylus's *Bassarai* (p. 9 Nauck), in which Dionysus punishes Orpheus with *sparagmos* for his devotion to Apollo. Other evidence (cited by Burkert 1972: 461 n. 69) might apply as well. Finally, ἀσέβεια might be thought to appear in the context of Hippasus in a papyrus fragment from Herculaneum (FGrHist 84 F 34), where Neanthes, who was a contemporary of Timaeus of Tauromenium (late fourth/early third century BCE) claims that someone had been (possibly) cast out of his homeland "on account of heresy" ([διὰ τὴν ἀ]σέβειαν αὐτοῦ). We know that Neanthes spoke about Hippasus within the context of the *Pseudo-Pythagorica* of his time: he did not consider Empedocles to be a pupil of Hippasus and Brontinus (FGrHist 84 F 26 = D.L. 8.55), as the "Telauges" letter he knew apparently attested. On this fragment and the other "Pythagoreans" discussed by Philodemus in the Herculaneum fragments, see Cavalieri 2002.

65. The term ἐκφέρειν and its cognates occur most frequently in Aristotle's biological works, where it tends to refer to an animal giving birth, that is, "to bring forward to full completion" (see *GA* 3.8, 748b30: ἐξενέγκειν εἰς τέλος; *HA* 6.24, 577b23 ἐξενέγκειν διὰ τέλος).

publishing a book perhaps called the "Mystic Speech" (Μυστικὸς λόγος) and referred to by Heraclides Lembus, the second-century BCE biographer and historical compiler of Aristotle's historical works?[66] That can remain only a remote possibility. Given the context of the use of the term ἐκφέρειν here, the place of this passage in the larger characterization of mathematical Pythagoreans as those who engage in demonstrations that provide the διότι, and grammar of the passage (διὰ δὲ τὸ ἐξενεγκεῖν καὶ γράψασθαι πρῶτος σφαῖραν τὴν ἐκ τῶν δώδεκα πενταγώνων), it appears most likely that what Iamblichus and his source (Aristotle?) are trying to communicate about Hippasus is that he was the first to present a written diagram that could be used in the *demonstration* of the construction of the sphere from twelve pentagons. Such a method of demonstration might anticipate Archytas's apparent employment of diagrams for making proofs concerning optics.[67] How precisely Hippasus's demonstration might have proceeded, and why doing so would have been considered an impious act for the acousmatic Pythagoreans, must nevertheless remain a bit of a mystery.[68]

The basic point remains, however: there is no evidence that *excludes* Aristotle as the authority behind passage (C) in Iamblichus's treatment of Hippasus in his work *On the General Mathematical Science* 25. On the contrary, Aristotle's peculiar employments of the term ἐξενεγκεῖν and its cognates suggests that Aristotle remains the ultimate source for the information in passage (C), even if we cannot be sure that the language and concepts deployed here are *uniquely and solely* Aristotelian.[69] Thus, with the advent of Hippasus of Metapontum on

66. Sotion F 24 Wehrli = D.L. 8.7, abridged by Heraclides Lembus: "[Heraclides, in his Epitome of Sotion] says that there was a "Mystic Speech" by Hippasus, written to slander Pythagoras." Also see Iambl. *VP* 258–260, 139.9–140.10, which is likely to be derived from Timaeus of Tauromenium, and on which see chapter 3.

67. On which, see Burnyeat 2005a: 40.

68. The best attempt to explain why Hippasus would have demonstrated the construction of a dodecahedron from twelve pentagons has to do with the problem of incommensurability, as von Fritz (1945: 257–260) argued. But Burkert is right to note that the evidence is flimsy and doesn't quite add up (1972: 457–463).

69. It should be noted that Nicomachus uses the term ἐκφέρειν twice (*Ar.* 1.19.20, 2.28.2), but of these usages, only the second even comes close to what we see in Aristotle and Iamblichus. In the second usage, Nicomachus refers to "setting forth" the fourth-through-sixth proportions "in an order based on their opposition to the three archetypes already described, since they are fashioned out of them and have the same order" (trans. D'Ooge 1926). This is dialectical organization according to τάξις, not the completion of a demonstration. Such linguistic evidence does not encourage us to see Nicomachus behind the philosophical tenor of this information, although it does not definitively count him out either. Zhmud (2007) offers evidence that there might be external influence on the passage after (C) (starting at p. 78.8 Festa) and the passage where Iamblichus begins to quote Aristotle's *Protrepticus* (*DCM* p. 79.1 Festa) derived from Nicomachus and Ptolemy. But see my assessment of this evidence above.

the scene, we have come into a place of deep disagreement both among current scholars of the history of philosophy and science, a consequence of the sometimes intractable confusions that mark ancient accounts of developments in philosophy and science.

Up to this point in this chapter, I have focused on the types of characterization of Pythagorean philosophy undertaken by Aristotle and his student Aristoxenus. We have seen that both Aristotle's and Aristoxenus's representations of Pythagoreanism feature a shared emphasis on the complication of metaphysical and political principles by appeal to the complex sense of the term *ἀρχή*: what is an "origin" or "first principle" in their presentation of Pythagorean philosophy is also implicated in political and moral ideals, and objects in the universe are placed in an axiological hierarchy according to what is "more honorable." It is thus clear that these early historians of philosophy invest Pythagoreanism in their own clothes, and consequently we are forced to pursue alternative sources for Pythagoreanism that might show important deviations from Aristotle's and Aristoxenus's accounts. As I will show in the next section, a small wealth of information on Hippasus of Metapontum can be culled from the doxography that traces chiefly from Aristotle's student Theophrastus of Eresus.

HIPPASUS OF METAPONTUM IN THEOPHRASTUS AND THE DOXOGRAPHICAL SOURCES

Scholars have tended to view the biographical and doxographical information concerning Hippasus of Metapontum with a reasonable degree of skepticism. I list here several of the main positions taken on the authenticity of the doctrine attributed to Hippasus and the influence he held over mathematical Pythagoreanism. Walter Burkert, whose position has become canonical, considers him a natural philosopher who undertook crude but significant experiments in harmonic and mathematical theory: for Burkert, Hippasus was "the oldest Pythagorean we know of who worked at mathematics and music theory, and also had something to say in the realm of natural philosophy—though, to be sure, not in terms of a theory of number, or of a philosophy of 'limit,' 'unlimited,' and 'harmony.'"[70] Leonid Zhmud, who is sensitive to the likelihood that Peripatetic categorization has distorted our understanding of Hippasus's philosophical position and influence, is skeptical about whether or not we can view the historical Hippasus as related to the split in Pythagoreanism that produces the acousmatic and the mathematical Pythagoreans;[71] but also he sees Hippasus as

70. Burkert 1972: 206.

71. Zhmud 2012: 124–126, 185–186.

an early student of Pythagoras "engaged in philosophy, harmonics, and mathematics."[72] Carl Huffman, too, sees Hippasus as a central figure in the development of Pythagorean activity of "scientific and mathematical analysis of music, which reaches its culmination in Archytas a century later," but he, too, is careful to distinguish this information from what we have of Hippasus's cosmology, which is colored by "Peripatetic attempts to classify him."[73] Unlike Zhmud, Huffman is more optimistic that the connections between Hippasus and the schism—most likely derived from Aristotle—are legitimately related.[74] Moreover, Huffman follows Burkert in concluding, against Kurt von Fritz and others, that the famous "discovery" of the irrational by means of analysis of the dodecahedron, attributed to Hippasus, is an invention of twentieth-century scholarship.[75] For my part, I am less interested in whether the historical Hippasus, who was probably active in the early to mid-fifth century BCE, *actually* "discovered" the irrational; what I am more concerned with is what historians of philosophy active over a century later were saying about him, that is, what importance Hippasus came to have for early histories of philosophy and science, starting from (apparently) the late fifth century BCE forward.[76] By assessing the classification and appropriation of Hippasus's thought by fourth-century BCE philosophers, we will be able to reconstruct the role he played in the minds of those philosophers who considered themselves to have been influenced by the legacy of his thought.[77] Such a study aims to provide a better sense of the ways Pythagoreanism came to be worth disputing among the dominant philosophical schools in

72. Zhmud 2012: 275.

73. Huffman 2010.

74. Huffman 2010.

75. Huffman 2010, contra von Fritz 1945.

76. The best analysis of the doxographical and historical evidence concerning Hippasus is Zhmud (2012: 274–277). I will depart from his interpretation in various ways, not least by attempting to analyze the evidence for the reception of his philosophical doctrines. The earliest historical evidence concerning Hippasus dates to the history of music of Glaucus of Rhegium (late fifth century BCE), which was consulted by Aristoxenus (F 90 Wehrli = DK 18 F 12). Interestingly, Glaucus also wrote that Democritus studied with the Pythagoreans (DK 68 A 1 = D.L. 9.38).

77. In addition to the accounts of Burkert, von Fritz, Zhmud, and Huffman, Riedweg (2005: 29) calls him "an important Pythagorean of the fifth century B.C.E." and regards him as an innovator in mathematics and musicology. Most recently, Creese (2010: 93–95) argues that we should accept the testimony that refers to Hippasus's modes of demonstration in harmonic theory. Kahn (2001: 15), who does not commit himself to any interpretation on these issues, considers him a "renegade Pythagorean" who is the only named Pythagorean before Philolaus about whom we know a good deal.

Athens during the second and third quarters of the fourth century BCE, namely the Academy and the Lyceum.

A remarkable account of the life and philosophical doctrines of Hippasus of Metapontum is a short description of his philosophy preserved in Diogenes Laertius's *Lives and Opinions of Eminent Philosophers* (third century CE):

> Hippasus of Metapontum was another Pythagorean, and (A) he claimed that the time for transformation of the cosmos was definite [*χρόνον ὡρισμένον εἶναι τῆς τοῦ κόσμου μεταβολῆς*] and that the universe is limited and always in motion [*πεπερασμένον τὸ πᾶν καὶ ἀεικίνητον*]. And (B) Demetrius [of Magnesia], in his *On Homonyms*, says that he left behind no writings. There were two men named Hippasus, one being our subject, and the other being a man who wrote a *Constitution of the Laconians* in five books. He, too, was a Laconian.
>
> (DK 18 F 1 = Diogenes Laertius 8.84)

Examination of this passage, I suggest, reveals that the information collected here most likely derives from two sources, namely Aristotle's student and successor to the Lyceum, Theophrastus of Eresos (A), and the first-century BCE biographer Demetrius of Magnesia (B).[78] The evidence derived from Theophrastus (ca. 371–ca. 287 BCE) is preserved by the Neoplatonist Simplicius of Cilicia (ca. 480–ca. 540 CE) in his commentary on Aristotle's *Physics*. It is likely that Diogenes's and Simplicius's information concerning the philosophical doctrine of Hippasus derives directly from a text ascribed to Theophrastus himself, rather than through an intermediate doxographical source.[79] Indeed, the origin of the information transmitted by source (A) appears to be book 1 of Theophrastus's *Physics*, since this book dealt both with general physical principles and motion and may have featured a description of the first principles.[80] At any

78. See Simpl. *in Phys.* pp. 23.21–24.12 Diels = Theophrastus F 225 FHS&G.

79. Simplicius seems to have access to the first book of the *Physics* (e.g. *in Phys.* p. 9.7 Diels = F 144B FHS&G and 20.20 = F 143 FHS&G). If Simplicius's source is doxographical, one would need to account for why he never quotes this doxographical source by title. See Sharples 1998: 5 and Baltussen 2000: 95–96, who follows Sharples more generally.

80. It is also possible that the information preserved by Diogenes and Simplicius (Theophrastus F 224–229 FHS&G = Simpl. *in Phys.* pp. 22.22–28.31 Diels) is owed to a work entitled the *Opinions of the Natural Philosophers*. Diogenes appears to have had access to both books (D.L. 9.21–22 = Theophrastus F 227D FHS&G, where he quotes from both a *Summary* and the *Physics* itself). For a useful discussion of the debate, see Sharples 1998: 4–5. Another position is that taken by Zhmud (2006: 159) who thinks that all this information derives from *On the Principles* and that Theophrastus follows Aristotle in seeing an affinity in the principles of Heraclitus and Hippasus.

rate, the information concerning Hippasus's cosmology traces back at least to Theophrastus, but this raises another troublesome issue: is it possible that Aristotle himself is the ultimate source for the information supplied by (A) in Diogenes's account of Hippasus's cosmological doctrine?

The evidence for Aristotle as the ultimate source for the information about Hippasus's cosmology in Diogenes Laertius's account is circumstantial but not thereby dismissible. Let's examine Theophrastus's description of Heraclitus's and Hippasus's doctrines of the chief element, persisting substrate and its alteration, and time:[81]

> Hippasus of Metapontum and Heraclitus also [said that] the principle was unified and in movement and limited, but they made it fire, and [they said that] things are made from fire by condensation and rarefaction, and are resolved into fire again, since this is the single underlying nature. For Heraclitus says that all things are an exchange for fire. And he [?] says that there is a certain order and definite fated time for the transformation of the cosmos, in accordance with some fated necessity [*ποιεῖ δὲ καὶ τάξιν τινὰ καὶ χρόνον ὡρισμένον τῆς τοῦ κόσμου μεταβολῆς κατά τινα εἱμαρμένην ἀνάγκην*]."
>
> (Theophrastus F 225 FHS&G = Simplicius, *On Aristotle's Physics* 23.21–24.12; translation after FHS&G)

81. The notion that the time for transformation of the cosmos is definite is here ascribed to Heraclitus, and not to Hippasus, which confuses matters more. KRS (1983: 200 n. 1) express some concern about the attribution of this doctrine to Heraclitus, given comparative evidence from Aristotle (*Cael.* 1.10, 279b14) and Plato (242d), which suggests that Heraclitus espoused a "*simultaneous* unity and plurality of the cosmos," whereas it is to Empedocles that we should attribute "separate *periods* of Love and Strife." Still, among the Peripatetics, it was possible to ascribe any of these theories of cosmos and time to Pythagoreans or to Heraclitus. The appeal to a doctrine of periodic eternal recurrence, in which "events exactly repeat themselves at fixed periods of time" (Huffman 2010), is also credited to Pythagoras by the Peripatetic Dicaearchus of Messana (Porph. *VP* 19 = F 40 Fortenbaugh and Schütrumpf) and to the "Pythagoreans" by Eudemus of Rhodes (Simpl. *in Phys.* p. 4.12 Diels = F 88 Wehrli). Compare Censorinus, *De Die Natali* 4.2–4 (= Dicaearchus F 53 Fortenbaugh and Schütrumpf), and Varro, *De Re Rustica* 2.1.3 (= Dicaearchus F 54 Fortenbaugh and Schütrumpf). For his part, Diogenes expands on the rather vague description of Heraclitus's theories of time and cosmos: "the universe is limited and the cosmos is one, and it is alternately born from fire and again resolved into fire according to certain periods in alternation for all eternity; this comes to be according to destiny." The most likely position on this matter is that of Zhmud (2006: 163), who claims that, like Aristotle, Simplicius "does not seem to have found in Theophrastus any trace of a specific teaching on principles by . . . Hippasus that would have been different from that of . . . Heraclitus." For what it's worth, the Suda (s.v. Heraclitus) credits Hippasus with being the teacher of Heraclitus.

First of all, apart from F 225 FHS&G, there is no reference to any other natural philosopher believing that "time" (*χρόνος*) was "definite" (*ὡρισμένος*) anywhere else in Theophrastus's surviving fragments.[82] This might be thought to imply that Simplicius has mixed Theophrastus's account with another, perhaps Stoic, treatment.[83] But there is a more likely scenario, in which Theophrastus has derived his account of Heraclitus's (and Hippasus's?) theory of time from a lost work of Aristotle, perhaps called *On Time*. This is suggested by a similarity in language and conceptualization. Indeed, one of the most interesting testimonies found in Simplicius's commentaries attributes to Theophrastus the claim that Plato held that "time" was "the motion and rotation of the universe" (*τὴν τοῦ ὅλου κίνησιν καὶ περιφορὰν τὸν χρόνον*).[84] This phrase represents a truncated form of Aristotle's summary of the theories of time ascribed to those who "came before" him at *Physics* 4.10, 218a30–b9, that is, Plato and the Pythagoreans, and the first portion of it is word for word.[85] Moreover, Aristotle in *Physics* 4 goes to extreme lengths to discuss precisely how time becomes definite (i.e. by means of motion, and vice versa).[86] The concept of "definite time," as ascribed to Heraclitus (and Hippasus?) by Theophrastus and as attributed to Hippasus by Diogenes Laertius may therefore have been derived from a work on time by Aristotle.[87]

82. This is not to say that Theophrastus was uninterested in assessing things that are "definite." But, in the extant fragments, Theophrastus makes reference to "definition" most commonly with reference to differentiation by means of "principles" and "natures" in a way that does not deviate significantly from Aristotle's method of definition. See van Raalte 1993: 69–70.

83. McDiarmid (1953: 137 n. 28) detected Stoic influence here with the term *εἱμαρμένη*. But the concept of "fated time" is certainly pre-Stoic and occurs three times in Plato's writing (*Phd.* 113a3, *Prt.* 320d1, *Tim.* 89c5), always in a myth; "fated necessity" also occurs at Plato's *Laws* (918e3). It is clear that the Stoics placed a certain amount of emphasis on the strong relationships between fate and necessity, even if they only saw the implication of these two concepts in certain contexts, but there might be Platonist influence here.

84. Theophrastus F 150 FHS&G = Simpl. *in Phys.* p. 700.16–19 Diels. Simplicius also attributes this information to Eudemus and Alexander of Aphrodisias.

85. "*οἱ μὲν γὰρ τὴν τοῦ ὅλου κίνησιν εἶναί φασιν* [apparently referring to Plato], *οἱ δὲ τὴν σφαῖραν αὐτήν* [apparently referring to the Pythagoreans]. *καίτοι τῆς περιφορᾶς καὶ τὸ μέρος χρόνος τίς ἐστι, περιφορὰ δὲ γε οὔ*" etc. The phrase *ἡ τοῦ ὅλου κίνησις* never appears anywhere in Greek philosophy other than in the writings of Aristotle and Theophrastus.

86. Arist. *Phys.* 4.12, 220b14–18. On this difficult passage, see Coope 2005: 104–107.

87. What could this work be? Given that the theory of eternal recurrence is ascribed to Pythagoras and the Pythagoreans by Aristotle's associates, it is possible that this discussion would have occurred in any of the works of Aristotle on the Pythagoreans or on the Platonists. It is also possible that this description would have appeared in the work *On Time*, which is attested in the *Vita Menagiana* along with the other work of the history of philosophy, the *Metaphysics;* neither of these appears in Diogenes's list.

This would be unsurprising: in general (although not always), it was Theophrastus's modus operandi in pursuing a history of philosophy to use the works of Aristotle as the basic foundation for pursuit of his own investigations.[88] It may also be the case that Theophrastus's accounts of Presocratic time were especially indebted to his teacher's writings. Simplicius claims that both Theophrastus and Eudemus "clearly thought and taught the same opinions as Aristotle concerning time."[89] That is to say, according to Simplicius, Theophrastus's investigation into time is not independent of Aristotle's.

Examination of the second doctrinal position attributed to Hippasus by Diogenes Laertius—that "the universe is limited and always in motion" (πεπερασμένον τὸ πᾶν καὶ ἀεικίνητον)—reveals further origins in the Peripatetic history of philosophy. In Simplicius's restatement of Theophrastus's dialectical presentation of the physics of his predecessors, we hear that Hippasus and Heraclitus believed that the first principle (ἀρχή) was "unified and both in motion and limited" (ἓν . . . καὶ κινούμενον καὶ πεπερασμένον).[90] The language used in the description here varies from that of Diogenes Laertius in minor but potentially significant ways: first, Diogenes does not refer to the status of Hippasus's first principle, that is, fire, as "unified" (ἕν) but instead refers to "the universe" (τὸ πᾶν). But this apparent distinction does not present us with actual difficulties, since Simplicius, when referring to Theophrastus's description of the doctrine of other natural philosophers in this and other passages, essentially understands "the universe" and "unity" to be coextensive.[91] This fact lends credence to the proposal that the doxographer Aëtius preserves the version of Hippasus's doctrine of physics closest to that actually given by Theophrastus:

> Hippasus of Metapontum and Heraclitus of Ephesus, son of Blyson, held that the universe, always in motion and limited, is unified, and that fire is

88. See Baltussen 2000: 56–60. It should be noted that Theophrastus certainly did not employ Aristotle *alone* as the source for his knowledge but probably used the writings of Speusippus and Xenocrates for his knowledge about the history of philosophy as well. See Horky: forthcoming.

89. Theophrastus F 151B FHS&G = Simpl. *in Phys.* pp. 788.34–789.4 Diels. See Sharples 1998: 60–61 and Baltussen 2000: 97–98.

90. Simpl. *in Phys.* pp. 23.33–24.6 Diels. I remain aporetic on the question of which precise work of Theophrastus this information derives from. Clement of Alexandria (*Protr.* 64.9 Marcovich = DK 18 F 8), in the light of Theophrastus's work, claims that Heraclitus and Hippasus both believed that god was fire.

91. When Simplicius appears to refer to Theophrastus directly as his source (F 224 FHS&G = *in Phys.* pp. 22.22–23.30 Diels), he establishes the relationship between the terms "first principle," "unified," "what is," and "the universe": "that the principle is unified—or that what is and the universe is unified, *etc.*"

> the first principle [ἓν εἶναι τὸ πᾶν ἀεικίνητον καὶ πεπερασμένον, ἀρχὴν δὲ τὸ πῦρ ἐσχηκέναι].[92]
>
> (DK 18 F 7 = Aëtius 1.5.5., from Theodoret)

The notion that the analysis of "the universe" (τὸ πᾶν) is of concern to Theophrastus in his classification of the doctrines of the natural philosophers follows from another passage, quoted in direct speech by Alexander of Aphrodisias, in which Theophrastus describes the philosophical doctrines of Parmenides and Xenophanes by appeal to description of "the universe."[93] In these doxographical accounts, appeal to "the universe" as the focal point of classification for Hippasus and the other natural philosophers seems to suggest an origin in Theophrastus's dialectical studies of the Presocratics.

It can be concluded, then, that these *testimonia* concerning the metaphysical and physical doctrines of Hippasus of Metapontum as preserved by Diogenes Laertius and Aëtius trace back ultimately to Theophrastus, and, in the case of the theory of time associated with Heraclitus and (possibly) Hippasus, possibly further back to Aristotle's lost work *On Time*. According to this doxographical strain, Hippasus's philosophical doctrine is characterized by the compatible ideas that the universe is unified, continually in motion, and limited, and that fire, the first principle and chief element of the universe, is also unified, in motion, and limited. What we have, then, is a formulation of Hippasus's philosophy *in which the universe and fire, its first principle, possess the same essential properties*. As Rachel Barney has recently argued, something similar happens in Aristotle's doxographical treatments of his predecessors' thought in *Metaphysics* A and Z, where the previous philosophers' pursuit of "what is" (τί ἐστιν)

92. Mansfeld and Runia (1997: 289–290) have argued that Theodoret's account often gives readings of Aëtius's *Placita* that are more accurate than those found in Ps.-Plutarch or Stobaeus.

93. Theophrastus F 227C FHS&G = Alex. *in Metaph.* p. 31.7–16 Hayduck: "Concerning Parmenides and his doctrine Theophrastus speaks as follows in the first book of his *On Natural Things*: 'Coming after this man'—he means Xenophanes—'Parmenides, son of Pyres, from Elea followed both routes. For he both declares that *the universe is eternal* [ὡς ἀΐδιόν ἐστι τὸ πᾶν], and also tries to give an account of the coming-to-be of the things that are. He does not hold the same opinion about both; rather he supposes that in truth the *universe is one and without beginning and spherical* [ἓν τὸ πᾶν καὶ ἀγένητον καὶ σφαιροειδὲς ὑπολαμβάνων].'" See Theophrastus F 229 FHS&G = Simpl. *in Phys.* p. 28.5–8 Diels: "however, he [Leucippus] did not follow the same path as Parmenides and Xenophanes concerning the things that are, but rather, as it seems, the opposite path. For they made the *universe one and unmoved and without origin and limited* [ἓν καὶ ἀκίνητον καὶ ἀγένητον καὶ πεπερασμένον ποιούντων τὸ πᾶν], and agreed not even to inquire into what is not." The combination of "unified" and "limited" with "unmoved" in this passage might compel us to follow Zeller in emending all accounts of Hippasus's philosophy that follow from Theophrastus from ἀεικίνητον to ἀκίνητον.

becomes confused (in Aristotle's eyes) with their pursuit of the first principles.[94] It is possible that Theophrastus would have attributed the same types of confusion to various predecessors as Aristotle had done. Indeed, in his own *Metaphysics* (11a27–b1), Theophrastus speaks of "Plato and the Pythagoreans" as "mak[ing] all things desire to imitate fully" (*ἐπιμιμεῖσθαι τ' ἐθέλειν ἅπαντα*) the first principles.[95] What does this mean for our investigation into the construction of mathematical Pythagoreanism? It suggests that, according to Theophrastus and the doxography that follows from his writings on the Presocratics, Hippasus's connection between the first principle and the universe may have been presented as mediated by a sort of "imitation" or "assimilation" of objects across the whole spectrum of existence. This is precisely the sort of metaphysics of imitation that Aristotle and Theophrastus criticized the "so-called" Pythagoreans and "Plato and the Pythagoreans," respectively, for employing.[96]

It does not appear to be the case, however, that the Peripatetics are the only sources whose description of the philosophy of Hippasus of Metapontum was passed down to later commentators. Other *testimonia*—preserved by Iamblichus in two passages—concerning Hippasus's doctrine of physics, number, and the soul, are striking and stimulating, but they, too, raise problems of authenticity and contamination:

> (I) But indeed some of the Pythagoreans assimilate [*συναρμόζουσιν*] this [i.e. number] to the soul simply, in this way: Xenocrates, insofar as it [i.e. the soul] is self-moving [*αὐτοκίνητον*]; Moderatus the Pythagorean, insofar as it contains ratios [*λόγους περιέχουσα*]; Hippasus, a mathematician among the Pythagoreans, insofar as it (A) is the discerning tool of God the world-maker [*κριτικὸν κοσμουργοῦ θεοῦ ὄργανον*].[97]
>
> (DK 18 F 11 = Iamblichus *On the Soul* 364, from Stobaeus, *Eclogae 1.49.32)*

> (II) The followers of Hippasus, the mathematicians, said that number is the (B) "first paradigm of the making of the world" and again (A) "the discerning tool of God the world-maker."

94. See Barney 2012: 76–85.

95. On this passage, see Horky: forthcoming.

96. On Aristotle's criticism of this procedure, see chapter 1. Also see Aristoxenus F 23 Wehrli (discussed above in the section entitled "Pythagoreanism and the Axiology of What Is 'Honorable'"). For Theophrastus's criticism of Plato and the Pythagoreans' presentation of a metaphysics of imitation (*Metaph.* 11a27–b7), see Dillon 2002: 185–186 and Horky: forthcoming.

97. Following Burkert (1972: 193 n. 8), I have emended the texts from *ἀκουσματικός* to *μαθηματικός*.

οἱ δὲ περὶ Ἵππασον μαθηματικοὶ ἀριθμὸν εἶπον παράδειγμα πρῶτον κοσμοποιίας καὶ πάλιν κριτικὸν κοσμουργοῦ θεοῦ ὄργανον.

(DK 18 F 11 = Iamblichus, *On the Introduction to Arithmetic of Nicomachus* 11, 10.20–23)

There is a variety of interesting interpretive problems here, and scholars have been hesitant to attribute any of this information to Hippasus at all (except, of course, for the possibility that he was a mathematical Pythagorean).[98] Ultimately, the challenges posed by these fragments have led to the abandonment of their content entirely, and it is unfortunate that scholars have neither taken seriously nor sufficiently discussed the doctrine or the language used to present it from these passages. But there is good reason, I suggest, to see their content as reflective of Peripatetic and early Platonist descriptions of Hippasus's doctrine, which is significant because it corroborates my hypothesis that, in the mid-fourth century BCE, philosophers in Athens were debating the role that Hippasus played in the development of philosophy. I will examine each doctrinal position separately, given the possibility that Iamblichus's doxographical source may have combined an authentic with an inauthentic doctrinal position.

The portion from Iamblichus's work *On the Soul* that refers to Hippasus (I) aims to distinguish between the various philosophers who believed that Soul was predicated of a "mathematical substance": some, such as the Platonists Seberus and Speusippus, thought that it was a figure (*σχῆμα*), while the Pythagoreans Xenocrates, Moderatus, and Hippasus held that it was a number (*ἀριθμός*). The reference to Moderatus suggests that Iamblichus's immediate doxographical source must be no earlier than the first century CE, and the word *κοσμουργός*, a term employed by later Neoplatonic commentators such as Proclus to refer to the tetrad, appears here for the first time.[99] Thus we might assume a late Hellenistic or early Imperial source for this particular *testimonium*, at the very earliest.[100] But, in this circumstance, source criticism does little to explain the significance of this fragment for Iamblichus's work *On the Soul*: if Soul-number is (A) the "discerning tool of God the world-maker," how would this explain the arithmeticization of the soul for Hippasus? A

98. Burkert calls this citation "apocryphal" (1972: 194 n. 9). Finamore and Dillon (2002: 94) speculate that it may be a late Hellenistic pseudepigraphon. Neither Huffman (2006) nor Zhmud (2012) discusses it.

99. Procl. *in Tim.* 3.232.17. See Finamore and Dillon 2002: 83–84.

100. It is possible that Moderatus himself is the immediate source here, since he appears to have developed a history of philosophy from the Pythagoreans to Plato and beyond (see Porph. F 236 Smith).

doxographical *testimonium* by Aëtius (4.3.4), which notes that Hippasus and Heraclitus thought that the soul was "fiery" (πυρώδη), offers little help.[101] One suggestion, that of Dillon and Finamore, is that Hippasus's Soul-number as κριτικὸν ὄργανον individuates particulars and arranges them into species and genera.[102] This explanation, which resonates with Proclus's description of the "world-making" power of the Demiurge in his *Commentary on the Timaeus*, seems somewhat anachronistic.[103] Generally, two mathematical uses of discernment are attested in earlier ancient philosophy, one Peripatetic and another Platonist. The Peripatetic use, which is more obviously mathematical, postulates that discernment functions as a mean that establishes relations between two extremes. In *On the Soul* 2.11, 424a5–12, Aristotle claims that sense-perception "discerns perceptibles" since "what is in the middle is discerning" (τὸ μέσον κριτικόν) and "relative to each extreme, it comes-to-be in the other," as in the case of the colors white and black. When discussing the particular types of sense-organs, Aristotle refers to "sight" as that which "discerns" (κρίνει) between white and black.[104] This description of "discernment" is the one commonly adhered to by late commentators such as Simplicius and Priscian, both of whom were influenced by Iamblichus himself. If a pseudepigrapher who was writing about Hippasus wished to Aristotelianize the Pythagorean's philosophy by invoking Peripatetic language and conceptual structures, as was common in the Hellenistic Pseudo-Pythagorean texts, for example, this would most likely produce a meaning for "discerning tool" that could be squared with the Aristotelian usage.[105] This does not, in fact, happen.

On the other hand, there is a tradition among the Middle Platonists that attributes "discernment" and the "discerning tool" to the soul itself in various ways.[106] Plutarch, in his work *On the Procreation of the Soul in the Timaeus*,

101. DK 18 F 9. See Tert. *de Anim*. 5.

102. Finamore and Dillon 2002: 84.

103. Compare the description of the soul's *logoi* (noetic v. organizational) at Procl. *in Tim*. 2.263.17.

104. Compare the system ascribed to Empedocles by Theophrastus (*DS* 7), in which vision, which is said to be unable to "discern" (κρίνειν) the objects of the other senses, can "recognize" (γνωρίζειν) the colors white and black by means of the pores that are either, respectively, fitted for fire or for water.

105. On the appropriation of Aristotelian and Platonic terms and concepts in the *Pseudo-Pythagorica*, see Huffman 2005: 91–100. It is surprising, in fact, that no extensive *Pseudo-Pythagorica* exist as such for Hippasus.

106. For a discussion of the Presocratic traditions concerning the discovery of the "arts of discernment," see chapter 6.

comments on Plato's description of the World-Soul's composition in the *Timaeus* (34b10):

> For each one proceeds from a different principle; the Same from the One, the Other from the Dyad; and they were first intermixed there, in the soul, and bound together [συνδεθέντα] by numbers, proportions, and harmonic means.[107] The Other, by coming-to-be in the Same, produces difference; the Same, by coming-to-be in the Other, produces order. And these things are clear in the primary powers of the soul [ἐν ταῖς πρώταις τῆς ψυχῆς δυνάμεσιν], which are discernment and motion [τὸ κριτικὸν καὶ τὸ κινητικόν]. . . . Discernment has two principles: mind from the Same, with regard to what is general, and sense-perception from the Other, with regard to what is particular. Reason is an admixture of both [mind and sense-perception], becoming thought in intelligible objects and opinion in perceptible objects and making use of the intermediary tools [ὀργάνοις μεταξύ] of imagination and memory.
>
> (PLUTARCH, *On the Procreation of the Soul in the Timaeus* 1032c3–d5)

Here, in one of the longest surviving Middle Platonist descriptions of τὸ κριτικόν, Plutarch defines it as one of the primary "powers" of the soul and as having two principles: mind (νοῦς) and sense-perception (αἴσθησις). The precise function of discernment, to be sure, is ambivalent: it appeals to mind when it distinguishes between things at the general or universal level, whereas it appeals to sense-perception when it distinguishes between particulars.[108] In this sense, Plutarch's description of the World-Soul's power of τὸ κριτικόν can function with appeal either to the higher, noetic stratum, or to the lower, bodily stratum.[109] Although this passage presents many challenges to interpreters, we can say with some plausibility that Plutarch considers the noetic type of discernment to be the tool of memory, whereas he sees the perceiving type of discernment to be the tool of imagination. Thus, in Plutarch's estimation, the World-Soul possesses two types of κριτικὸν ὄργανον: (1) memory, whose principle is mind and whose objects are intelligible, and

107. See Pl. *Tim.* 37a2–4: "Since the soul was comprised of three portions, namely the nature of the Same, the nature of the Other, and Essence, and since it was divided and bound together [συνδεθεῖσα] according to proportion."

108. It should be noted that Plato's cosmos-organizing figures (the Demiurge in the *Timaeus* and Mind in the *Philebus*) do not explicitly undertake diacritical activities of these sorts.

109. It is difficult to correlate this description with Plutarch's own of human souls, as described by Dillon (1996: 211–213).

(2) imagination, whose principle is sense-perception and whose objects are perceptible.

Among Middle Platonists of the first and second centuries CE, Alcinous, too, describes something like the *κριτικὸν ὄργανον* as illustrated by Plutarch and attributed to Hippasus by Iamblichus. As Dillon suggests, Alcinous's employment of the term *κριτήριον* effectively Platonizes an otherwise Stoic term by proposing that it is an activity subordinate but related to Platonic dialectic:[110]

> Since there is something that discerns, and there is something that is discerned, there must also be something that results from these, and that may be called discernment [*κρίσις*]. In the strictest sense, one might declare discernment to be the act of discernment, but more broadly that which discerns. This may be taken in two ways: (1) that *by the agency of which* what is discerned is discerned, and (2) that *by means of which* it is discerned. Of these the former would be the mind [*νοῦς*] in us, while that "by means of which" is the natural discerning tool [*ὄργανον φυσικὸν κριτικόν*]—primarily truth, but consequently also falsehood; and this is none other than natural reason.
>
> (ALCINOUS, *Handbook of Platonism* 154.10–18; translation after Dillon 1993)

As George Boys-Stones has convincingly argued, what makes Alcinous's account of *κριτήριον* distinctive in the history of ancient philosophy is the way it focuses on "what is required of the agent judging" in the context of epistemology; the focus, then, is not on the quality of perceptible information with which we make judgments.[111] Alcinous develops this agent-centered epistemology by going on to say that "the discerning agent" could either be the philosopher or reason, which, he reiterates, is a "tool" (*ὄργανον*) of the agent.[112] He then proceeds to describe two types of reason: (1) that which is ungraspable and unerring, and (2) that which is only free from error when it attempts to cognize reality.[113] The former type of reason, which is a tool, can only be employed by God, while humans are capable of attaining the latter. Finally, human reason may be further divided into two types: (1^1) that which concerns intellectual objects, and (1^2) that which concerns perceptibles.[114] A proper diaeretic schema

110. Dillon 1993: 61–64.

111. Boys-Stones 2005: 208–209.

112. Alcin. *Intr.* 154.18–21.

113. Alcin. *Intr.* 154.21–25.

114. Alcin. *Intr.* 154.25.

for these types is difficult to draw, since Alcinous is analogizing the ὄργανον across all the schematic divisions: the "tool" is reason, which, as in Plutarch, can deal with intelligibles and perceptibles alike; but what Alcinous introduces here is the notion that there is one type of reason-tool, namely that of God, which is distinct from that of human beings.[115] While both these dogmatic Middle Platonists conceive of "discerning tools," only Alcinous conceives of a divine tool that is the *explicit* provenance of God. Thus, among Middle Platonists, Alcinous's appeal to the "discerning tool of God" coordinates most closely with the description of Soul-number attributed to Hippasus of Metapontum by Iamblichus.

If it is the case that both Alcinous and Plutarch emphasize a κριτικὸν ὄργανον in the context of Platonist epistemology, and that this term helps them to articulate an agent-centered doctrine for Middle Platonism, the question arises: how far back in the history of the Academy does the κριτικὸν ὄργανον reach? Of course, Plato himself had made gestures in this direction, but without being systematic: in *Republic* 9 (582a3–d9), Socrates distinguishes the philosopher from the honor-lover and the profit-lover on the grounds that one must "make distinctions" (κρίνεσθαι) through the philosopher's "tool" (ὄργανον), that is, through "arguments" (λόγοι).[116] Moreover, discernment appears in connection with vision in the *Timaeus* (67d5–e8) and is thus related to perception: the "dilating" (διακριτικόν) type of vision is called "white," and the "contracting" (συγκριτικόν) type of vision is called "black." The eye, of course, is compared with the "tool in the soul" with which we learn in *Republic* 7, but it is not explicitly allotted any discernment-functions there.[117] We should be wary to attribute the concept of the κριτικὸν ὄργανον to Plato's unwritten teachings: Socrates's description in the *Republic* seems not to participate in an established debate over discernment-functions as we see in Plutarch's and Alcinous's treatments, and it holds little in common with the description of Soul-number as the κριτικὸν ὄργανον attributed to Hippasus of Metapontum, which is, importantly, a tool *of God*.[118] The ascription of such a tool to a divine being smacks of the type of Middle Platonism that sought to recover the lost limbs of

115. See Boys-Stones 2005: 209–210, with n. 8, and Sedley 1996: 303–306.

116. See the discussion of this passage at Dillon 1993: 61. See Sedley 1996: 302, who also cites *Theaetetus* 184–185 as a possible source. For "discernment" in the heurematographical contexts of Greek philosophy from Heraclitus to Plato, see chapter 6.

117. Pl. *R.* 7, 527d6–e3. See *R.* 7, 518c5–10 and 6, 508b3–4.

118. Clement of Alexandria (DK 18 F 8 = *Protr.* 64.9 Marcovich), deriving his information from a doxographic source, states that Hippasus and Heraclitus thought fire *is* god.

metaphysics and cosmology in Plato's philosophy, as against Academic Scepticism and others.[119] For my purposes, it is important to note that the definition of a *κριτικὸν ὄργανον* as an instrument to be used in Platonic dialectical procedure is likely not Plato's but is derived from one of the Platonist exegetes. So if it is not to Plato that we refer such a description of the *κριτικὸν ὄργανον*, to whom should we look?

My investigation into the source for the claim that Hippasus's philosophy believed that number is the "discerning tool of God the world-maker" (*κριτικὸν κοσμουργοῦ θεοῦ ὄργανον*) has gained some traction thanks to the accounts of the Middle Platonists Plutarch and Alcinous, but it is also worth bringing to bear on the investigation another testimony, that of the Anonymous Commentator on Plato's *Theaetetus* (perhaps second century CE), who contrasts his own approach to Plato's epistemology with that of other Platonists:

> Some of the Platonists have thought that the dialogue [the *Theaetetus*] was on the topic of the criterion, in view of the considerable space it also devotes to investigation of this. That is wrong. Rather, the declared aim is to speak about simple uncompounded knowledge, and it is for this purpose that he necessarily investigates the criterion. By "criterion" in the present context I mean the criterion through which we judge, as an instrument [*λέγω δὲ νῦν κριτήριον τὸ [δ]ι'οὗ κρίνομεν ὡς ὀρ[γ]άνου*]; for it is necessary to have that whereby we will judge things; then, whenever this is accurate, the permanent acceptance of the things which we have judged properly becomes knowledge.
>
> (Anonymous Commentator on Plato's *Theaetetus* 2.11–32 ed. Bastianini and Sedley; translation by Sedley)

Sedley is surely right to see figures such as Alcinous in the crosshairs of this account, but we cannot assume that the debate began with him. Dillon speculates about the possibility of Antiochus of Ascalon (second to first century BCE). This works for Alcinous's account of the human *κριτήριον*, although neither Antiochus nor his mouthpiece Varro in Cicero's *Academica* refers to anything like the "discerning tool" of God, as we find in the *Handbook* of Alcinous and in Iamblichus's doxographic account of the cosmology of Hippasus.[120] We ought to look, instead, for a Platonist before Antiochus who, while still holding some interest in matters of logic and dialectic, contextualizes these with cosmology, mathematics, and organization of the universe. It also might

119. See Boys-Stones 2001: 138–142, with reference to Numenius of Gerasa.

120. Dillon 1996: 63–69, 273–274.

be worth considering Platonist thinkers who focused, as Hippasus is said to have done, on "divine" epistemology. One alluring candidate, I suggest, would be Xenocrates of Chalcedon, who held doctrinal positions on metaphysics and psychology quite similar to those ascribed to Hippasus by Iamblichus, and who is associated with him in the doxographical grouping. Indeed, as I will show, two important fragments attributed to Xenocrates coordinate, in striking ways, with the doctrine ascribed to Hippasus of Metapontum by Iamblichus. Examination of these *testimonia* ultimately suggests that, even in the absence of direct evidence, there is good reason to speculate that the Early Platonists wrote about Hippasus and may have been the first to assimilate the doctrine of Hippasus to Pythagorean ideals.

Concerning the tradition that attributes to Hippasus (A) the notion that Soul-number is the discerning tool of God, it is worth examining the account of Nemesius of Emesa (fourth century CE), who compares the psychological theory of Pythagoras with that of Xenocrates:

> Pythagoras, who, in all cases, was accustomed to assimilate symbolically both God and everything [else] to numbers [συμβολικῶς εἰκάζειν ἀεὶ καὶ τὸν θεὸν καὶ πάντα τοῖς ἀριθμοῖς εἰωθώς], also defined the soul as "number that moves itself." And even Xenocrates followed him, not because the soul is number, but because it is among things that can be numbered as well as among things that are multiple, and because the soul is that which distinguishes things by assigning shapes and characters to each [ὅτι ἡ ψυχή ἐστιν ἡ διακρίνουσα τὰ πράγματα, τῷ μορφὰς καὶ τύπους ἑκάστοις ἐπιβάλλειν]. For the soul is that which separates out Forms from [other] Forms [αὕτη γάρ ἐστιν, ἡ τὰ εἴδη ἀπὸ τῶν εἰδῶν χωρίζουσα] and brings to light their differences, both in terms of otherness of Forms and in terms of numerical count, and thereby it makes things numbered. For this reason things are not altogether divorced from their sharing in numbers. And he also gave witness to the soul's being self-changing.
>
> (NEMESIUS, *On the Nature of Man* 2.102 = Xenocrates F 190 IP)

This testimony of Xenocrates's theory of the soul's activity of judgment is remarkable and unique among similar doxographical accounts for its detailed expression. While it is likely that Nemesius is deriving this description of Xenocrates's and Pythagoras's soul as self-moving number ultimately from an intermediary source, the other doxographical accounts that refer to this relationship in Aëtius's collection are skeletal and far less descriptive: in light of the proposition that Soul is "number that moves itself," Pseudo-Plutarch does not mention Xenocrates at all, and Stobaeus only mentions that Xenocrates is in agreement with Pythagoras concerning nature of the soul.[121] Neither source, though, preserves the description that follows in Nemesius, which goes into far greater detail than might be expected: it is especially striking to see that Nemesius suggests that Pythagoras and Xenocrates believed that Soul had a *discerning-function*, specifically the function of discerning things by assigning "shapes and characters" (μορφὰς καὶ τύπους) to each thing. This doctrine and the language used to describe this intellective activity are unparalleled in Aëtius and in earlier doxographers and historians of philosophy, so we might be on safest ground if we assume that his source taps into an alternative tradition, perhaps Middle Platonist.[122] While it is possible that the association of the soul and number with a discerning-function is genuine Xenocratean doctrine, the doxographical reports and later commentaries infuse this principle with Middle Platonist and Neoplatonist language and concepts, and as such we cannot safely conclude, *on this evidence alone*, that the attribution of the doctrine of number as the (A) "discerning tool of God the world-maker" to Hippasus predates the first century BCE.[123] It may remain a possibility, but we would need to have some sort of corroborating evidence that figures earlier than the Middle Platonists were deliberately Platonizing Hippasus's doctrines.

On the other hand, some evidence survives that suggests that the second doctrinal position preserved by Iamblichus, namely (B) that number is the "first paradigm of the making of the cosmos" (παράδειγμα πρῶτον κοσμοποιίας) can be traced back to the Early Academy, and, moreover, that this position might be associated with Xenocrates's and/or Speusippus's characterizations of the Pythagoreans. Information presented by Proclus constitutes a stimulating *testimonium* about Xenocrates's theory of the Forms:

> For this reason [Plato] ascended to these [i.e. the Forms] as first principles and made the whole of creation depend on them, in accord with what Xenocrates, says, who defines the Form [ἰδέα] as the paradigmatic cause

121. See Aët. 4.2.3–4. Isnardi Parente (1982: 389–390) hypothesizes that the ultimate source here is Plutarch.

122. The unusual description of soul as that which "assigns shapes and characters" (μορφὰς καὶ τύπους ἑκάστοις ἐπιβάλλειν) to everything can only be paralleled in one passage that I can find, in Hermias's *Commentary on Plato's Phaedrus*, which is substantially derived from Syrianus's commentary. There (p. 193.1 Couvreur) Hermias describes how Plato (*Phdr.* 253c) gives attributes and characteristics to the various figures that are composite in the soul, namely the two horses and the charioteer.

123. The additional doxographical material presented in Nemesius's description of the psychology of Xenocrates and Pythagoras would thus likely represent what Mansfeld and Runia (1997: 297) describe as an "intervention" of Nemesius or an intermediate source.

> of whatever is composed continually in accordance with nature [αἰτία παραδειγματικὴ τῶν κατὰ φύσιν διεστώτων] . . . even if we say that it creates by reason of its very essence, and that becoming like to it is an end for all generated things, nevertheless the final cause of all things in the strict sense and that for the sake of which all things are is superior to the Forms, and the efficient cause in the strict sense is inferior to them, looking to the paradigm as a criterion and rule [πρὸς κριτήριον βλέπων καὶ κανόνα τὸ παράδειγμα].
>
> (PROCLUS, *On Plato's Parmenides* p. 691 Stallbaum = Xenocrates F 94 IP)

This passage affords us numerous difficulties in assessing what precisely may be attributed to Xenocrates and what is modified by later doxographers. Most scholars agree that the definition of the Form as "paradigmatic cause of whatever is composed continually in accordance with nature" is genuine Xenocratean doctrine, although there are interesting debates about what actual words to attribute to Xenocrates.[124] What follows in the *testimonium* is an attempt to explain the "paradigmatic cause" (αἰτία παραδειγματική) in Aristotelian terms, i.e., as both the final and efficient cause. These are sure signs of distortion, as Tarán and Dillon have noted.[125] Still, their analyses have not accounted for the reiteration and further elaboration of the concept of the "paradigm" (παράδειγμα), understood to be to that which the "efficient cause" (i.e. the prime mover) looks as the "criterion and rule" (κριτήριον καὶ κανών). As Long and Sedley note, the technical phrase κριτήριον καὶ κανών appears frequently in Hellenistic philosophy after Epicurus, so it would appear likely that this description cannot be older than the end of the fourth century BCE.[126] But the appeal to the "rule" (κανών) and its applications in a wide philosophical sense is, as Huffman has recently argued, genuinely Archytan and thus traces back to

124. For the debate about this fragment before 1982, see Isnardi Parente 1982: 325–326. Since then, to my knowledge, it has been discussed extensively only by Tarán (2001: 564–622, originally published in 1987) and Dancy (2011), who both agree that this doctrine is actually Xenocratean. Remarkably, it is not treated in Thiel's monograph on Xenocrates (2006).

125. Tarán 2001: 587 and, more generally on this fragment and its relation to a corresponding description in Alcinous's *Handbook of Platonism* (*Intr.* 163.11–14), see Dillon 1993: 96–97.

126. Long and Sedley 1987, vol. 1: 88–90. One therefore wonders if the testimony of Sextus Empiricus on Philolaus (*Adv. Math.* 7.92 = T A 29 Huffman), which claims that Philolaus thought that the reason that arises from the mathematical sciences is a κριτήριον, is an invention of the third century BCE.

the mathematical Pythagoreans of the first half of the fourth century BCE.[127] This concept also may have been linked by Anticleides of Athens, a late fourth-century BCE historian, to Pythagoras.[128] While it is highly likely that Xenocrates did indeed refer to the ἰδέα as a παράδειγμα, as Tarán and others have argued,[129] we cannot be absolutely sure that Xenocrates assimilated the "Form-paradigm" to the κριτήριον καὶ κανών or that Xenocrates was adapting some sort of Archytan metaphysics for his own proposition of a formal theory.[130]

If Xenocrates were to consider the "Forms" as "paradigms," he would not be innovating dramatically: a strong association of these concepts follows almost naturally from a reading of Plato's *Timaeus*, and more important, it might have already been circulating in the Early Academy after Plato's death. Xenocrates's predecessor in the Academy Speusippus may have had a more comprehensive doctrine of the efficient capacities of the "forms": if we are to trust the account in the Iamblichean *Theology of Arithmetic* (83.1–5 = F 28 Tarán), Speusippus, in a work entitled *On the Pythagorean Numbers*, ascribed to the Pythagoreans the notion that the Decad was

> the most natural and perfective of entities, a sort of productive form in itself (and not owing to our thoughts or to luck) for cosmic things that have come to completion, a preexistent foundation set before the god, maker of the universe, as the most complete paradigm.
>
> φυσικωτάτην . . . καὶ τελεστικωτάτην τῶν ὄντων, οἷον εἶδος τι τοῖς κοσμικοῖς ἀποτελέσμασι τεχνικὸν ἐφ᾽ἑαυτῆς (ἀλλ᾽ οὐχ ἡμῶν νομισάντων ἢ ὡς ἔτυχε) θεμέλιον ὑπάρχουσαν καὶ παράδειγμα παντελέστατον τῷ τοῦ παντὸς ποιητῇ θεῷ προεκκειμένην.
>
> (Ps.-Iamblichus, *Theology of Arithmetic* p. 83.1–5 = Speusippus F 28 Tarán)

127. In Archytas F 3 Huffman, Archytas refers to calculation (λογισμός) as the "standard and hindrance" (κανὼν καὶ κωλυτήρ) that prevents types of injustice. See Huffman 2005: 218–220. Democritus, whose connections with Pythagoreanism cannot be doubted (see Zhmud 2012: 45) wrote a work on logic in three books κανών (see DK 68 B 10b and 11).

128. See FGrHist 140 F 1 (= D.L. 8.12). See Burkert 1972: 375 n. 22 and 407 n. 37.

129. See Dancy 2011, Dillon 2003: 118–121, and Tarán 2001: 590. On Plato's appeal to the Form-number, see chapter 5, section entitled "What Is Wisest? Three."

130. In this context, it is worth mentioning Xenocrates F 83 IP (= Sext. Emp. *Adv. Log.* 1.147–49), where Sextus ascribes to Xenocrates the doctrine of three essences (the noetic beyond heaven, the mixed, and the sensible under heaven), each of which has an epistemological criterion (κριτήριον) assigned to it: knowledge for the noetic, opinion for the mixed, and perception for the sensible.

As Tarán notes in his commentary, the language here might have been doctored by the excerptor or the immediate source (Nicomachus of Gerasa?),[131] but he doubts this hypothesis on the grounds that most of the terms were meant to recall the vocabulary of Plato, especially in the *Timaeus*.[132] Relative to the doxography of Hippasus of Metapontum, at least, we might note that the language here is more baroque (e.g. superlatives, adjectives ending in *-ικος*) and less representative of what we would expect of a plain doxographic report. Be that as it may, the description of the Pythagorean Decad attributed to Speusippus's work *On Pythagorean Numbers* represents an extended version of what is otherwise, in the doxographic tradition of Hippasus, a rather simply and obscurely rendered definition of number. In other words, *both Speusippus's Decad and Hippasus's number are what God employs as the paradigm in order to impart order onto things in the universe.*[133] Given the affinities between these descriptions, we might consider two possibilities. First, it is possible that Speusippus himself wrote about Hippasus and implicitly compared his cosmology with Pythagoreanism. In this case, we would have the earliest example of a philosopher who claimed Hippasus of Metapontum as a Pythagorean, in contradistinction to the Peripatetic description of him as a natural philosopher and material monist.[134] Second, it is possible that the doctrinal position ascribed to Hippasus, namely that Soul-number is the "first paradigm of the making of the world," might be thought to derive from the same intermediary source as the description of the Decad by Speusippus—possibly Nicomachus of Gerasa, or perhaps another Middle Platonist or Neopythagorean who was seeking to approximate Pythagorean numerology to Platonic metaphysics.[135]

131. Also see Huffman 1993: 361.

132. Tarán 1981: 270–271.

133. Contrast this with Proclus's association of the Tetrad with the Paradigm, and the Decad with the Demiurge, both of which he associates with "the Pythagoreans" (*in Tim.* 2.316.17–317.5).

134. Of course, the musician and historian of music Glaucus of Rhegium was the first to write about Hippasus (F 90 Wehrli = DK 18 F 12) and one of the earliest figures to classify people as Pythagorean (such as Democritus: D.L. 9.38).

135. Compare *Ar.* 1.4.2, where Nicomachus says that the "arithmetic method . . . existed before all the others in the mind of the craftsman god [*τεχνίτης θεός*] like some cosmic and paradigmatic proportion [*λόγον τινὰ κοσμικὸν καὶ παραδειγματικόν*], relying on which as a design and archetypal model [*ἀρχέτυπον παράδειγμα*] the craftsman of the universe sets in order the products made from matter [*ἐκ τῆς ὕλης ἀποτελέσματα*] and makes them attain to their proper ends" (trans. after D'Ooge 1926). On this passage, see Dillon 1993: 354–358. The juxtaposition of "paradigm" and "archetype" also recalls Philo (e.g. *Opif.* 78.13–14), whose lost *On Numbers*, were it to have survived, might have been a valuable point of reference. See

This hypothesis encounters some difficulties: at least among extant texts, Nicomachus never names Hippasus, although he does have intimate and reliable knowledge of Archytas's fragments and speculates about those who are responsible for discovery of the various mathematical means.[136] Given the reliable evidence of Hippasus's discoveries in harmonics given by Aristoxenus and those discoveries in the mathematical means probably given by Eudemus, both of which Nicomachus seems to have had access to,[137] we would expect Nicomachus to discuss Hippasus somewhere either in the *Introduction to Arithmetic* or the *Enchiridion*, but he does not obviously do so.[138] Moreover, no other surviving Middle Platonists or Neopythagoreans refer to Hippasus: the sources for the philosophical doctrines of Hippasus appear to trace from Aristotle, Theophrastus, and Aristoxenus into the doxographical traditions, disappear for several centuries, and resurface in the writings of Clement of Alexandria and Iamblichus. There is no sign of significant interest in Hippasus, or anything like the doctrines under (A) or (B) in Iamblichus's account from *On the Introduction of Arithmetic of Nicomachus*, in the Hellenistic Pseudo-Pythagorean writings. It thus appears more likely, I suggest, that Iamblichus has retained an original description of "Soul-number" that both traces back to the Early Academy and was passed down in the doxographical tradition to him without any direct contamination by Middle Platonism or by the Pseudo-Pythagorean traditions.

Dillon 1996: 158–159. One other option would be Eudorus of Alexandria, who was interested not only in the mathematical means (see F 7 Mazzarelli = Plut. *An. Proc.* 1019e–f) but also in earlier theories of divine *κοσμοποιία* (e.g. F 25 Mazzarelli = Stob. *Ecl.* 2.7.3).

136. See Huffman 2005: 112–124 and 1993: 168–169. Zhmud (2012: 265–266) thinks that it was Eudemus who associated Hippasus with discoveries in the mathematical means. Both Archytas and Hippasus are credited with changing the name of the third proportion from subcontrary to harmonic (Iambl. *in Nic.* 141–142, 100.19–101.1 Pistelli = DK 18 F 15), and Archytas links himself to Hippasus specifically by speaking of the third proportion as "what we call harmonic" (F 2 Huffman).

137. See Zhmud 2012: 173–174 and Barker 2012: 322.

138. Instead, Nicomachus (*Ar.* 2.22.1, 122.11–20 Hoche) attributes knowledge of the first three means to "Pythagoras, Plato, and Aristotle," does not attribute discovery to anyone for means 4–6; he goes on to claim that "the more recent" (*οἱ νεώτεροι*) were the ones who "filled out [*συμπληροῦντες*] the tenth number [i.e. the Decad] as the 'most complete' [*τελειότατον*], according to the thinking of the Pythagoreans." Apparently, he is distinguishing the "more recent" philosophers (e.g. Speusippus) from the Pythagoreans. This account contradicts the systematization of Iamblichus, who claims (*in Nic.* 163, 116.1–7 Pistelli = Archytas F 2 Text C Huffman) that the final four means were discovered by the Pythagoreans Myonides, Euphranor, and their followers. Where Iamblichus obtains this information is unclear.

One might wish to entertain the possibility that the Peripatetics may have ascribed the doctrine of Soul-number as (B) "first paradigm of the making of the world" to Hippasus of Metapontum. The term *κοσμοποιία* appears either in its nominal or verbal form in Aristotle five times, and always in reference to the Presocratic natural philosophers.[139] Given this fact, one might generalize that Aristotle's discussion of *κοσμοποιία* would not normally come to be associated with a Pythagorean or Platonist. But in one instance, Aristotle, while criticizing the physics of the Pythagoreans and Platonists, expresses confusion about physical systems he is describing. He claims that the Pythagoreans believe that eternal things are generated, which leads to an absurdity that, consequently, emphasizes the physical assumptions of the Pythagoreans and the Platonists (by association):

> They [i.e. the Pythagoreans] clearly state that when the one had been constituted [*συσταθέντος*]—whether out of planes or superficies or seed or out of things they are at a loss to say—immediately the nearest part of the unlimited began to be drawn in and to be limited by the limit. But since they are making the world [*κοσμοποιοῦσι*] and intend to speak in terms of nature [*φυσικῶς βούλονται λέγειν*], while it is rather just to criticize their physical theories, we should leave them be for our present inquiry. For we are pursuing the first principles in unchangeable things, with the result that we ought to examine the generation of these sorts of numbers.
>
> So they deny that there is generation of the odd, which evidently implies that there is generation of the even; and *some* hold that the even is first [*πρῶτον*] constructed out of unequals—the great and the small—when they are equalized.
>
> (ARISTOTLE, *Metaphysics* 14.3, 1091a13–26)

Commentators and translators since Ross have distinguished the Pythagoreans from the "they" who "deny that there is generation of the odd," which obscures what is clear and obvious in Aristotle's mind: that the natural philosophers who posit a "world-making" are, or at least claim to be, "Pythagoreans," on Aristotle's estimation.[140] Here, it is important to recognize, with Huffman, that Aristotle

139. Arist. *Phys.* 8.1, 250b16 (Anaxagoras, Empedocles). Simplicius, in his *Commentary on Aristotle's Physics* (p. 1121.6 Diels), refers to Anaximander, Leucippus, and Democritus; Aëtius (2.1.3) adds Anaximenes, Archelaus, Xenophanes, and Diogenes of Apollonia. See *Cael.* 3.2, 301a13 (Anaxagoras, Empedocles), *Phys.* 2.4, 196a22 (Empedocles), *Metaph.* 1.4, 985a19 (Anaxagoras) and 14.3, 1091a18, on which see below.

140. It is difficult to assess why Aristotle is uncomfortable about describing the physics of the "Pythagoreans." He apparently wishes to relegate discussion of the issue of *κοσμοποιία* to the

did indeed describe certain Pythagoreans as dealing, in their philosophical pursuits, with the natural sciences. As Huffman has suggested, the best source for the information regarding those Pythagoreans who "make the world" is Philolaus of Croton, who believed (F 7 Huffman) that "the first (*τὸ πρᾶτον*) thing fitted together, the one in the center of the sphere, is called the hearth."[141] What Aristotle means when he claims that once the "one" was "constituted" the "nearest part of the unlimited began to be drawn in and limited by the limit," has been the subject of a great debate among scholars that need not concern this study at this point, since we are focusing chiefly on Aristotle's classification of various Pythagoreans according to his own schemes.[142] A bifurcation within the "Pythagoreans" as a group occurs when Aristotle describes how "some" (*τινές*) of them believe that "the even is first constructed out of unequals" (*τὸν ἄρτιον πρῶτον ἐξ ἀνίσων*), that is, the great and the small. As Tarán acutely notes in his edition of Speusippus's fragments, the "some" here likely refers to the Platonist Xenocrates, especially given the fact that Xenocrates presented his own metaphysical doctrines as those of Plato himself.[143] Apparently, Aristotle understands *each* of these positions as a type of "making of the world" that is fit more for discussions of natural science; the emphasis on what happens "first" (*πρῶτον*) in the cosmogonies attributable to Philolaus, Hippasus, and Xenocrates might imply that each of these "Pythagoreans" (as they were called) were held by Aristotle and his school to have posited theories of *κοσμοποιία*.[144] In essence, then, the approaches of those who posit theories of cosmology lay at the threshold that distinguishes natural from theoretical philosophy, in Aristotle's estimation.[145] Aristotle, of course,

---

Physics, but there he only mentions the Pythagoreans' physical theories involving the constitution of the universe briefly (4.6, 213b22–29), in a passage that is full of textual problems, and with no apparent relation to what Aristotle says here in the *Metaphysics*. See Annas 1976: 210–212.

141. Philolaus F 7 Huffman = Stob. *Ecl.* 1.21.8.

142. In general, I agree with the assessment of Huffman (1993: 203–211).

143. Tarán 1981: 333–334. Also see Horky: forthcoming. This was also the practice of another Platonist, Hermodorus of Syracuse, on whose metaphysics see Horky 2009: 85–90.

144. See Thphr. *Metaph.* 6b7–9, where Theophrastus claims that Xenocrates distinguishes himself from those who do not "generate" the universe, e.g. Speusippus, by "assigning a place to all things in the cosmos."

145. In the *Parts of Animals* (1.1, 640a1–10) Aristotle seems to lay out a strong bifurcation between approaches to these various sciences, as natural science has as its starting point "what will be," whereas theoretical science starts from "what is"; but, as Gotthelf has argued (1987: 170–171 and 197–198), Aristotle believes that both admit of demonstrations, so that, while it is true that the types of demonstration that can be used in mathematics and natural science differ, we should not overstate the case for their difference.

found it difficult to fit in the Pythagoreans between the Presocratics and Plato in his teleological history of philosophy, and it remains ambiguous whether Hippasus of Metapontum, Philolaus of Croton, or Archytas of Tarentum would have been considered chiefly as natural scientists or Pythagoreans by Aristotle and Theophrastus. Be that as it may, it is likely that—whatever the ultimate doxographical source was for the notion (A) that Hippasus believed that Soul-number was the "discerning tool of God the world-maker" (probably Middle Platonist, but possibly as early as Xenocrates)—the (B) analogy between Soul-number and "first paradigm of the making of the world" (*παράδειγμα πρῶτον κοσμοποιίας*) derives from the Early Academy, most likely from Xenocrates or Speusippus, whose philosophy was generally correlated with that of the "Pythagoreans" by Aristotle, and who, like Aristotle, wrote books of various sorts about the Pythagoreans and their philosophical ideas.[146]

Even if we accept the proposal that the Early Academy is the ultimate source for the doctrinal information regarding the theories of physics, number, and soul of Hippasus of Metapontum, we must consider this question: how might this information have been passed down to the doxographers? It is difficult to know whether they had access to the original writings of Speusippus or Xenocrates. In the absence of proof, we must assume that they did not, and that much of this information was transmitted by an intermediary. Since what survives of Aristotle's *Physics* does not refer to Hippasus at all, one candidate would be Theophrastus's dialectical writings on physics and mathematics; on the one hand, they do not survive complete, which would still force us to make an argument ex silentio; on the other hand, as I have argued elsewhere, Theophrastus's information concerning the Pythagoreans was not uniquely derived from Aristotle but also seems to arise out of information he garnered from the Early Academy, such as the writings of Xenocrates.[147] While we cannot confirm that the fifth-century BCE natural philosopher Hippasus of Metapontum *actually* held any of the doctrinal positions concerning physics, metaphysics, time, the soul, and number attributed to him by (possibly) Aristotle, (very likely) the early Academy, and (definitely) Theophrastus, we can say with more confidence that the attribution of these positions to Hippasus's philosophy was established by the last quarter of the fourth century BCE in the Lyceum, and it likely reaches

146. In addition to Speusippus's work *On Pythagorean Numbers*, Xenocrates wrote several books on mathematics (*On Numbers*, *Speculation of Numbers*, *On Geometry*, and *On Dimensions*) and a book *On Pythagorean Things*. On Xenocrates's appropriation of Pythagoreanism to Platonism, see most recently Horky: forthcoming.

147. As originally proposed by Burkert (1972: 62–64), accepted by Huffman (1993: 22–23, with n. 7), and argued in Horky: forthcoming.

back to the mid-fourth century in the Academy, when Speusippus and Xenocrates undertook in a systematic way the commensuration of the philosophy of their teacher Plato with the ancient Pythagoreans.

CONCLUSIONS

Aristotle and his students Theophrastus and Aristoxenus on one side and the Platonists Speusippus and Xenocrates on the other appear to be responsible for the bifurcation of Hippasus's classification as a philosopher and doctrines of physics, time, mathematics, soul, and knowledge that fascinated Middle and Neoplatonist commentators such as Iamblichus and Proclus. Aristotle and Theophrastus in their dialectical works probably classified Hippasus as a natural philosopher, a material monist with Heraclitean leanings who, in Theophrastus's opinion, understood the universe and fire, its first principle, to be unified, continually in motion, and limited. In this sense, Theophrastus may have believed that Hippasus conceived of the universe and its first principle as having the same characteristics, that is, as being in a relationship of imitation. Aristotle, who ultimately may be responsible for *some* (but not necessarily all) of the basic material found in Theophrastus's doxography, seems to have had trouble inscribing Hippasus into his own history of philosophy: Aristotle may have seen Hippasus as a natural philosopher who put forward theories of the "making of the world" (κοσμοποιία), but he also seems to be the source for the claim that Hippasus was the progenitor of the "so-called" or mathematical Pythagorean school, whose employment of theoretical demonstrations provided an early version of speculation into the superordinate sciences. Indeed, as I argued in chapter 1, Hippasus may have been the first Pythagorean to perform demonstrations of the sort that produce a "reason why" (διότι), especially involving mathematics. The trouble that the mathematical Pythagoreans presented to Aristotle's systematization of philosophy arises out of their collapsing of intelligible and sensible objects without a clear sense of stratification: they thereby "correct" reality in a way that flies in the face of experience. Another influential aspect of Aristotle's characterization of mathematical Pythagorean activity was arguments from what is "honorable," which Aristoxenus of Tarentum takes over from Aristotle and employs in his own descriptions of Pythagorean philosophical method and first principles. By appeal to what is "honorable," Aristotle and Aristoxenus configure the Pythagorean *pragmateia* as something that reaches beyond metaphysics and religion into "practical" aspects of daily existence, including ethics and politics.

In the Early Academy, both Speusippus of Athens and Xenocrates of Chalcedon may have spoken about Hippasus, too. Speusippus, like the mathematical Pythagoreans with whom Aristotle associates him, employed arguments from what is more "honorable." Moreover, the doctrines of Speusippus concerning the Decad and of Hippasus concerning Number are strikingly similar: in both, God is said to employ a numerical object as a paradigm in order to give order to objects in the universe. Such a striking correlation suggests that the doctrine ascribed to Hippasus, that he believed that "Soul-number is the first paradigm of the making of the world," is owed to Speusippus's or possibly Xenocrates's writings on the Pythagoreans, in an attempt to align Hippasus's supposed ancient doctrine with their own (which, subsequently, has been derived in various fashions from Plato's *Timaeus*). None of the evidence adduced in this chapter allows us to conclude anything about what the historical Hippasus, who seems to have lived in the first half of the fifth century BCE, *really* thought; rather it suggests to us the ways the Platonists, especially, were invested in seeing in Hippasus of Metapontum a precursor whose philosophy lent credence to their own, or their teacher's, philosophical tenets. It remains an almost impossible challenge to infer from this complex of doxographical sources anything about the real person Hippasus of Metapontum or his philosophical activities. In order to gain some traction on this problem, I will turn in chapter 3 to sources other than those listed above in order to see whether a historical Hippasus indeed can be found anywhere among the fractured pieces that remain.

3

Exoterism and the History of Pythagorean Politics

Most scholarly treatments of Pythagorean political history begin with an apology. This chapter will deviate from the norm in this and other ways. Pythagorean political history has been variously described as "contradictory" (Riedweg), "too partisan and too incomplete for us to reconstruct . . . with any confidence" (Kahn), and a "*punctum dolens* of archaic Greek history" (Musti).[1] The many problems in source criticism concern both the Pythagoreans as a community of wisdom-practitioners and (especially) Pythagoras himself. Referring to Pythagoras's political activities in Croton, Burkert elegantly notes that "the Master himself can be discerned, primarily, not by the clear light of history but in the misty twilight between religious veneration and the distorting light of hostile polemic."[2] I, too, am hesitant to admit that we can with any certainty attribute *particular* political activities to the historical Pythagoras. Broad sketches of Pythagoras's historical activities can be and have been cautiously attempted recently by Kahn and Riedweg, so I will point the curious reader in the direction of those useful scholarly works.[3] Be that as it may, I submit that we should not simply throw our hands up at the prospect of reconstructing the political history of the Pythagoreans who lived after him. In this chapter, I seek to shed some light on the political history of those figures, especially those political actors who seem to have established a heterodox Pythagorean philosophy in the wake of democratic revolutions throughout Southern Italy and Sicily in the early to middle fifth century BCE (roughly 473–450 BCE). I will trace what we might call

1. Riedweg 2005: 18; Musti 2005: 161; Kahn 2001: 7.

2. Burkert 1972: 120.

3. See Riedweg 2005: 17–20 and Kahn 2001: 7–9, both of which follow the authoritative study of Burkert (1972). More positivistic studies of Pythagorean political history (e.g. Lévêcque and Vidal-Naquet 1996: 69–72, de Vogel 1966: 218–231, van der Waerden 1979: 202–222) often *acknowledge* the historiographical challenges but also synthesize the historical material without attempting to *account for* the local biases of the historians.

the "reception-history" of Pythagoras's and the Pythagoreans' political activities from the early fourth century BCE in Athens, where the associates of Socrates (Antisthenes, Aeschines of Sphettus, Isocrates, and Plato) were the first witnesses on record to associate political activities with Pythagoras and the Pythagoreans, through an often understudied account that describes a schism in the Pythagorean brotherhood, preserved by a certain Apollonius in Iamblichus's work *On the Pythagorean Way of Life* (254–264) and derived ultimately from the *Italian and Sicilian Histories* of Timaeus of Tauromenium (ca. 350–ca. 260 BCE). While I will examine the accounts of Pythagorean political history that come down to us through the fourth-century BCE Peripatetics Dicaearchus of Messana and Aristoxenus of Tarentum, as well as later accounts of figures such as the Middle Platonist Nicomachus of Gerasa (first to second century CE), the reader will quickly discern that I place a greater emphasis on the account of Timaeus. One might ask: surely the historical account of Aristoxenus, who was from Tarentum and a student of Aristotle, and who claims to have known certain Pythagoreans personally, should be preferred over that of the Hellenistic historian Timaeus, who lived too late to have been familiar with any of those known as the "last of the Pythagoreans" or any reliable external critic who may have had knowledge of their philosophical and political activities, such as Aristotle?

There are two reasons I emphasize Timaeus's version. First, methodologically, the account of Timaeus of Tauromenium represents, in the words of the historian Domenico Musti, a *lectio difficilior* for the political history of Pythagoreanism.[4] This means that it provides an account otherwise generally unattested that does not share much in common with the Peripatetic accounts, which focus chiefly on the characteristic excellence of Pythagoras and degeneracy of those Pythagorean pretenders who revolted against his authority. Indeed, Timaeus's account diverges in at least two ways from the accounts of Dicaearchus and Aristoxenus: neither is it obviously biased in favor of Pythagoras, nor does it take any clear ethical stance against those who revolted within the brotherhood. Timaeus thus emerges as an important external witness. That does not mean, however, that his account can be considered "objective," at least in any simple way. As I will show, Timaeus associates the Pythagorean revolts of the mid-fifth century BCE with a democratizing ideology, of which we can identify two special attributes: (1) the publication of "secret" Pythagorean knowledge otherwise shared only among a small few within a ἑταίρια, and (2) the promotion of political rights for the citizens of the city-states of Magna Graecia. These democratizing values were of special interest to Timaeus, who himself seems to have written his history with the purpose of celebrating democracy in Western Greece. Generally, to be sure, the "misty twilight between religious

4. Musti 2005: 164–165.

veneration and the distorting light of hostile polemic" that worried Burkert is not to be found in the account of Timaeus of Tauromenium.

The second reason why I will emphasize Timaeus's account in this study of Pythagorean political history is not novel, but it, too, has not been discussed by scholars in the English-speaking world often enough: Timaeus of Tauromenium distinguishes his own account of the history of Southern Italy by appealing to *documentary evidence*, namely, the "registers of the Crotonians" (τὰ τῶν Κροτωνιατῶν ὑπομνήματα).[5] Such written records appear to be the source for details of Timaeus's retelling of the political conspiracy against the Pythagoreans. Importantly, in keeping with Timaeus's historiographical methodology, the account of the Cylonian conspiracy in sections 254–264 of Iamblichus's work *On the Pythagorean Way of Life* retains traces of civic agreements that were made binding on inscriptions. Again, this aspect of Timaeus's historiography reinforces his own polemic against historical accounts that sensationalize the past and its chief political actors, as we sometimes see in the early biographical historiography of Pythagoras and the Pythagoreans. Indeed, as I will show, Timaeus denies the mythographic approach to history specifically by appealing to documentary evidence, which he claims to examine locally. Of course, even the written records of the Crotonians, if they did indeed exist, are subject to the biases that attend all historical accounts, and it is no coincidence that the existence itself of civic registers has been associated with democratization in the historiography of Western Greece. For this very reason, they would have appealed as source evidence to Timaeus of Tauromenium. And, for this reason, we need to consider them and all accounts that may have followed from them within the larger "reception-history" of Pythagorean political activity. I will first trace the beginnings of this "reception-history" with the traditions of Pythagorean politics written by the Socratics, which were formative for the representation of Pythagoreans and their philosophical-political activities in all subsequent historical accounts, even in the alternative version of Timaeus. By contrast with antecedent historiographical traditions concerning the political history of the Pythagoreans, Timaeus manages to construct what I am calling a history of the Pythagorean "exoterics,"[6] a group of philosophers, including Hippasus of Metapontum,

5. Iambl. *VP* 262, not discussed by Huffman (2005 and 1993), Riedweg (2005), Kahn (2001), or van der Waerden (1979), but emphasized by Musti (1988: 27–28), Morrison (1956: 149), and von Fritz (1940: 65–67). Burkert (1972: 117 n. 50) calls them "suspect" following Delatte (1922: 218). Neither, however, attempts a detailed contextualization of the passage with Timaeus of Tauromenium's other fragments, which I undertake below.

6. The terms "esoterics" (ἐσωτερικοί) and "exoterics" (ἐξωτερικοί) were certainly used by the third-century CE heresiologist Hippolytus of Rome (*Ref.* 1.2.4) to refer to the two branches of Pythagoreans. The question is whether these terms are original with Hippolytus,

Empedocles of Agrigentum, Epicharmus of Syracuse, Philolaus of Croton, Archytas of Tarentum, and (possibly) Plato himself, who were credited with a sustained heterodox strand of Pythagorean thought distinguished by its democratizing tendencies. In particular, these democratizing tendencies take the shape of both upheavals in political organization in the Pythagorean city-states of Magna Graecia and the publication of secret knowledge, which was supposed to be kept in the possession of a small group of intimates. The majority of this chapter will deal with a reconstruction of this "exoteric" Pythagorean history, as told in the *Italian and Sicilian History* of Timaeus of Tauromenium.

PYTHAGORAS AMONG THE ATHENIAN PHILOSOPHERS IN THE FOURTH CENTURY BCE

Stories that tell the mythological origins of Pythagorean political philosophy first arise in the late fifth to early fourth century BCE and assign speeches to Pythagoras. In Athens, interest in Pythagoras and Pythagoreanism more generally arises in the wake of the influence of Socrates, even though some scholars have detected indirect references to Pythagoras and/or Pythagoreans in earlier Athenian literature.[7] Pythagoras's political speeches are first attested in a fragment of the Socratic philosopher Antisthenes of Athens (ca. 445–365 BCE), wherein Antisthenes[8] demonstrates how Pythagoras, like Odysseus, could be

go back to an intermediary such as Nicomachus of Gerasa or Plutarch, or are original with Timaeus of Tauromenium. Timaeus seems to be the source behind the term "inside the curtain" at Iambl. *VP* 72, 41.15, where we also see that Iamblichus employs the term "esoteric" (ἐσωτερικοί). For this reason, I speculate that the terms "esoteric" and "exoteric" probably derive from Timaeus, but I cannot be absolutely certain. In this chapter, to be sure, I use the terms "esoteric" and "exoteric" heuristically, in order for us to construct an alternative to Aristotle's division into "acousmatic" and "mathematical" Pythagoreans.

7. The Aeschylean *Prometheus Bound* (459) preserves a version of the Pythagorean *acusma* "what is wisest? Number." On the supposed Pythagoreanism of this passage, see the skeptical analysis of Griffith 1978: 109–111. But also see chapter 6, on the literary *topos* of the "first discoverer."

8. The provenance of this fragment has been under dispute since Schrader's edition of the *Homeric Questions* (1880–82), in great part due to the fact that Porphyry, in his *Life of Pythagoras* (18), attributes the same information to the Peripatetic Dicaearchus. A useful survey of the positions taken on its provenance is Giannantoni (1985: 308). More recently, Luzzatto (1996) has argued that the passage reflects Porphyrian patchwork. But arguments in favor of the unity of Antisthenes's claims within the text have been expressed by Susan Prince (email to the author, June 16, 2010): "The arguments [against authenticity] would be the seams in the argument (which I think are slight—the medical analogy is certainly germane to Antisthenes, and the change from *oratio obliqua* to *oratio recta*, which has been used as an argument against continuity, can also be explained as a choice for clarity) and the fit between

considered an orator πολύτροπος ("of many turns") or, in Antisthenes's usage, capable of appropriating types of speech to particular audiences (i.e. one who can "turn the many" toward a single goal):

> So too Pythagoras, who was considered worthy of making speeches, is said to have composed playful speeches for children, speeches appropriate to women for women, archontic speeches for the archons, and ephebic speeches for the young men. For the discovery of what manner of wisdom is suitable to each person is characteristic of wisdom [τὸν γὰρ ἑκάστοις πρόσφορον τρόπον τῆς σοφίας ἐξευρίσκειν σοφίας ἐστίν]. But it is characteristic of ignorance to employ a single manner of speech [χρῆσθαι μονοτρόπῳ] for different people. Medicine also depends on the correct use of its art, in that the "turning of the many" for an experienced treatment is what happens through the variegated constitution of those being cared for.
>
> (SCHOLIUM on *Porphyry's Homeric Questions of the* Odyssey α 1 = Giannantoni V A F 187)

Several aspects of Antisthenes's argument suggest how Socratics other than Plato might have understood Pythagoreanism. Antisthenes argues that Pythagoras, like Odysseus, exemplifies the importance of making speeches appropriate to the listener, in a way that is comparable with the art of medicine, in which a doctor must be attentive to the varieties of constitution that patients could possess. Implicit here is the principle that wisdom (σοφία) demands attention to the peculiar qualities of each object that forms one's interests, whether in politics or in medicine. Such a conceit appears among intellectuals of the early fourth century BCE: the author of the Hippocratic *Regimen* claims that a doctor must be attentive to the variety of constitutions that men could have;[9] likewise, Gorgias in the eponymous Platonic dialogue (456b1–c7) emphasizes the power that the rhetorical art

---

the exemplum and Antisthenes's main point about Odysseus, which I think is also clear—both speakers could say the same thing in many ways, determined by the audience, and the goal was unity of message. There is apparent evidence also for Peripatetic intervention into Antisthenes's interpretation of Homer in F 189 [Giannantoni], and in this case I think it's clear the intervention would be pre-Porphyry, not by Porphyry (details are difficult), so one could also imagine the situation . . . that Dicaearchus received Antisthenes's interpretation of Odysseus and added the comparison, then it went down to Porphyry all assembled." I think, given Prince's cogent reply to Luzatto and other critics' skepticism, that we can admit of two possibilities: that most or all of the passage quoted here goes back to Antisthenes's study of how rhetoricians make use of τρόπος, or that Dicaearchus has paired Antisthenes's description of the τρόπος of Odysseus with one of Pythagoras. See White 2001: 211 n. 38 and Zhmud 2012: 46, who take this formulation to be original with Antisthenes.

9. Hippoc. *Reg*. 28.

can have, namely the medical art, a portrait that cannot be far from the mark, given parallels to Gorgias's own *Encomium of Helen*;[10] and Socrates, too, in the *Phaedrus* (277b6–c6) baldly states that the rhetorician must have knowledge of the souls he hopes to persuade. In Plato's *Statesman* (304b11–e1, 305e2–311c7), the Eleatic Stranger extends this principle of psychic knowledge beyond rhetoric, which is subordinated to the political (also called "arbitrational": ἐπιτροπεύουσα) art of the statesman, who is expected to understand the difference between courageous and temperate souls in order to interweave them into the fabric of the city-state. Finally, in the *Laws* (720b8–e2, 722b5–c2), the lawgiver is exhorted to work like a free doctor who, by employing persuasion, makes free citizens of the state more favorably disposed to learning; this is especially the case for the young, who are expected to respond favorably to preludic addresses, which persuade by providing easily accessible versions of the law.[11]

The implication of rhetoric and medicine in politics—as associated with Pythagoras by Antisthenes—raises the important chicken-and-egg question: did the association of Pythagorean "wisdom" (σοφία) with rhetoric and medicine in Athens arise out of Socrates's influence, or was it Socrates who was influenced by Pythagoreanism?[12] That question may be impossible to answer, although we are on surer ground to posit a real interest in Pythagorean political and philosophical activities on the part of Socrates's students. The Socratic Aeschines of Sphettus (d. after 356 BCE), in a comedy that may have staged a discussion between Socrates and a Pythagorean named Telauges, apparently ridiculed Telauges on two related grounds: that he, an ascetic, dressed shabbily, and that he was a poor orator.[13] Isocrates (436–338 BCE) in the epideictic speech *Busiris*, likely crafted in the early or middle portion of Isocrates's career (late 390s–early 370s BCE),[14] speaks of Pythagoras as the first to bring the "other philosophy"[15] (ἡ ἀλλὴ φιλοσοφία) to the Greeks; but he also implicitly criticizes

10. See Holmes 2009: 211–216 and Horky 2006: 375–378.

11. On this subject, see especially Bobonich 2002: 97–105 and 113, with reference to *Leg.* 823c1–824a21.

12. As queried by Riedweg (2005: 13), who suggests the latter by appeal to evidence from Aristophanes's *Clouds* (performed the first time in 423 BCE) that portrays Socrates as adopting Orphic-Pythagorean tenets.

13. Athen. 5, 220a = Giannantoni VI A F 84. See Kahn 2001: 49.

14. See Livingstone 2001: 40–47 and Eucken 1983: 173–183.

15. It is remarkable how little discussion of the "otherness" associated with Pythagoras's philosophy has occurred. Perhaps this is in part because scholars don't actually translate ἀλλή, e.g. "was first to introduce high culture" (Kahn); "was first to bring to the Greeks all philosophy" (van Hook). What does Isocrates mean by "other" here? An answer does not

the philosophical activities of Pythagoras and those followers who engage in such activities:

> More conspicuously than others, Pythagoras concerned himself with sacrifices and ritual activity in shrines, since he believed that even if he would not gain advantage from the gods through these activities, still he would be of the greatest repute among men on account of them [μάλιστ' εὐδοκιμήσεν]. And this indeed did happen to him. For he excelled all others in reputation so greatly that not only did all the younger men desire to be his students, but also the older men were more pleased to see their children in his company rather than caring for their own affairs. And one cannot disbelieve these reports: for even now people admire more those who fashion themselves [προσποιουμένους] as his silent pupils than [they admire] people who have the greatest renown for eloquence.
>
> (ISOCRATES, *Busiris* 28–29)

When reading this description of Pythagoras—one of the earliest that survives from Athens—we need to keep in mind the lightheartedness of Isocrates's playful encomium. Isocrates's characterization of Pythagoras's philosophy associates it strongly with piety (ὁσιοτής), and those pretenders who "fashion themselves" (προσποιουμένους)[16] his followers practice such philosophy by being silent rather than producing public speeches.[17] We cannot take Isocrates seriously in his claim that by ignoring private affairs, the pretenders who follow Pythagoras's philosophy are to be admired; the statement that "one cannot disbelieve these reports" is tinged with irony.[18] Pythagoras's philosophy, which fails to provide the youth with the means to be effective political agents, must be understood to be in contrast to the valuation of rhetorical eloquence that Isocrates and apparently Aeschines of Sphettus shared.[19] The "other philosophy"

---

obviously present itself, but we might imagine that he means a philosophy that is "other" than his own. Now Isocrates's *philosophia* means something like "education in speech-making," and, if Burkert is right (1972: 216), what Pythagoras's philosophy chiefly dealt with was issues of "religious" piety, not speech-making. But what "piety" is associated with in Isocrates's *Busiris* is not simply relegated to activities involving prayers and sacrifices for the gods. See below, where I suggest that the "other" philosophy is that of the Egyptians.

16. The most common meaning for this term is something like "pretend." See *LSJ* s.v. προσποιέω 2.3 and 2.4.

17. See Zhmud 2012: 48–50.

18. See Kahn 2001: 12.

19. See Morrison 1956: 138.

ascribed to Pythagoras, to be sure, appears to refer back to a type of philosophy that the Egyptian priests introduced and that Busiris practiced, namely, "wisdom" (φρόνησις):[20]

> And then the Egyptian priests introduced training in philosophy for souls, which has the power not only to establish laws but also to pursue the nature of the universe [τὴν φύσιν τῶν ὄντων ζητῆσαι]. Busiris assigned the older men to the most important matters [ἐπὶ τὰ μέγιστα], but he persuaded the younger men to neglect pleasures and devote themselves to astronomy, arithmetic, and geometry, the capacities of which some praise on the grounds that they are useful in certain ways, whereas others attempt to show that they are most conducive to virtue.
>
> (ISOCRATES, *Busiris* 22–23)

Busiris's application of philosophy functions as a precursor to Pythagoras's "other philosophy": Busiris assigns the task of taking care of the affairs of the city-state, including lawgiving, to the older men, while he persuades the youth to shun the public life—associated with "pleasures"[21]—and to study in private the natural sciences by learning astronomy, arithmetic, and geometry, three of the four parts of the quadrivium. So-called Egyptian philosophy, writ large, also implicates medicine in the activities of "piety" that are inherited by Pythagoras and his followers.[22] There is a hint of criticism of Pythagoras and Busiris for employing speeches to render the youth politically useless while improving their own fame and popularity in the city-state. It is, of course, not impossible that Isocrates could be referring to Plato when he mentions that some believe that those mathematical sciences may have the potential to conduce to virtue, and one might detect some playful polemic going on between Isocrates and Plato here.[23] Be that as it may, it is clear that Isocrates conceives of this Pythagorean-Egyptian "other" philosophy as an activity that distinguishes between activities proper to youth and elders, with an emphasis on educating the youth in mathematical sciences. That is to say, regardless of whether Isocrates was implicating Plato in this discourse, he was associating Pythagoreanism with Egyptian culture—an association at least as old as Herodotus[24]—and characterized the

20. As explicitly identified at *Bus.* 21. See Zhmud 2012: 48–49.

21. Also see Isocrates's *On the Peace* 5, on which see Morgan 2004: 132.

22. Isoc. *Bus.* 22.

23. See Livingstone 2001: 48–55, and Eucken 1983: 173–195.

24. Hdt. 2.81.2.

educational curriculum of the philosophy that was associated with these religious doctrines as based in the mathematical sciences.[25] Moreover—and this point is especially important for the characterization of mathematical Pythagoreanism that we find a half century later in the *Metaphysics* of Aristotle—*Isocrates understands this Pythagorean-Egyptian philosophical education in mathematics (astronomy, arithmetic, and geometry) as a type of inquiry into the nature of things in the universe* (τὴν φύσιν τῶν ὄντων ζητῆσαι). Study of the celestial phenomena along with the abstract mathematics afforded by (possibly) arithmetic and (definitely) geometry is understood to constitute what Isocrates considers the *philosophia* that was said to have been invented by the legendary Busiris and brought to the Greeks by Pythagoras.[26]

If, indeed, it is the case that the account of Pythagoras's speeches originates with Antisthenes of Athens, then the evidence from Aeschines of Sphettus's *Telauges* and Isocrates's *Busiris* takes on a new and significant import for this study: it suggests that as early as the 390s BCE, Athenian philosophers were debating the role that Pythagoras and his followers played in public discourse, especially with regard to the value of the religious and ethical teachings for citizens that have been transmitted through rhetoric. Antisthenes characterizes Pythagoras's activities in Croton as fundamentally political in nature, whereas Isocrates considers them impractical and implies that the pretenders "who fashion themselves" (προσποιουμένους) the younger "silent" followers of Pythagoras, suffer from political inertia because they concern themselves too much with mathematics in their scientific inquiry. By the time of Aristotle, at least, mathematics came to be understood as an essentializing axiomatic discourse opposed to the adaptive qualities of political speeches, which can respond to social and political challenges in an ad hoc manner and are associated with on the one hand Sophists and on the other orators and legislators in Athenian democracy. In both of the latter cases, the goal was compulsion by persuasion, not successful deduction by appeal to axioms.[27] Tempting as it might be to see Plato and his associates in the Academy as "those who fashion themselves" the followers of Pythagoras, we cannot be absolutely sure (1) that they are definitively intended to be the referents, or (2), even if we do admit that possibility, whether Plato and his associates in the Academy are the *only* objects

25. Pace Kahn 2001: 13, who suggests that it is with Plato that Pythagoras becomes associated with mathematics.

26. On the vexed problem of whether Pythagoras invented the word *philosophia*, see most recently Riedweg 2005: 90–98. In the context of this passage, Zhmud (2012: 241) adduces Democritus's claim (DK 68 B 299) to excellence in the construction of lines by means of demonstrations, comparing himself with the Egyptian land surveyors.

27. See Lloyd 1979: 110–116.

of Isocrates's playful banter. For however much we might think that Isocrates is poking fun at the ideal constitution of Plato's *Republic*, as Niall Livingstone has fittingly observed, the presentation of Pythagorean-Egyptian philosophy and its attendant political constitution applies equally to any idealizing state of the Laconizing tradition, such as that expressed in the Xenophontic *Constitution of Sparta* or implied as an alternative to Athenian democracy by the Old Oligarch.[28] The nature of Isocrates's epideictic speech makes it virtually impossible to pin down a simple object of ridicule here.

It also might be tempting to see Aristotle's "so-called" Pythagoreans (*οἱ καλούμενοι Πυθαγόρειοι*),[29] whom I argued in chapter 1 were one and the same as the mathematical Pythagoreans, as tied to Isocrates's terminology for those "who fashion themselves" (*προσποιουμένους*) after Pythagoras's philosophy. Aristotle seems to have adapted Isocrates's description of Pythagorean pretenders in accordance with his own notion of the division of sciences. As Aristotle argues in *Metaphysics* A, while objects of the philosophical inquiry of the "so-called" Pythagoreans are mathematical, their *pragmateia* deals especially with nature.[30] What for Isocrates is a relatively straightforward relationship between mathematics (arithmetic and geometry) and natural science (astronomy) becomes, in Aristotle's estimation, a confounding of superordinate and subordinate sciences. Moreover, while Aristotle does not highlight the religious piety of the mathematical Pythagoreans—that is, of course, a defining characteristic of the *pragmateia* of the acousmatic Pythagoreans—he does seem to grant the possibility that their employment of the superordinate sciences constitutes a sort of theological pursuit.[31] The overall point stands, though: the description of certain Pythagoreans interested in mathematics as pretenders is not original with Aristotle but represents an adaptation of a characterization already present in the writings of Isocrates.

Aristotle also avails himself of the same kind of description of Pythagoras's speech-making found in the fragments that seem to trace back to the Socratic Antisthenes. In a fragment from his lost works on the Pythagoreans and quoted in direct speech by Iamblichus, Aristotle adapts both accounts by describing

28. Livingstone 2001: 50–51.

29. Of course, the participle could be middle rather than passive voice: "those who call themselves Pythagoreans."

30. Arist. *Metaph.* 1.8, 989b29–990a5, discussed in chapter 1, section entitled "On the 'So-called' and Mathematical Pythagoreans."

31. The problem of how first philosophy could be related to theology is one that cannot be sufficiently resolved here. Suffice it to say that there might be good reasons for Aristotle to describe first philosophy as theology, as Nightingale (2004: 236–240) has argued.

how Pythagoras discoursed with the elder and younger men and elaborates further on the medical-political analogy:

> Pythagoras came from Ionia, more precisely from Samos, at the time of the tyranny of Polycrates, when Italy was at its height, and the first men of the city-states became his associates. The older of these [men] he addressed in a simple style, since they, who had little leisure on account of their being occupied in political affairs, had trouble when he conversed with them in terms of sciences and demonstrations. He thought that they would fare no worse if they knew *what* they ought to do [εἰδότας τί δεῖ πράττειν], even if they lacked the explanation [ἄνευ τῆς αἰτίας] for it, just as people under medical care fare no worse when they do not additionally hear *the reason why* they ought to do [διὰ τὶ πρακτέον] each thing in their treatment. The younger of these [men], however, who had the ability to endure the education, he conversed with in terms of demonstrations and sciences. So, then, these men [i.e. the mathematicians] are descended from the latter group, as are the others [i.e. the acousmatics] from the former group.
>
> (Iamblichus, *On the General Mathematical Science* 25, 77.4–18)

As I argued extensively in chapter 1, the appeal to various kinds of methodologies that are associated with the mathematical and acousmatic Pythagoreans here represents Aristotle's concerted attempt to map his own philosophical assumptions about the sciences onto their activities.[32] Overall, this represents Aristotle's peculiar approach to distinguishing the peculiar attributes of the *vita activa* from the *vita contemplativa*. But what the testimonies of Isocrates, Aeschines of Sphettus, and Antisthenes on Pythagoras and his students reveal is that Aristotle's formulation of the *pragmateiai* of the Pythagoreans is arrived at by way of descriptions of Pythagorean philosophy that were *already available in Athens a generation before Aristotle was writing, and in Socratic intellectual circles*.[33] Even among surviving authors and works in the Athenian context of the early fourth century BCE, as we can see, there was a debate about whether Pythagoras and those who called themselves Pythagoreans could be considered

32. Aristotle might not be the only figure in the mid-fourth century BCE who represented Pythagoreans as engaging in demonstrations. The obscure dialogue between Pythagoras and Phalaris of Agrigentum, preserved by Iamblichus (*VP* 215–219, 116.22–119.3), highlights Pythagoras's argumentative abilities, especially his capacity to make reasoned arguments. The text might ultimately go back to Heraclides of Pontus's *Abaris*, but it must have been significantly modified by the time Iamblichus recorded it. See Dillon and Hershbell 1991: 215 n. 2 and, more generally, Gottschalk 1980: 123–126.

33. See Riedweg 2005: 94–95.

effective political agents or Sophistical nuisances, hallowed priests or spirited charlatans, technologically advanced natural scientists or abstruse geeks. Most of us will be familiar with this paradigm, as would any Athenian who went to the Greater Dionysia in 423 BCE and saw Aristophanes's *Clouds*.[34] This raises an important question for the present study of early Pythagoreanism: is it possible to identify the exact historical circumstances in which the split between the mathematical and the acousmatic Pythagoreans occurred? In order to advance on this very difficult question, we need to take account of the evidence that seems to derive from the historiography of Western Greece.

PYTHAGOREAN EXOTERICS IN THE FIFTH CENTURY BCE? THE HISTORICAL EVIDENCE OF TIMAEUS OF TAUROMENIUM

The aforementioned descriptions of Pythagoras and his followers that are associated with the students of Socrates focus the philosophical activity of the Pythagoreans around political effectiveness, especially in public oratory. In each case I have considered (Antisthenes, Aeschines of Sphettus, and Isocrates)—regardless of whether the figures are ridiculed for their absent-minded idealization, absurd piety, crackpot medical expertise, or phoniness—a chief criterion for evaluation of their philosophical activities is their application to the sphere of politics. We should be suspicious of the historical value of this representation of Pythagoras and his followers, especially given the association of similar philosophical activities with negative attributes in Aristophanes's picture of Socrates and his followers in the *Clouds*. It could easily be a literary *topos* applied to Pythagoras and the Pythagoreans in order to make them relevant to Athenians. In order to evaluate the historical value of the evidence of Pythagorean political activity given by these Socratics, we would need to look elsewhere. In the comic tradition from Southern Italy in the fourth century BCE, we see mockery of Pythagoreans made explicit. The comic poet Alexis of Thurii, who, given the location of his birth, may have been familiar with fourth-century Pythagoreanism, satirizes the Pythagoreans in a play called *The Tarentines*. In particular, he singles out the speciousness of their λόγοι (Πυθαγορισμοὶ καὶ λόγοι / λεπτοὶ διεσμιλευμέναι τε φροντίδες).[35] What is also interesting

34. As suggested by Demand (1982: 183–184), although I cannot thereby agree that the speeches attributed to Pythagoras and preserved by Iamblichus (*VP* 38–57, 22.9–31.16) antedate Aristophanes's *Clouds*. More likely, I suspect, they are products of Timaeus of Tauromenium's historical accounts, on which see Burkert 1972: 104 n. 37.

35. F 223 K.-A. = Ath. 4, 161b. Apparently, Timaeus of Tauromenium (FGrHist 566 F 132 = Schol. Eur. *Hek*. 131) defended Pythagoras against the charge, leveled against him by Heraclitus, of inventing Sophistic arguments (εὑρετὴς γενόμενος τῶν ἀληθινῶν κοπίδων). On Timaeus's account of Heraclitus's criticism of Pythagoras, see Musti 1988: 30 n. 28.

about this play, moreover, is the fact that it presents Tarentine Pythagoreanism as something foreign and eccentric no later than the last quarter of the fourth century BCE.[36] At one point, Athenian life is explained to one character; and elsewhere in the play, a character explains what those who "Pythagoreanize" do, according to what people say (*ὡς ἀκούμεν*): allegedly, Pythagoreans aren't supposed to eat fish or anything else that has a soul, but the profligate Athenian Epicharides, who fashions himself a Pythagorean, eats dog.[37] Mendicant Athenian Pythagoreanizers, who pretend to live the Pythagorean life, are contrasted against genuine Pythagoreans, to full comic effect.[38]

Still, in order to advance beyond Athenian representations of Pythagorean political activities, we need to look for evidence from outside Athenian circles, to Western Greek historiography. In the social conditions of Magna Graecia in the fifth century BCE we find a potentially viable account of Pythagorean political organization. As Burkert has argued, the basic social unit of the Pythagoreans was a special type of Pythagorean *ἑταιρία*, a brotherhood that was organized around a particular lifestyle, what we might call the *βίος Πυθαγόρειος*.[39] Typically, throughout Greece during the sixth and fifth centuries BCE, the *ἑταιρίαι* and the closely related *συνωμοσίαι* (clubs) functioned as associations of friends with shared aims, especially in politics, and often bound by a common oath among the participating *ἑταίροι*. Such oaths are often associated with particular affinities between private citizens (*ἴδιοι*), by contrast with the broader oath promised to a civic community associated with the people (*δῆμος*).[40] Sartori distinguishes between *ἑταιρίαι* and the *συνωμοσίαι*—at least before the oligarchic coup in Athens of 404 BCE—chiefly on the grounds that the *ἑταιρία* is

36. Burkert (1972: 200–201, with nn. 41 and 51) allows for the possibility that the Alexis/ Cratinus the Younger plays called *Tarentines* were composed either around 360 or 330–320 BCE.

37. See Olson 2007: 244. The identity of Epicharides is difficult to pin down—assuming this is a real person at all—but in addition to the reference to Epicharides's profligacy in Alexis's *Phaedrus* (F 248 K.-A. = Ath. 4, 165e), it should be noted that a certain Athenian Epicharides dedicated a boat called *Boetheia* in 358/7 BCE. We might recall that the earliest evidence for Pythagoras's belief in the transmigration of the soul, in Xenophanes's satire (DK 21 B 7), characterizes Pythagoras as defending a puppy from being whipped because he recognized the "soul of a friend" when it was barking.

38. See Zhmud 2012: 181–182. Similar tendencies can be noted in the Tarentines of Cratinus the Younger, which portrays the Pythagoreans as employing various tricks in their speeches.

39. Burkert (1981: 14–15, with nn. 61–63) notes that the Pythagoreans were called *filoi, hetairoi, gnôrimoi,* and *homilêtai*. For another thorough treatment of the Pythagorean *ἑταίριαι*, see Minar 1942: 19–29.

40. At least by Plato, on which see *R.* 4, 443a3–7, where Socrates implicitly distinguishes between the two sorts of oaths (civic and private).

more often associated with political groups that seek to overthrow a current governmental or constitutional order, whereas the *συνωμοσία* tends to function with the legitimate approval of a civic governing body, even if that governing body is a tyrant.[41] It is chiefly a matter of perspective, in a world of constant political revolt and general instability such as Southern Italy evinced in the fifth century BCE, whether one is to brand a group of friends a politically legitimate *ἑταιρία* or a (potentially) subversive *συνωμοσία*.

There is good reason to adopt Burkert's proposal that the Pythagoreans' central communal grouping was the *ἑταιρία*. As I will show, there is a small series of early and reliable figures who testify to it, both in Athens and in Western Greece. The earliest surviving witness to this is Plato himself, who, in *Republic* 10, has Socrates articulate with some precision what the influence of the great charismatic sages of the past was on the development of human culture. Chiefly in his crosshairs is Homer, who was allegedly famed for the knowledge he possessed in various public arts, such as warfare, generalship, city government, and education of the people. Socrates and Glaucon dismiss this possibility, on the grounds that there is no evidence of the product of such knowledge—in the form of, for example, a Homeric political constitution, by contrast with the well-known Lycurgan constitution of Sparta, Charondan constitution of Southern Italy and Sicily, and Solonian constitution of Athens. It is at this point that Plato explicitly describes the activities of Pythagoras and his followers:

> Then, if there's nothing of a public nature [*δημοσίᾳ*], are we told that, when Homer was alive, he was a leader in the education of certain people [*ἰδίᾳ τισίν*] who took pleasure in associating with him in private [*ἐπὶ συνουσίᾳ*] and that he passed on a Homeric way of life to those who came after him, just as Pythagoras did? Pythagoras is particularly loved for this [*διαφερόντως ἠγαπήθη*], and even today his followers are conspicuous for what they call the Pythagorean way of life [*οἱ ὕστεροι ἔτι καὶ νῦν Πυθαγόρειον τρόπον ἐπονομάζοντες τοῦ βίου διαφανεῖς*].
>
> (PLATO, *Republic* 10, 600a8–b4; translation by Grube and Reeve in Cooper and Hutchinson 1997)

After Glaucon agrees, Socrates continues:

> But Glaucon, if Homer had really been able to educate people and make them better, if he'd known about these things and not merely about how to imitate them, wouldn't he have made himself many companions and been loved and honored by them [*οὐκ ἄρ' ἂν πολλοὺς ἑταίρους ἐποιήσατο καὶ ἐτιμᾶτο*

41. See Sartori 1967: 17–33.

> καὶ ἠγαπᾶτο ὑπ' αὐτῶν]? Protagoras of Abdera, Prodicus of Ceos, and a great many others are able to convince anyone who associates with them in private [ἰδίᾳ συγγιγνόμενοι] that he wouldn't be able to manage his household or city unless they themselves supervise his education, and they are so intensely loved because of this wisdom of theirs that their companions [οἱ ἑταῖροι] do everything but carry them around on their shoulders.
>
> (PLATO, *Republic* 10, 600c3–d5; translation after Grube and Reeve in Cooper and Hutchinson 1997)

This is tremendously important evidence for the present study, in part because it is the only place in his oeuvre where Plato directly names Pythagoras, as well as the only one where he explicitly inscribes the activities of the Pythagoreans into the larger project of establishing the best means to pursue the good life.[42] Socrates highlights several important aspects of Pythagoreanism—as Plato reconstructs it—that will continue to be influential for the present study of the ways Plato inherited and adapted its tenets. First of all, the "Pythagorean way of life" (Πυθαγόρειος τρόπος τοῦ βίου) secures relationships between private individuals, rather than "the people" at large. It is unclear whether Plato means that Pythagoras's social reforms were rather benignly directed at a small group of acolytes rather than city-states at large or whether a certain antidemocratic hostility toward "the people" should be inferred. Interpretation of this issue is made more difficult by the fact that Plato's larger concern here is not with representation of Pythagoras but with the monolithic normative status Homer holds over Greek culture, in both the private and public realms. Setting aside this issue, however, we can also see that Pythagoras's activities, and the "way of life" that resulted from his teachings, are marked as charismatic: in Socrates's opinion, Pythagoras's followers, "have affection" for and "honor" him in ways that go far beyond even what the Homerides feel for Homer. Finally, and most important for my current analysis, Pythagoras's activities are compared with the models of private education provided by Sophists such as Protagoras and Prodicus, whose activities are said to influence their companions' (οἱ ἑταῖροι) private and public conduct. The assumption is that the Pythagorean community, with its distinctive "way of life," resembles the sorts of ἑταρίαι that rally around a charismatic wisdom-practitioner, such as Protagoras or Prodicus.

Later on, in chapter 6, I will revisit the issue of the relationship between the Pythagoreans and the Sophists at greater length; here, it is important to note that as early as Plato's *Republic* 10 (ca. 380s BCE), both Sophists and Pythagoreans are

42. Plato also refers to the Pythagoreans (οἱ Πυθαγόρειοι) in *Republic* 7 (530d6–531c7), when he criticizes Archytas of Tarentum's approach to harmonic theory. Plato nowhere else refers to the Pythagoreans or to Pythagoras by name.

said to associate with one another in groups of ἑταῖροι. Other evidence for Pythagorean ἑταιρίαι, this time at a civic level, is preserved by Plutarch (*On the Daemonion of Socrates* 13, 583a–c), who stages what is likely to have been a fictional dialogue between the Pythagorean Theanor of Croton and Epaminondas of Thebes, in which Epaminondas begins his account of the expulsion of the Pythagoreans from Southern Italy by referring to the "brotherhoods of Pythagoreans throughout the cities" (αἱ κατὰ πόλεις ἑταιρίαι τῶν Πυθαγορικῶν) and then a smaller subgroup of "fellows"[43] who met, apparently for political reasons, at a house in Metapontum.[44] All versions of this famous episode, which scholars tend to refer to as the "Cylonian Conspiracy" (named eponymously after its author, Cylon, who allegedly set the house on fire and led the group of revolutionaries), describe such a political council (βουλή), in various levels of detail.[45] It is virtually impossible to separate out the various confused strands that we encounter in the historical traditions and to reconstruct a single narrative account; consequently, developing a synthetic account of the Cylonian Conspiracy that elides out the differences would be historically dubious. Apart from some linguistic similarities, Plutarch's account diverges from those of the Peripatetics Aristoxenus of Tarentum and Dicaearchus of Messana from the late fourth century BCE. In particular, neither Aristoxenus nor Dicaearchus seeks to describe the larger brotherhoods of Pythagoreans in any detail. This may be due to the emphasis each places on the preservation of unified Pythagorean ethical doctrines, at the cost of historical information pertaining to the political activities or social organization of the Pythagoreans.[46] Or it might be because Plutarch was deriving his information from a source that told a similar story about the Cylonian Conspiracy but focused on the descriptions of the political and social institutions of the city-states of Magna Graecia. Regardless, we are prompted to consider whence Plutarch might have come up with the idea that the Pythagorean communities were constituted of ἑταιρίαι.

43. Variously called συνεδρεύοντες (by Aristoxenus [F 18 Wehrli = Iambl. *VP* 249, 134.5–6] and Plutarch) and ἑταίροι παρεδρεύοντες (Dicaearchus F 41A Fortenbaugh and Schütrumpf = Porph. *VP* 56–57).

44. Evidence for this subgroup also occurs in Justin's epitome of Pompeius Trogus, where it is referred to as a *sodalicium* (Justin 20.4.14) among the youth. Von Fritz (1940: 41–42) argued that Justin used a single source, Timaeus of Tauromenium, for his account of Pythagoreanism. It is therefore likely that at least some portions of Plutarch's account also trace from Timaeus's tradition.

45. Aristoxenus (F 18 Wehrli = Iambl. *VP* 249, 134.6–7) refers to the men "deliberating about civic affairs" (βουλευόμενοι περὶ πολιτικῶν πραγμάτων) in Milo's house.

46. For a useful if brief discussion of Aristoxenus's and Dicaearchus's representations of Pythagoreanism, see Kahn 2001: 68–71.

Plutarch's brief description of the Pythagorean brotherhoods should be compared closely, I suggest, with other sources that derive ultimately from Western Greek historiography. It is chiefly in the account of a certain Apollonius,[47] as preserved by Iamblichus in his work *On the Pythagorean Way of Life* (254–264), that we get the most extensive treatment of the social organization and political activities of the Pythagorean *ἑταιρίαι* in Southern Italy during the fifth century BCE.[48] Indeed, Iamblichus himself differentiates Apollonius's account from those that have come before in his narrative (Aristoxenus's and Nicomachus's) by emphasizing the important additions it makes to those otherwise skeletal versions.[49] Iamblichus was attentive to the way Apollonius's account, which focuses on historical elements not explored in the accounts of Aristoxenus and Nicomachus, diverges from tradition. As Minar noted, this claim at least partially explains why the account of Apollonius that Iamblichus presents, after he introduces it, is chronologically confused: Iamblichus is likely to have selected only the parts of Apollonius's account that were of value to him in filling in the gaps left by Aristoxenus's and Nicomachus's accounts.[50] As I will show, the confusions of evidence and omissions of material have catalyzed a great deal of

47. Who this Apollonius was is unfortunately impossible to determine. One candidate is Apollonius of Tyana (FGrHist 1064), who is said (Suda s.v. Ἀπολλώνιος, Τυανεύς = FGrHist 1064 T 9) to have written a *Life of Pythagoras* or books *On Pythagoras* (Porph. *VP* 2 = FGrHist 1064 F 1). But also see the arguments of Gorman 1985 against this hypothesis. The best studies of the account of this Apollonius remain Morrison 1956: 147–149, von Fritz 1940: 44–67, Lévy 1926: 105–111, and Rostagni 1913–14: 382–395.

48. Iamblichus's version of Apollonius's account appears to begin after p. 136.17 of Deubner-Klein's Teubner edition (*φέρε δὴ καὶ τὴν τούτου παραθώμεθα διήγησιν περὶ τῆς εἰς τοὺς Πυθαγορείους ἐπιβουλῆς. λέγει . . .*) and conclude just at p. 142.7 Deubner-Klein (*περὶ μὲν οὖν τῆς κατὰ τῶν Πυθαγορείων γενομένης ἐπιθέσεως τοσαῦτα εἰρήσθω*). It shares several qualities with a brief mention in Polybius's *Histories* (2.39) of the burning of the houses of the Pythagoreans and, subsequently, a "widespread political revolution" (*γενομένου κινήματος ὁλεσχεροῦς περὶ τὰς πολιτείας*) that resulted in the death of the leading men in each city. Both Apollonius and Polybius were probably using the same source, Timaeus of Tauromenium. See Walbank 1959: 223.

49. Iambl. *VP* 254, 136.14–17: "But since Apollonius differs in some places about the same events, and adds many things not said by these authorities, let us also give his account of the plot against the Pythagoreans." Burkert (1972: 101) evaluates the relationship between Nicomachus's and Apollonius's accounts most reasonably: "[Rohde's negative evaluation of Apollonius's account] can only be reckoned as true a parte potiori. Apollonius too used good sources, and it is precisely his material that has provided most opportunities for those who, from time to time, have tried to discover really ancient lore. Nicomachus, on the other hand, who calls himself a Pythagorean, is so intimately concerned in his narrative that, at least in selection, arrangement, and interpretation, his personal contribution must not be underestimated."

50. Minar 1942: 60.

unfortunate speculation on the part of scholars of Pythagorean political history but also encourage us to admit various possibilities in our evaluation of the material.

The account of Apollonius is largely, if not wholly, based on the history of Southern Italy that is presented in books 9 and 10 of the *Italian and Sicilian History*[51] of Timaeus of Tauromenium (ca. 350–ca. 260 BCE).[52] It describes the political history of the Pythagorean city-states in the fifth century BCE, with special attention focused on the nature of political revolutions there, in particular those caused by democratic revolt.[53] Apollonius describes the envy of those in the city-state of Croton who felt disenfranchised as inferiors to the smaller group of Pythagoras's intimates, who began to "gain prominence privately" (*ἐν τοῖς ἰδίοις βίοις πρωτεύειν*) and "govern the city publicly" (*τὸ κοινῇ τὴν πόλιν οἰκονομεῖν*) due to their proximity to Pythagoras and their parental lineage and inheritance. Apprehensions about the intermingling of public and private activity in a *ἑταιρία* go back at least to Thucydides, in his account of the attempted overthrow of the Athenian democracy by the oligarchs in 411 BCE.[54]

51. As it is called in the Suda (s.v. *Τίμαιος* = FGrHist 566 T 1), but alternative titles include *Sicilian History* and *Histories*. For accessible recent discussions of Timaeus, see Vattuone 2007 and Baron 2012.

52. On Timaeus as the chief source for Apollonius's account, see Burkert 1972: 104–105, with n. 37, which covers the relevant bibliography before 1972. There is one important exclusion, however: Walbank (1957: 223) refers to Apollonius's account as "ultimately Timaeus, but . . . Timaeus in a much worked-over and distorted form," but does not go on to explain what he means by this claim, or how he might justify it. Since 1972, very little has been written on the relationship between Iamblichus, Apollonius, and Timaeus at Iambl. *VP* 254–264, 136.17–142.7. Gorman (1985) argued that the Apollonius in question was not Apollonius of Tyana, although he identified the language of "Apollonius" at *VP* 255–256 as Hellenistic. Pearson (1987: 113–115), in his discussion of the life of Pythagoras, admits the possibility that Timaeus is a source for Apollonius, but he does not cite or discuss any of the studies that evaluate Apollonius's account (e.g. Burkert 1972, Minar 1942, Rostagni 1913–14). Vattuone (1991: 213, with n. 23) accepts the conclusions of Burkert concerning Apollonius without further elaboration. Zhmud (2012: 68–70) argues that "it would be wrong to accept a minimalist line and reject what has been successfully reconstructed" of Timaeus's history of the Pythagoreanism. Baron (2012: chap. 7) is critical of Rohde's assumption that the speeches of "Pythagoras" given by Apollonius preserve Timaeus's own wording (*VP* 37–57), but he has little to say about the passages of Apollonius's account of significance to this study (*VP* 254–264, 136.17–142.7).

53. Notably, Timaeus was concerned with establishing a historical account of Locri, Croton, and Sybaris that corrected the account of Aristotle, which, according to Polybius (12.8 = FGrHist 566 F 156), Timaeus complained about excessively. See Pearson 1987: 41.

54. On which see Sartori 1967: 115–126. Radicke (1999: 156) plausibly suggests that the "political terminology and the sociological model of class struggle [in Apollonius's account] . . . also point to a fourth-century author, well versed in Athenian history, as a source."

Apollonius, following Timaeus, describes the Pythagorean brotherhood in some detail, explaining that the Pythagoreans "formed a large *ἑταιρία*—for they were three-hundred—but they were just a small part of the city, since it was not yet governed by their customs and practices."[55] The combination of proximity to Pythagoras and familial inheritance fuels the growing dissent among the citizens of Croton. Another fragment associated with Timaeus's account of the political history of the Pythagoreans, this time preserved by Diogenes Laertius (8.3, not in Jacoby's collection), explains that Pythagoras became famous along with his followers (*ἐδοξάσθη σὺν τοῖς μαθηταῖς*) for the laws that they had enacted among the Italians, a curious combination of the emphases on the fame of the Pythagoreans found in the Athenian traditions of Isocrates and Plato and on the political activity that marks the later Pythagorean historical traditions from the Platonist Speusippus and the Peripatetics Dicaearchus and Aristoxenus forward.[56] After the destruction of Sybaris—most likely the one that occurred in 454/3 BCE, when Croton took control of its conquered land and its leaders chose not to divide it according to a system of equal distribution—what appears to be a democratic revolt against the Pythagoreans ensued.[57] The contours of this revolt are especially rich and complex, and it is worth devoting some attention to them.

Apollonius describes the revolution as fueled by the consternation of the people (*τὸ πλῆθος*) concerning the distribution of conquered Sybarite land, but led by the Crotonians who "stood closest in ties of kinship and friendship to the Pythagoreans" (*οἱ ταῖς συγγενείαις <καὶ> ταῖς οἰκειότησιν ἐγγύτατα καθεστηκότες τῶν Πυθαγορείων*).[58] Ostensibly, we are dealing with a democratic revolt against a *ἑταιρία* with oligarchic tendencies. This group of leaders and the people protested against the particularity (*ἰδιασμός*) of the life that the Pythagoreans practiced, which was exclusive and consisted of various daily

55. Iambl. *VP* 254, 137.1–4: *μεγάλην μὲν ἑταιρείαν συναγηοχόσιν* (*ἦσαν <γὰρ> ὑπὲρ τριακοσίους*), *μικρὸν δὲ μέρος τῆς πόλεως οὖσι, τῆς οὐκ ἐν τοῖς αὐτοῖς ἔθεσιν οὐδ' ἐπιτηδεύμασιν ἐκείνοις πολιτευομένης*. Translated after Dillon and Hershbell 1991. Justin (20.4.1) also knows this story from Timaeus. Von Fritz (1940: 108) correctly, in my opinion, argues that this account suggests indirect rule by Pythagoreans.

56. For the political activities of Pythagoras, see Dicaearchus F 40A–B Fortenbaugh and Schütrumpf (= Porph. *VP* 18–19) and F 41 (= Porph. *VP* 56–57). Speusippus, too (F 3 Tarán = D.L. 9.23), apparently in a work called *The Philosopher*, referred to Parmenides as giving laws to the city-states.

57. See Walbank 1957: 223–224 and Minar 1942: 77–78, contra von Fritz 1940: 78–79.

58. As Dillon and Hershbell 1991 translate this difficult phrase. The land of Sybaris conquered by the Crotonians was massive, and its distribution would have tested any ancient community's ideological barometer. See Musti 2005: 104–105.

rules, regulations, and prohibitions that were formed from *acusmata* of the *τί πρακτέον* variety.[59] The ties that constituted this particularity, according to Apollonius, were grounded in their private education (*ἰδιάζοντας ἐν αὐτοῖς τοὺς συμπεπαιδευμένους*), which set them apart from their relatives (excepting their parents) and from the "people" of Croton. Moreover, these Pythagoreans practiced the sharing of property with one another, in accordance with the famous Pythagorean *acusma* "things among friends are shared" (*κοινὰ τὰ φιλῶν*) but were not allowed to share their property with their own relatives.[60] The nuanced ethical practices of the Pythagorean *ἑταιρία* thus appear to exercise some influence over the political actions of its members.

At this point in Apollonius's account, something very important for the present broader study of the role of Hippasus of Metapontum in Pythagorean history arises: Apollonius gives the names of the various leaders of the pro- and anti-Pythagorean groups in Croton and includes Hippasus among them. Here is Apollonius's account of what happened after the Pythagoreans' relatives began to dissent:

> And when, from the council of the Thousand [*ἐξ αὐτῶν τῶν χιλίων*], Hippasus, Diodorus, and Theages spoke in behalf of all citizens having a share in the political offices and the assembly [*κοινωνεῖν τῶν ἀρχῶν καὶ τῆς ἐκκλησίας*], and of having public officials give accounts of the conduct to those who had been elected by lot from all citizens, the Pythagoreans Alcimachus, Deinarchus, Meton, and Democedes opposed this proposal and sought to prevent the ancestral constitution [*πάτριος πολιτεία*] from being abolished. Those who were champions of the

59. The same sort of behavior mocked by the comedians Alexis of Thurii and Cratinus the Younger. See Vattuone 1991: 216 n. 31. Note that the Pythagoreans must "determine what must be done" (*προχειρίζεσθαι τί πρακτέον*) when they awake in the morning. The Pythagorean intimates are further prohibited from calling Pythagoras by name, rising from bed later than sunrise, wearing a signet ring with a divine image, and blaspheming.

60. First attributed to Pythagoras by Timaeus of Tauromenium (FGrHist 566 F 13), along with the phrase "friendship is equality" (*φιλία ἰσότης*) by Diogenes Laertius (8.10). The Scholiast to Plato's *Phaedrus* (279c) quotes Timaeus directly: "When the young men came to visit him and wished to practice his way of life with him [*συνδιατρίβειν*], he did not admit them immediately, but said that all who shared his company must also share their property, holding it in common [*δεῖν καὶ τὰς οὐσίας κοινὰς εἶναι τῶν ἐντυγχανόντων*] . . . and it was because of these men that the saying 'things among friends are shared' came to be used in Italy" (trans. after Pearson 1987). This passage bears a very strong resemblance to another passage that might originate with Apollonius in Iamblichus's *VP* (71–74, 40.15–42.22), on which see Burkert 1972: 192 n. 1. Also notable is the use of similar language in Apollonius's account at *VP* 256, 138.6–8, in which we see an explanation of what it means for someone to "practice" the Pythagorean communal way of life (*τις τῶν κοινωνούντων τῆς διατριβῆς*).

> common people prevailed. Thereupon, when the people assembled, the politicians Cylon and Ninon, apportioning between themselves the thrust of their speeches, launched an attack [on the Pythagoreans].
>
> (Apollonius FGrHist 1064 F 2 = Iamblichus, *On the Pythagorean Way of Life* 257–258, 138.22–139.8; translation after Dillon and Hershbell 1991)

What do we learn from this passage about the Pythagoreans? Notably, Apollonius stages a typical[61] democratic revolution in atypical historical circumstances: the ancestral constitution (*πάτριος πολιτεία*) of Croton seems to have provided for two active bodies of government: the magistrates of the city and the council of the Thousand.[62] This is consonant with, and expands on, Diogenes Laertius's statement (8.3), also derived from Timaeus, that explains how Pythagoras and his followers governed Croton so nobly (*ᾠκονόμουν ἄριστα τὰ πολιτικά*) that the constitution itself was a true aristocracy (*ὥστε σχεδὸν ἀριστοκρατίαν εἶναι τὴν πολιτείαν*).[63] But Apollonius's account in particular focuses on the effect that the democratic revolt, perpetuated by several political actors, including Hippasus of Metapontum, had on the constitution of Croton.[64] At issue in Apollonius's account is not only the esoteric lifestyle of the Pythagoreans, which interested

61. See, inter alia, Arist. *Pol.* 6.2, 1318a3–10.

62. What was the precise nature of the Crotonian ancestral constitution (*πάτριος πολιτεία*) that underwent change? We cannot be sure. It is likely that it was originally some sort of oligarchy, on the grounds that all sources that discuss the revolution in political terms (Apollonius at *VP* 265, Strabo at 8.7.1, and Polybius at 2.39.1) refer to it as democratic. We should also add the testimony of Dicaearchus (F 40 Fortenbaugh and Schütrumpf), who refers to a "council of elders" (*τὸ τῶν γερόντων ἀρχεῖον*) to which Pythagoras appealed, and Diodorus Siculus (12.9, possibly from Ephorus) refers to a council and an assembly (*ἡ σύγκλητος καὶ ὁ δῆμος*) who deliberate on the issue of whether to surrender suppliants to the Sybarites before the battle on the Traeis in 511/10 BCE.

63. See Minar 1942: 15–16. This account of Timaeus shows correspondences with that of Dicaearchus of Messana (F 40 Fortenbaugh and Schütrumpf = Porph. *VP* 18–19), on which see White 2001: 211–212. The nature of this "governance" is under debate, and I am inclined to imagine the Pythagoreans as being an influential group who participated in the larger civic council while retaining their own private rites and ability to withdraw from public engagements, that is, a *ἑταιρία*. I would also note that Musti's interpretation (2005: 159–160), which sees Nicomachus as a source intermingled with what is attributed to Dicaearchus, does not account for the Athenian tradition of Pythagoras as a public orator, as directly illustrated by Isocrates and Antisthenes, and as implied by Plato.

64. We know of no other Hippasuses in South Italy during the fifth century BCE. Iamblichus (*VP* 81, 47.3–4) claims that "some say" that Hippasus was originally from Croton, "others" that he was from Metapontum. The list of Pythagoreans at the end of Iamblichus's work *On the Pythagorean Way of Life* (267, 144.20), which traces back to Aristoxenus, lists him as coming from Sybaris.

Aristotle, Aristoxenus, and those who followed them, but also and especially the *political history of the city-states of Magna Graecia.* All of this continues to point to Timaeus of Tauromenium as the chief source for specific details in Apollonius's version of the story, especially given the fact that Timaeus pursued an account of the history of the city-states of Western Greece that celebrated those "democratic" political revolutionaries who represented the cause of the people and maligned anyone who had designs on oligarchy or tyranny, in accordance with their proclivity to luxury (τρυφή).[65]

The specificity of the details found in Apollonius's account prompts us to evaluate the evidence from which Timaeus of Tauromenium, his chief source, derived his own history of those city-states whose legislation and conduct were influenced by Pythagoreans. It is almost certain that Timaeus, who spent a great deal of his lifetime in Athens,[66] had access to the Aristotelian *Constitutions* (Πολιτεῖαι), likely collected in the Lyceum sometime during the 320s BCE.[67] This might lead us to speculate that Timaeus obtained his information about Magna Graecia as well as the Pythagoreans from the Aristotelian *Constitutions* or from Aristotle's lost works on the Pythagoreans. But we also need to account for Timaeus's polemic against Aristotle concerning various historical and methodological issues, including how each discusses the origins of the colony of Epizephyrian Locri and the significance of the famed Zaleucus as lawgiver there.[68] In particular, Timaeus leveled against the philosopher, whom he characterized as employing arguments from probability (καθ' εἰκὸν λόγον) in his

65. See Justin 20.4.5 and Iambl. *VP* 255, 137.4–17, with reference to FGrHist 566 F 44–45 (= Athen. 12, 522a and c). On Timaeus's criticism of tyrannical luxury, see especially Vattuone 1991: 222–236, who also argues not implausibly that the democratic political theory of Archytas is relevant to Timaeus's ethics. In this light, it is notable that Timaeus champions Empedocles (FGrHist 566 F 2 = D.L. 8.66; see 566 F 134 = D.L. 8.63) as a political figure both wealthy and democratically minded (ἦν . . . τῶν τὰ δημοτικὰ φρονούντων), just like Cylon. Timaeus credits Empedocles with abolishing a group called "the Thousand" (τὸ χιλίων ἄθροισμα) in Agrigentum, which might, in Timaeus's mind, have been comparable with the "Thousand" at seats of Pythagorean conservatism in Croton and Rhegium in the same period (see Heraclides Lembus F 55 Dilts). Note that Heraclides, who is basing his information probably on Aristotle, describes the council of the "Thousand" in Rhegium as "aristocratic" (ἀριστοκρατικῆς). It is not obvious from the evidence that Timaeus would have considered Empedocles a Pythagorean, except in a limited sense. See Pearson 1987: 127–128.

66. For a good summary of Timaeus's life and writings, see Brown 1958: 1–20.

67. Generally, on the Aristotelian collection of *Constitutions* as well as Peripatetic historiography, see Rhodes 1991: 58–63.

68. On Timaeus's dispute with Aristotle about Locrian history and Zaleucus, see especially Pearson 1987: 98–108. It should be noted that Timaeus (at Iambl. *VP* 255, 137.4–11) has Pythagoras depart for Metapontum before the anti-Pythagorean revolt in Croton, in agreement with Aristotle (F 191 Rose = Apollon. *Mirab.* 6) and Aristoxenus (F 18 Wehrli = Iambl. *VP* 249, 133.23–25), and against all other sources (pace Minar 1942: 67). See Burkert 1972: 117 n. 46.

historiography,[69] the charge of ignorance of local customs and law (ἔθη; νόμος) in Southern Italian city-states; and while it is likely that Timaeus used the texts of Aristotle and his students in his own research, he appears to have done so chiefly in order to refute them.[70]

Indeed, we have evidence that points to two sources outside Aristotle for Timaeus's knowledge of political affairs in Southern Italy. First, conversations between Timaeus and a certain Echecrates[71] are attested by Polybius (12.10.7–9 = FGrHist 566 T 10).[72] According to Polybius, Timaeus cites Echecrates as the authority behind his information about the history of Epizephyrian Locri and reinforces this information with the statement that Echecrates's father was an envoy of Dionysius II,[73] implying that the father of Echecrates, given his status and connections, was a reliable source of knowledge about Epizephyrian Locri during the first half of the fifth century BCE.[74] Who was this Echecrates? We are most familiar with Echecrates of Phlius, to whom Phaedo tells the story of Socrates's death in Plato's *Phaedo* but about whom we know very little outside Plato's work. From the portrayal in the *Phaedo*, we gather scraps of knowledge: Echecrates of Phlius was very curious, somewhat of an amateur historian who inquired about the previous generation of philosophers, and he may have known many of them by name, if not personally.[75] Plato also represents him as

69. FGrHist 566 F 12 = Polyb. 6.9.2.

70. FGrHist 566 F 11a = Athen. 6, 264c–d, quoting Timaeus directly. Timaeus thus explains that the Locrians themselves criticized Aristotle's treatment of their laws (probably in the *Constitution of the Locrians*): "καθόλου δὲ ᾐτιῶντο τὸν Ἀριστοτέλη διημαρτηκέναι τῶν Λοκρικῶν ἐθῶν· οὐδὲ γὰρ κεκτῆσθαι νόμον εἶναι τοῖς Λοκροῖς." Regardless of this historicity of this statement, it is a deft rhetorical move: not only does it cast doubt on Aristotle's skill as a historian for ignoring local historical sources, it also places the historical authority in the local community itself. See Pearson 1987: 98–100.

71. That Timaeus spoke with this Echecrates is explicit in Polybius's summary: ποιήσασθαι τοὺς λόγους.

72. Brown (1958: 48–50), following Oldfather, speculates that the Pythagorean Echecrates of Phlius was Timaeus's source for most of his knowledge of the early philosophers of Magna Graecia. More cautious is Pearson (1987: 100–101).

73. Polybius only says "Dionysius," but, given Dionysius II meddling in Locrian affairs (especially the annulment of the ancestral constitution of the Locrians in the 350s BCE, which would have been of interest to Timaeus, as it was to Aristotle, *Pol.* 5.1, 1307a34–40), we might speculate that it was the younger tyrant.

74. See Vattuone 1991: 50–51.

75. For example, at *Phd.* 59c2, Echecrates, in the manner of a busybody, asks whether Aristippus and Cleombrotus were present at the death of Socrates. It is to be assumed that Echecrates at least knew of these two men, even if we cannot be sure whether he knew them personally. Aristippus and Cleombrotus had been represented as teenagers who met with Socrates in Plato's *Lysis*.

believing that the soul is not immortal but rather a material harmony perishable at death (*Phaedo* 88d3–6). Aristoxenus (F 19 = Diogenes Laertius 8.46; see Iamblichus, *On the Pythagorean Way of Life* 251, 135.3–8), whose information often differs considerably from that of Timaeus, considers Echecrates one of the famous "last of the Pythagoreans" (τελευταῖοι τῶν Πυθαγορείων) whom Aristoxenus claims to have seen personally. This may have happened either when he was a young man and in the company of his "father" Spintharus, who was associated with the Tarentine Pythagoreans, or when he spent time with Xenophilus of Chalcidice.[76] The combined traditions of Plato and Aristoxenus corroborate Burkert's claim that Echecrates would have been considered some sort of mathematical Pythagorean, along with Philolaus, who is said to have been his teacher, and Archytas of Tarentum.[77] Another later tradition known to Cicero (*On Ends* 5.29.87) and Valerius Maximus (8.7.ext.3) makes Echecrates—this time of Epizephyrian Locri—one of the Pythagoreans whom Plato met possibly on his second visit to Southern Italy around 367 BCE, in addition to Archytas of Tarentum.[78] There is also indirect evidence that the Phliasians sought to preserve their Pythagorean history through genealogical record, probably oral.[79] It is impossible, however, to evaluate the evidence from this tradition independent of further context.

There are some problems, however, with identifying the source of Timaeus's knowledge about Locri with Echecrates of Phlius. The chief problem is chronology: if the Platonic and Aristoxenian Echecrates is the same figure—and there is no obvious reason why this could not be the case—and this figure was a contemporary with Archytas and the other "last of the Pythagoreans," when would Echecrates have met Timaeus of Tauromenium, who was likely born only in the 350s? The most probable scenario would be that the two men met when Echecrates was at a very advanced age and Timaeus himself was in his teens or early twenties. This is not chronologically impossible, in the absence of further corroborating evidence from the fourth century BCE, we should be at

76. On Aristoxenus's intimate knowledge of Pythagoreanism, see Huffman 2006: 107 and Burkert 1972: 198.

77. Ibid. In the catalogue of Pythagoreans at the end of Iamblichus's work *On the Pythagorean Way of Life* (267, 144.15 and 146.6), Echecrates is listed both as a Tarentine and as a Phliasian. We might speculate that the double origin might be explained by Echecrates's emigration from Tarentum to Phlius.

78. That this occurred on the second visit is suggested by Apuleius's reference (*Plat.* 1.3) to Plato studying with Archytas specifically on his second voyage to Italy.

79. See D.L. 8.1 and Paus. 2.13.2, both of whom mention Hippasus as well. Aristoxenus may have been influenced specifically by the Phliasian traditions (D.L. 8.46 = F 19 Wehrli). More generally on the Pythagoreans of Phlius, see Huffman 2006: 107 and Burkert 1972: 206 n. 77.

least skeptical of the argument, advanced by, for example, T. S. Brown, that Echecrates of Phlius was Timaeus's source for his knowledge about Pythagorean matters.[80] Another problem is the fact that the Echecrates to whom Polybius refers as Timaeus's source for information about Epizephyrian Locri is not named as a Pythagorean, which Polybius could have easily done, as he has referred to Pythagoreans by reference to Timaeus's fragments earlier in his work.[81] Neither of these problems definitively counts out the possibility that Echecrates of Phlius was Timaeus's interlocutor and source for knowledge about Magna Graecia and Pythagoreanism, but caution should be exercised.

The other non-Aristotelian source for Timaeus's knowledge about political affairs in Southern Italy is of greater value to us, as it surely was to him in his desire to undermine the authority of Aristotle's probabilistic account of South Italian history: documentary evidence. Regarding the early history of Epizephyrian Locri, which, in particular, provided him with reason to criticize Aristotle, Timaeus distinguishes his own historiographical method by appealing to written sources preserved locally. Polybius is our best ancient source on Timaeus's method. He explains (12.9.1–12.10.9 = FGrHist 566 F 12) that rather than accept Aristotle's treatment of the founding of Locri, which Timaeus characterized as constructed according to "argument from probability" (*καθ' εἰκὸν λόγον*), Timaeus went personally to the mother-city and consulted a "written treaty" (*συνθηκότες ἐγγράπτοι*) preserved between the Opuntian and Epizephyrian Locrians. As proof, Timaeus quotes the first line of the treaty.[82] Polybius goes on to suggest that Timaeus subsequently traveled to Epizephyrian Locri, where he may have examined a law code that provided him with evidence against Aristotle's account of the colony. Polybius, to be sure, has reason to doubt Timaeus's testimony on this particular issue, but as Polybius later admits, it is a universally accepted fact (*πάντες γιγνώσκομεν*) that Timaeus employed public records with precision in his historical inquiries.[83] Indeed, Polybius's celebration of Timaeus's discovery of "inscribed columns in the inner chambers of temples and the proxeny decrees inscribed on their doorways"[84] may be the greatest praise the later Hellenistic historian was willing to pay his intellectual forebear and competitor.

80. See Pearson 1987: 100–101.

81. Plb. 2.39.1. On Timaeus as the most likely source for this passage, see Walbank 1957: 223.

82. The first line was "as parents to children" (*ὡς γονεῦσι πρὸς τέκνα*), understood with something like ἔδοξε. The formula is accepted as genuine by Pearson (1987: 104).

83. Plb. 12.10.4: "*λέγω δὲ κατὰ τὴν ἐν τοῖς χρόνοις καὶ ταῖς ἀναγραφαῖς ἐπίφασιν τῆς ἀκριβείας*." See Diod. 5.1.3.

84. Plb. 12.11.1 = FGrHist 566 T 10, translated by Pearson: "*καὶ μὴν ὁ τὰς ὀπισθοδόμους στήλας καὶ τὰς ἐν ταῖς φλιαῖς τῶν νεῶν προξενίας ἐξευρηκὼς* Τίμαιός ἐστιν."

It will not be surprising, given Polybius's characterization of Timaeus as a historian with an "almost maniacal care for documents,"[85] that we find traces of documentary evidence sprinkled throughout Apollonius's account of the story of the expulsion of the Pythagoreans from Croton. The relevant portion of Apollonius's account, as preserved by Iamblichus in his work *On the Pythagorean Way of Life* (257–262), features references to various sorts of binding agreements: rewards given for fulfillment of the decrees of the assembly in Croton, treaties agreed on by various city-states of Magna Graecia, and oaths ratified by foreign ambassadors.[86] The sources for this historical information are given explicitly: we hear of the "record-books of the Crotonians" (τὰ τῶν Κροτωνιατῶν ὑπομνήματα) and inscriptions at Delphi which may have been consulted at the source.[87] Once again, we can detect the characteristic stamp of Timaeus's historiographical method on Apollonius's account of the revolt against the Pythagoreans. Moreover, as would be expected of Timaeus's approach, various historical details concerning the Pythagoreans are incorporated into a larger constitutional history of Croton, chiefly for the sake of exposing the dangers that luxury brings to a civic community.[88] As I mentioned earlier, the account first describes a meeting of the council of the "Thousand," in which Hippasus and two other figures, Diodorus and Theages, convince the

85. In the words of Vattuone (2007: 199).

86. Iambl. *VP* 261–263, 140.23–141.19: the Crotonian assembly votes to reward whomever kills the would-be tyrant Democedes with 3 talents, and Theages, who overcomes Democedes, is awarded the 3 talents; power of jurisdiction over the case against the fugitives is granted not to Crotonians but to Tarantines, Metapontines, and Caulonians; in order to reconcile the Pythagorean exiles with the Crotonians, Achaean ambassadors come and ratify the oaths between them. Note that Polybius, in a passage most likely derived from Timaeus (2.39.1), refers to a confederation between the Crotonians, Sybarites, and Caulonians, which probably took root after the Pythagoreans had been recalled from exile, sometime after the founding of Thurii (in 443 BCE) but likely before Thurii declared war on Croton in the 420s BCE. Polybius suggests that the city-states of Magna Graecia adopted the political ideals of the Achaean league twice: when they expelled the aristocratic Pythagoreans, and when they formed their own league in the threat of Lucanian invaders. See Walbank 1957: 225–226.

87. The importance of the Crotonian registers has been emphasized by von Fritz (1940: 65–67) and Musti (1988: 28). It is also possible that Timaeus obtained his evidence from registers kept in Heraclea Italica, which held the treasury of the Italiote league after 393 BCE (on which see Wuilleumier 1987: 70–71). There is no reason to confuse these civic ὑπομνήματα with the Pythagorean ὑπομνήματα of Alexander Polyhistor (D.L. 8.25–33 = FGrHist 273 F 93), which postdate Timaeus and show no relevant connections to the political account as preserved by Apollonius.

88. Compare Justin's account of the same episode (20.4.18), with the emphasis on the dangers that *luxuria* presented to the Pythagoreans of Croton.

council to pass significant changes to the ancestral constitution of Croton. These changes are democratic in design: the extension of the right to participate in the assembly as well as in all magistracies to all citizens, and the requirement that magistrates be accountable to officials selected by lot from the entire citizen body.[89] The measure is passed, despite an attempt to derail it by the defenders of the ancestral constitution, the Pythagoreans Alcimachus, Deinarchus, Meton, and Democedes. Subsequently, the *ἐκκλησία* assembles, and two Crotonian orators, Ninon and Cylon, attack the Pythagoreans in public. Cylon's speech is unfortunately passed over in the account, and this omission stimulates us to speculate that there might be a lacuna in the transcription of Timaeus's original version.[90] Then, in Apollonius's account, something quite bizarre happens that is noteworthy for this study.

Someone is said to have written an account or pamphlet (*βιβλίον*) intended to discredit the Pythagoreans and given it to the recorder to read out to the assembly of Crotonians. Scholars have generally assumed that the author of this text is the aforementioned Ninon:

> When the speeches were given [*τοιούτων δὲ λόγων*]—the longer of them spoken by Cylon [*μακροτέρων δὲ παρὰ τοῦ Κύλωνος ῥηθέντων*]—the other one goaded the assembly on [*ἐπῆγεν ἅτερος*], pretending to have inquired after the secrets of the Pythagoreans [*προσποιούμενος μὲν ἐζητηκέναι τὰ τῶν Πυθαγορείων ἀπόρρητα*] but fabricating in writing things through which he might slander them [*πεπλακὼς δὲ καὶ γεγραφὼς ἐξ ὧν μάλιστα αὐτοὺς ἤμελλε διαβάλλειν*], he gave the book to the recorder and ordered him to read it aloud. Its title was *Sacred Discourse* [*Λόγος ἱερός*], and the following is an outline of its contents:
>
> "They [i.e. the Pythagoreans] revere their friends as if they were gods, but subdue others as if they were beasts. This very opinion the disciples recollect in verse, in reference to Pythagoras, when they say: "his companions he held equal to the gods / the rest neither so in speech nor in value." They praise Homer especially for the verses in which he speaks of the shepherd

89. In Iamblichus's words (*VP* 257, 138.23–139.2), "*ὑπὲρ τοῦ πάντας κοινωνεῖν τῶν ἀρχῶν καὶ τῆς ἐκκλησίας καὶ διδόναι τὰς εὐθύνας τοὺς ἄρχοντας ἐν τοῖς ἐκ πάντων λαχοῦσιν*." Aristotle (*Pol.* 6.5, 1320b11–14) describes democratic innovations in Tarentum, which are to be dated probably to the expulsion of the king Aristophilides in 473 BCE, in similar terms: "they divide all their offices into two classes, some of them being elected by vote, the others by lot; the latter, so that the people may participate in them [*ὅπως ὁ δῆμος αὐτῶν μετέχῃ*], and the former, so that the state may be better administered" (trans. Jowett). See Berger 1989: 308–309.

90. See Minar 1942: 56 with n. 19.

> of the people. For, as a supporter of oligarchy, Homer represented the rest of men as cattle. They declare war against beans, since beans are source of the lot and of the policy of placing those chosen by lot in their charges. They incite to tyranny by declaring that it is better to be a bull for one day than a cow for a whole lifetime. While they praise the lawful behavior of the others, they order them to obey what they themselves have decreed."
>
> In short, he declared that their philosophy was a conspiracy against the people and exhorted them not even to tolerate the expression of their advice, but to keep in mind that they would not even have come together in the assembly at all if they [i.e. the Pythagoreans] had persuaded the Thousand to accept their advice.
>
> (APOLLONIUS FGrHist 1064 F 2 = IAMBLICHUS, *On the Pythagorean Way of Life* 258–260, 139.9–140.10; translation after Dillon and Hershbell 1991, with significant modifications)

Apollonius provides a few more details regarding the book read out by the recorder and finally concludes his outline of this speech by bookending the summary with the aforementioned "slander" (*τῇ διαβολῇ*) that was employed by its author to incite the people to attack the Pythagoreans on the day of the festival of the muses. There is a problem with the content of the *Sacred Discourse*, since Apollonius furnishes us with a summary of the contents but does not preserve the actual speech itself. Contextual evidence for this *Sacred Discourse* is also lacking from the ancient world, and the only evidence that corresponds with this account of the ironic *Sacred Discourse* refers to Hippasus of Metapontum, whose *Mystic Discourse* (Μυστικὸς λόγος) looks to have been a parody of Pythagoras's *Sacred Discourse* (Ἱερὸς λόγος).[91] The immediate source of this information is Heraclides Lembus, the excerptor of the Aristotelian *Constitutions*, who first attributes to Pythagoras a *Sacred Discourse* (from which Heraclides quotes the first lines) and then, by contrast, to Hippasus a *Mystic Discourse* "written to slander Pythagoras" (*γεγραμμένον ἐπὶ διαβολῇ Πυθαγόρου*).[92] Of course, Heraclides Lembus is not the only figure to have claimed that Hippasus wrote. As I argued in chapter 2, Aristotle seems to have been the source for Iamblichus's information that Hippasus was declared a heretic and drowned at sea for being the "first to publish in written form the sphere" (*διὰ δὲ τὸ ἐξενεγκεῖν καὶ γράψασθαι πρῶτος σφαῖραν*).[93] It would be unsurprising if Heraclides Lembus were to

91. As noted by Burkert (1972: 207 n. 78).

92. D.L. 8.7 = Heraclides Lembus F 9 Müller (vol. 3: 169–170).

93. Iambl. *DCM* 25, 77.18–24.

have taken the information concerning Hippasus's *Mystic Discourse* from some sort of Peripatetic account, especially given the fact that he excerpted extensively from works that circulated under the name of Aristotle.[94] Did Timaeus, who is certainly lying somewhere in the shadows behind Apollonius, have at his disposal Peripatetic evidence concerning the political activities of the democratic revolutionaries? This is possible, especially given that Aristotle, in his lost works on the Pythagoreans, likewise refers to Pythagoras's classification of human beings in the same breath as the Pythagorean "secrets" (ἀπόρρητα)[95] and apparently recognized Hippasus as the progenitor of the mathematical Pythagoreans, who, as pretenders (in the eyes of the acousmatic Pythagoreans as well as Isocrates), called themselves Pythagorean.[96] We are on safer ground, however, if we conclude that Timaeus seems to adapt the dramatic characteristics of the Pythagorean pretenders illustrated in various Athenian literary and philosophical circles by ascribing similar qualities to Ninon: he is an effective political speaker (ἐπῆγεν) who excels in the art of fiction (προσποιούμενος; πεπλακώς).[97]

When working with fragmentary historical sources, it is very often difficult to achieve certainty. In order to venture any plausible reconstruction of the political history of the Pythagoreans, we are required to navigate between various fragmentary historical traditions. As I have already shown in chapters 1 and 2, the influence of Aristotle over the philosophical history of the Pythagoreans is paramount and cannot be underestimated; but my investigations into the account attributed to Apollonius by Iamblichus near the end of his work *On the Pythagorean Way of Life* (257–262) reveal a tradition that, at the very least, counterposes Aristotle's history of the South Italian city-states, especially

94. Rhodes (1991: 65) notes that the information concerning Athenian political history and institutions in Heraclides's *Epitome* is entirely derived from the Aristotelian *Athenian Constitution*. It is not clear where Heraclides obtained information concerning Pythagoras and Hippasus. It is possible that Heraclides derived this information from the work *On Pythagoras* by Hermippus of Smyrna (FGrHist 1026 F 21–27), cited in a papyrus fragment of Heraclides's *Epitome* of a work by Hermippus of the same name (P.Oxy. 1367 = FGrHist 1026 T 5). But it is also important to note that Heraclides is cited along with the Peripatetic Dicaearchus of Messana by Diogenes Laertius (8.40 = F 41B Fortenbaugh and Schütrumpf) as describing Pythagoras's death from starvation in Metapontum, by contrast to the rationalizing account of Hermippus (ibid.), which described Pythagoras's death in a war between the Agrigentines and Syracusans.

95. F 192 Rose = Iambl. *VP* 31, 18.12–16.

96. See chapter 2, section entitled "Aristotle on Hippasus of Metapontum."

97. Compare Isoc. *Bus.* 29. Timaeus elsewhere (FGrHist 566 F 16 = Athen. 4, 163e–f) also describes Diodorus of Aspendus in similar terms, as someone who "pretended to resemble the Pythagoreans" (τοῖς Πυθαγορείοις πεπλησιακέναι προσποιηθέντος) through his special attire.

with regard to the city-states where Pythagoreans seem to have held a great deal of political influence in the fifth century BCE. It is chiefly to the third-century BCE Sicilian historian Timaeus of Tauromenium that we owe this alternative Western Greek account of Pythagorean politics, although there are traces of local histories of communities near Croton sprinkled throughout later Greek historiography that might derive from other sources.[98] By the middle of the third century BCE at the very latest, charismatic lore had come to be deeply intermingled with Pythagorean history, to such an extent that it becomes virtually impossible for us to extract actual historical evidence of Pythagorean politics in the writings of historiographers such as Neanthes of Cyzicus and the mysterious Hippobotus, with whom Neanthes is associated in the traditions.[99] By contrast, the historical writings of Timaeus of Tauromenium, allegedly rooted in documentary evidence, focused on the role that the Pythagoreans played in civic rule and subsumed the individual histories of various important Pythagoreans under the larger umbrella of the constitutional history of ancient Italy.

Timaeus's account thus raises many important questions about the history of the Pythagoreans. As I have argued, Hippasus was seen by Aristotle to have been the author of the methodological split between acousmatic and mathematical Pythagoreans. Scholars have tended to accept this evidence as legitimate, but they have not attempted to make sense of the evidence from Timaeus that makes Hippasus one of the leaders of the democratic revolution in Croton. Are these two stories in any way related? It is initially difficult to say for sure. On the one hand, Aristotle, adapting a characterization given by Isocrates, does describe the Pythagoreans interested in mathematics as pretenders ("so-called").[100] On the other hand, it is not clear that he associates the mathematical Pythagoreans with the grubby pedants, such as Diodorus of Aspendus, as he is described by Timaeus.[101] From what we can gather of the evidence, Aristotle was not especially concerned about the habitus of the mathematical Pythagoreans but focused on the ways their philosophical advancements anticipated his

98. For a comprehensive but ultimately speculative synthetic account of the history of the Pythagorean city-states in Southern Italy and Sicily, see Minar 1942: 36–49.

99. See Burkert 1972: 102. Neanthes's account seems to have carried over and influenced Nicomachus of Gerasa's *Life of Pythagoras* (FGrHist 1063, second century CE), which focuses on the miracles of Pythagoras (see Radicke 1999: 124–130). Zhmud (2012: 68 n. 30) thinks that Timaeus made use of Neanthes's fragments, but it is equally possible (1) that Neanthes used Timaeus, or (2) that both were using a lost source. More generally, on Neanthes's history of Pythagoreanism, see Riedweg 2005: 37–39.

100. See chapter 1, section entitled "On the 'So-Called' and Mathematical Pythagoreans."

101. FGrHist 566 F 16 = Ath. 4, 163e–f.

own philosophical system. If, however, we ask the question in the light of the evidence that Aristotle provides—that Hippasus was charged with impiety specifically because he *publicized* the sphere in writing—some aspects of the problem regarding Diodorus of Aspendus in the account of Timaeus disappear. Timaeus, for his own part, seems to distinguish two groups in the Pythagorean *ἑταιρία* not based on their personal ethics and comportment but on their proximity to the master: those who were the real or "esoteric" (*οἱ ἐσωτερικοί*) Pythagoreans, who received their name because they were "inside the curtain" (*ἐντὸς σινδόνος*), and those who were "outside" (*ἐκτός*; *ἔξωθεν*), who were rejected by Pythagoras and publicized the secrets of the Pythagoreans.[102] The "esoteric" Pythagoreans, those who belonged to the inner circle, have stimulated the imaginations of ancient and modern readers, but it remains very difficult to know what their precise activities might have been.[103] The story of the heretical Pythagoreans, as I am attempting to reconstruct it, starting from the original apostate Hippasus of Metapontum, is the history of a group of intellectuals who sought to give noninitiates access to the wisdom that had been passed down from Pythagoras.[104] That is to say, the larger project of attempting to secure the relationship between the democratic ideology of figures such as Hippasus and the scientific innovations of the mathematical Pythagoreans appears to be the reconstruction of the history of a group of exoterics. Indeed, we might describe this exoteric activity as the *democratization of scientific knowledge*,

102. See Iambl. *VP* 72, 41.15; 89, 52.14–18; 86, 50.6–17; and 266, 143.10–15 (on which see below). Iamblichus also discusses "esoteric" and "exoteric" teachings in the light of levels of education at *DCM* 18, 63.1–6, a passage that, I suspect, might have been drawn from the writings of Nicomachus. There, Iamblichus distinguishes between the "esoterics," who are taught "with the knowledge of reality" (*μὲν μετ᾿ ἐπιστήμης τῶν πραγμάτων*), and the "exoterics," who are taught "the same thing, [but] only mathematically" (*δὲ αὐτὸ τοῦτο μόνον μαθηματικῶς*). See, recently, Brisson 2012: 46–48, who translates *μόνον* unpersuasively, I think, as "exclusively." It is doubtful that Iamblichus would see the mathematicians as having a more exclusive learning than those who learn "with knowledge of reality." For my analysis of *VP* 86, see chapter 1, section entitled "Aristotle on the *Pragmateiai* of the Pythagoreans."

103. It is possible that Nicomachus and Hippolytus followed Timaeus in positing "esoteric" and "exoteric" Pythagoreans, which complicates the situation considerably. For example, in Porphyry's *On the Pythagorean Way of Life* 57, Nicomachus seems to be the source (see Iambl. *VP* 252, 135.18–24, but Porphyry offers better evidence; see Burkert 1972: 219 n. 4) for the following claims: "With the death of the Pythagoreans came the loss of their knowledge, since they had guarded it close in their hearts until that point, except for a few things committed to memory by those outside the group [*παρὰ τοῖς ἔξω διαμνημονευομένων*]. Pythagoras himself left no writings; but a few sparks of his philosophy, difficult to understand and unexplained, were preserved by those who were exiled, such as Lysis and Archippus." On Iamblichus and Nicomachus, O'Meara (1989: 14) remains central.

104. See Burkert 1972: 205.

which was part and parcel of larger political revolutions against oligarchies in the ancient world.[105]

We might at this juncture ask: what does Timaeus focus on when he refers to the pretenders who were not given access to the most intimate circles of the Pythagorean community? Consider the two other exoterics to whom Iamblichus refers in an account that also appears to derive from Timaeus's histories (Iamblichus, *On the Pythagorean Way of Life* 266): Diodorus of Aspendus, an early fourth-century BCE Pythagorean pretender who is said to have distributed throughout Greece the Pythagorean "sayings" (διέδωκε τὰς Πυθαγορείους φωνάς), which Aristotle apparently consulted for his own writings on the Pythagoreans, and Epicharmus of Syracuse, the early fifth-century BCE tragedian, is said to have "*published* the hidden doctrines" (ἐκφέροντα κρύφα δόγματα) of Pythagoras in meter.[106] Also mentioned in this context are Cleinias and Philolaus—here said to have been from Heracleia Italica (which would date them, correctly, to after the Tarentine colony's foundation in 433 BCE)—as well as Theorides and Eurytus of Metapontum, and finally Archytas of Tarentum, each of whom is said to have "devoted himself *to writing*" (ζηλωτὰς γράφειν) about those Pythagoreans who led the school before them.[107] I have already discussed Cylon, who, along with an otherwise unknown figure, Perillus of Thurii, is said to have been rejected from the inner part of the Pythagorean school (ὁμακοεῖον) by Iamblichus.[108] Then there are the very curious cases of Empedocles and Plato as exoteric Pythagoreans: Timaeus of Tauromenium (FGrHist 566 F 14 = Diogenes Laertius 8.54) elsewhere claims that Empedocles was prevented from participating in the Pythagorean discussions (τῶν λόγων ἐκωλύθη μετέχειν) because he was convicted of "stealing the arguments"

105. Musti (1988: 24–28) implicates the development of historiographical method in Western Greece within the larger competitive political ideologies of the fifth century BCE. One must be careful, though, not to claim that all historiography is meant to democratize knowledge without qualification. For example, Thucydides marks his study as critical both of democracy and of democratization of knowledge by selecting carefully what he considers most *worthy* to preserve for future generations. See Ober 1998: 60–63.

106. That is, if I am right in taking the entire list of successors at Iambl. *VP* 265–266, 142.15–143.15 (starting from λέγεται) as derived from Timaeus of Tauromenium (compare FGrHist 566 F 16 = Athen. 4, 163e–f). See Burkert 1972: 203–204.

107. For complications involved in interpreting this text, see chapter 4, section entitled "Growth and Being: Mathematical Pythagorean Philosophy before Plato."

108. Iambl. *VP* 74, 42.4–22, a passage that might go back to Timaeus or Nicomachus (for the term ὁμακοεῖον, see Iambl. *VP* 30, 17.17–19). Aristoxenus, too (F 18 Wehrli), described Cylon as rejected from sharing in the Pythagorean approach to life, although it must be noted that Aristoxenus's characterization of Cylon's character as "ill-tempered, violent, turbulent, and tyrannical" far exceeds the simple description of Timaeus, which exhibits no obvious hostility to Cylon.

(λογοκλοπία), "just like Plato" (καθὰ καὶ Πλάτων).[109] What λογοκλοπία might have meant to Timaeus remains difficult to infer from the passage. Within a generation of Timaeus, however, Neanthes of Cyzicus (fl. early third century BCE?), who seems to have collected information from his predecessors (perhaps the elusive Hippobotus and probably Timaeus) in order to write a work *On Illustrious Men*,[110] develops an account that corresponds with that of Timaeus on these basic points and general tenor,[111] but with some striking additions:

> Neanthes states that down to the time of Philolaus and Empedocles, all Pythagoreans participated in the discussions. But when Empedocles himself made them public [ἐδημοσίωσεν] in his poetry, they passed a law that no poet should be granted access to them. He says that Plato received the same treatment; for he too was barred. He did not say which of the Pythagoreans Empedocles studied under, since, as he claims, the letter that circulates as Telauges's, as well as the story that he studied with Hippasus and Brontinus, are unreliable.
>
> (NEANTHES, FGrHist 84 F 26 = Diogenes Laertius 8.55; translation after Inwood)

It is virtually impossible to be certain about where Neanthes got his information concerning Empedocles and Plato, and it is also difficult to deduce whether

109. See Burkert 1982: 18. Did Timaeus include Plato's name? Jacoby is doubtful, and Burkert makes no mention of Plato in his reference, but there is no obvious reason to exclude Plato from Timaeus's account. After all, the charge that Plato stole his philosophical tenets from Pythagoreans was leveled by Hermippus of Smyrna (FGrHist 1026 F 69 = D.L. 8.85) a generation later and hinted at (though not explicitly) by Timaeus's contemporary Timon of Phlius (F 54 Diels), who suggested that Plato had adapted Philolaus's book. For good discussions of the problem, see Riginos 1976: 169–174 and Brisson 2000: 25–41. Hermippus, for his part, probably had access to Timaeus's works (see Vattuone 1991: 210–227). It should also be noted that Timaeus denies Empedocles (FGrHist 566 F 6 = D.L. 8.67) the fame that had accrued from the legendary story of his death (by jumping into Mount Etna), declaring instead that Empedocles died in the Peloponnesus and without a tomb, "just as is the case for many men." The implicit suggestion is that Empedocles did not elevate himself above other people and that figures such as Heraclides of Pontus, who may have made Empedocles look like he was a miracle-worker and proto-tyrant, had misrepresented him (see Pearson 1987: 126–128; contra Kingsley 1995: 234, who misrepresents Timaeus's historical proclivities by referring to him as "motivated . . . by a strong element of polemic" that he never defines). For Timaeus's account of Empedocles as "democratic" and hostile to luxury (τρυφή)—which makes him a hero to Timaeus—see FGrHist 566 F 2 (= D.L. 8.66) and 566 F 134 (= D.L. 8.63).

110. Diogenes Laertius (8.72–73 = FGrHist 84 F 28) preserves a title for Neanthes's work *On the Pythagoreans*, but we cannot be sure whether this constituted a subsection of the larger work or a different book altogether.

111. Neanthes (ibid.) also seems to have agreed with Timaeus by writing that Empedocles promoted democratic political ideology in Agrigentum.

this account is chronologically earlier or later than the accounts of Timaeus of Taurominum.[112] Still, the strong relationship between Timaeus's and Neanthes's accounts suggests some connection—perhaps Neanthes had Timaeus's account at hand, or possibly both were deriving their information from a common source. What is notable in Neanthes's version for my purposes, to be sure, is the criticism of a tradition in which Empedocles was considered to have studied with Hippasus and Brontinus, attributed to a letter written by Telauges, the son of Pythagoras.[113] As Zhmud notes, Neanthes presents the earliest evidence of Pseudo-Pythagorean writing, already in the late fourth century BCE.[114] Obviously, we should not discount the historical value of the contents of this letter simply because Neanthes does so. But what else can we say about the letter of Telauges? Who wrote it? Diogenes Laertius refers to it twice: first (8.53) as a letter written to Philolaus (thus keeping within the tradition of exoteric Pythagoreans), where he tells us that Telauges claimed that Empedocles's father was Archinomus, and second (8.74) as a "small letter" (*ἐπιστολίον*), noting that it described Empedocles's death. According to the Telauges letter, Empedocles fell into the sea and died, apparently owing to his old age.[115] This seems to be an apocryphal account, in keeping with the other rationalizing accounts that emphasize the banality of Empedocles's death, such as those of Timaeus of Tauromenium and Strabo (6.2.8). For his own part, Iamblichus (*VP* 146, 82.3–12) knew of a *Sacred Discourse* (Ἱερὸς λόγος) that was considered by "some of the famous and reliable members of Pythagoras's school" (*ἔνιοι τοῦ διδασκαλείου ἐλλόγιμοι καὶ ἀξιόπιστοι*) not to be the writing (*σύγγραμμα*) of Pythagoras, as many others maintained, but to have been passed down by Telauges.[116] It is

112. Zhmud (2012: 67–68) thinks that Neanthes was personally affiliated with Philip of Opus, which is doubtful.

113. On the Telauges letter, see Burkert 1972: 289 n. 59. Aeschines wrote a dialogue called *Telauges* (Giannantoni IV A F 84 = Athen. 5, 220a) in which Socrates spoke with the eponymous figure, who is characterized, perhaps paradigmatically, by his ragged clothing. For Brontinus, see Riedweg 2005: 109.

114. Zhmud 2012: 68.

115. Death by drowning in the sea is often associated with Pythagorean apostates such as Hippasus who published the secrets (e.g. the rebuke of "Hipparchus" by "Lysis" at *VP* 75, 42.23–43.14; Hippasus at *VP* 88, 52.5; and an unnamed figure at *VP* 247, 132.17–20). On this tradition, see Burkert 1972: 459, with n. 63.

116. The story in Iamblichus is hopelessly confused, and he himself is inconclusive about the precise authorship of this text. The *Sacred Discourse*, if it is actually the same text as the letter of Telauges, seems to have included an explanation of its origins, in which Telauges claimed to have received the *ὑπομνήματα* of Pythagoras from his sister on her death. I cannot agree, however, with Burkert (1972: 100 n. 15) that "doubtless Nicomachus is to be understood" as one of the "famous and reliable members" of Pythagoras's school who attributed the *Sacred Discourse* to Telauges.

not clear whether this document is the same as the letter of Telauges to Philolaus. It is equally unclear whether Timaeus of Tauromenium knew about this tradition and might have included Telauges among the names of those who were original exoterics in the first generations after Pythagoras.

What can be discerned from the tangle of evidence discussed here, however, is that the early heresiological traditions that derive from Timaeus of Tauromenium and Neanthes describe Pythagorean exoterics as those who made available the unwritten teachings of Pythagoras to the wider public and, in certain cases, incurred punishment for having done so: Perillus of Thurii, Cylon and Ninon of Croton, Empedocles of Agrigentum, Diodorus of Aspendus, Epicharmus of Syracuse, Cleinias and Philolaus of Heracleia, Theorides and Eurytus of Metapontum, Archytas of Tarentum, and possibly Plato of Athens.[117] But the most notorious of the Pythagorean exoterics was, of course, Hippasus of Metapontum, here described by Iamblichus in a passage that seems to derive ultimately from Aristotle:

> And concerning Hippasus, they say while he was one of the Pythagoreans, he was drowned at sea for committing heresy, on account of being the first to publish, in written form [διὰ τὸ ἐξενεγκεῖν καὶ γράψασθαι], the sphere, which was constructed from twelve pentagons. He acquired fame for making his discovery, but all discoveries were really from "that man" (as they called Pythagoras; they do not call him by name] . . . well, then, such are basically the characteristic differences between each philosophical system and its particular science. The Pythagoreans devoted themselves to mathematics. They both admired the accuracy of its arguments, because it alone among things that humans practice contains demonstrations [εἶχεν ἀποδείξεις ὧν μετεχειρίζοντο], and they saw that general agreement is given in equal measure to theorems concerning attunement, because they are [established] through numbers, and to mathematical studies that deal with vision, because they are [established] through diagrams.
>
> (Iamblichus, *On the General Mathematical Science* 25, 77.18–78.18)

Thus we have come full circle and returned to Aristotle's account of mathematical Pythagoreanism, in which any historical evidence that pertains to Hippasus is subordinated to the larger project, in Aristotle's mind, of establishing a teleological history of the development of the sciences.[118] In essence, the

117. Or Hipparchus, according to the letter of "Lysis."

118. Zhmud (2006: 125–128) suggests that a unified project of the historiography of the sciences was distributed by Aristotle to his students Theophrastus (physics), Eudemus (mathematics), and Meno (medicine), although we should be wary of a top-down assignment of responsibilities, especially since Peripatetics such as Theophrastus did not simply accept Aristotle's word as law.

early history of the schism in the Pythagorean brotherhood in the fifth century BCE can be reconstructed along two main lines: Timaeus of Tauromenium's story of a political division that is associated with revolutions throughout various city-states in Southern Italy and Sicily during the period of (roughly) 473–453 BCE and Aristotle's story of a methodological departure that is associated with the public demonstration of mathematical secrets in writing.[119] Timaeus lays claim to documentary evidence in his pursuit of a chronological history of the South Italian city-states, whereas he characterizes Aristotle as employing arguments from probability in the project of classifying the *Constitutions* of the city-states of Magna Graecia. Both men seem to be treating the same historical objects, but from diverging points of view that reflect their own projects of historiography. Yet the account of Timaeus is not likely to have been written without some access to Aristotle's texts themselves. The situation of Polybius's criticism of Timaeus is analogous, where Polybius's universal history is unthinkable without Timaeus's work informing its content and methodology, even though Polybius's work was employed chiefly in order to refute that of Timaeus. We might ask: did Aristotle speak about the political divisions among the Pythagoreans? In the fragments of his lost works on the Pythagoreans, we hear stimulating suggestions of it. He knew about Cylon, whom he considered a rival of Pythagoras.[120] Aristotle also claimed that Pythagoras foretold the future political strife among the Pythagoreans before withdrawing to Metapontum, in secrecy.[121] His associates Aristoxenus and Dicaearchus knew some version of the story involving Cylon, even if it confused an earlier uprising against Pythagoras himself and a later uprising against the Pythagoreans of Croton.[122] Dicaearchus, for his part, did not claim

119. Musti (2005: 166) also identifies democratic revolutions at Agrigentum and Himera (472 BCE), Syracuse (466/465 BCE), and Naxos, Catania, and Rhegium (461/460 BCE). To this list should be added what appears to be the first democratic revolution in Southern Italy, in Tarentum around 473 BCE. See Berger 1989: 308–309.

120. Arist. F 75 Rose = D.L. 2.46.

121. Arist. F 191 Rose = Apollon. *Mirab*. 6.

122. The relevance of the divisions is suggested in Dicaearchus's story of Pythagoras's expulsion (F 41A Fortenbaugh and Schütrumpf = Porph. *VP* 56–57), whereupon he is welcomed by Caulonia, but then sent away from Locri and Tarentum to Metapontum, where he dies unable to gain access to food. All historical accounts preserve strong relations between Caulonia and Croton, but the relationships with Locri and Tarentum are more complicated. Did Dicaearchus retroject mid-fifth-century BCE history involving the expulsion of the Pythagoreans back onto Pythagoras? If so, his version would preserve a coalition of "democratic" city-states: Croton, Locri, Tarentum, and Metapontum, with Caulonia as an intermediary.

to rely on Aristotle for his own information but asserted of the "great [Pythagorean] revolutions everywhere" that they were public knowledge: "even now people call them to mind and discuss them, calling them 'the revolutions in the time of the Pythagoreans.'"[123] It is difficult to know, given the scarcity of evidence, where Dicaearchus obtained his information. Aristoxenus's version, in spite of the fact that it synthesizes the two uprisings into one story, preserves the vestiges of the split in the brotherhood: Aristoxenus claims that the ambition of the "so-called Cylonians" (*οἱ Κυλώνειοι λεγόμενοι*) was so excessive that it "extended to the last Pythagoreans" (*διατεῖναι μέχρι τῶν τελευταίων Πυθαγορείων*).[124] There is no mention here of Hippasus, Ninon, or any of the other exoteric figures associated with the democratic uprising against the Pythagoreans in Croton, as in Apollonius's account.[125] But Aristoxenus does confirm that Archytas of Tarentum was the only Pythagorean to have remained in Italy after the expulsion. All the rest of the Pythagoreans fled, first to Rhegium and then to Thebes (Lysis), Phlius (Phanton, Echecrates, Polymnastus, and Diocles), and Chalcidice in Thrace (Xenophilus).[126] Aristoxenus does not describe a return and reconciliation, although a lacuna in the narrative makes it at least possible that he did so in the original text.[127] Xenophilus of Chalcidice was considered in antiquity to have been Aristoxenus's teacher (F 20a–b Wehrli), which would imply his probable influence over Aristoxenus's version of the story.[128] We cannot be sure that Aristoxenus adapted anything about Pythagorean *history* directly from Aristotle's accounts, even though, as I argued in chapter 2, his presentation of Pythagorean philosophy was influenced by Aristotle's formulations; and the focus on character traits in Aristoxenus's biographical sketches of figures such as Cylon implies an engagement with the scientific projects of Dicaearchus and Theophrastus.[129] Even so, we are forced to take care not to accept without justification Aristoxenus's character descriptions of Cylon and his followers, on the grounds that they demonstrate

123. Dicaearchus F 41A Fortenbaugh and Schütrumpf = Porph. *VP* 56–57.

124. F 18 Wehrli = Iambl. *VP* 249, 133.18–134.1.

125. Of course, Hippasus plays a very important role in Aristoxenus's history of musicology. See chapter 6, section entitled "The Mathematical Pythagoreans and the Heurematographical Tradition."

126. F 18 Wehrli = Iambl. *VP* 250–251, 134.22–135.6.

127. At Iambl. *VP* 251, 135.3.

128. See Huffman 2006: 107.

129. See Zhmud 2006: 139.

an obvious bias against these figures that is likely to have been carried over by Xenophilus.[130] Thus the early accounts of the fissure among the Pythagoreans, given by the Peripatetics Aristoxenus of Tarentum and Dicaearchus of Messana and by Timaeus of Tauromenium, all offer valuable (and sometimes contradictory) evidence for our reconstruction of the political history of the Pythagoreans; but the most "objective" of these accounts, from a modern historiographical perspective, would have to be the *lectio difficilior* of Timaeus, who took a critical stance against the (lost) Aristotelian history of the Pythagoreans by reconstructing the political history of Italy, with the aid of documentary evidence.

CONCLUSIONS

In this chapter, I have traced some of the important early evidence that presents Pythagoras, his followers, and a rival splinter group in the context of fifth- and fourth-century BCE political thought who seem to be referred to by most authorities as "pretenders." At every turn, it is important to try to figure out what their main traits might have been. The political historiography of Pythagoras and his followers goes back at least to the early fourth century BCE among the Socratics in Athens, with the parody of Pythagoras and the Pythagorean pretenders in Isocrates's *Busiris* and the more flattering portrayal of Pythagoras in the fragments of Antisthenes. Another Socratic, Aeschines of Sphettus, apparently illustrated the poor rhetorical abilities and grubby clothing of a Pythagorean who spoke with Socrates in a dialogue called *Telauges*. In the *Republic*, Plato, too, has something to say about the lifestyle of Pythagoras and his followers and focuses on the role that Pythagoras had in the education of private citizens, thus associating Pythagoras with the Sophists. The sum total of these early but oblique references to Pythagoras and his followers among Athenian intellectuals suggests that the fifth-century BCE Pythagoreans organized themselves in a political *ἑταιρία*. Isocrates and Plato preserve traces of a mendicant Pythagorean lifestyle, casting doubt on the values of the pretender Pythagoreans by saying that they practiced their way of life in order to achieve fame. A generation later, Aristotle systematizes a division in the Pythagorean brotherhood while underdetermining the ethical component emphasized by the students of Socrates—which continues to be influential well into the Hellenistic

130. Following a lacuna in the text, Aristoxenus refers to Phanton, Echecrates, Polymnastus, Diocles, and Xenophilus as the "most important" (*οἱ σπουδαιότατοι*) of the Pythagoreans who fled, implying a preference for this group. He goes on to explain that these men "kept safe the original customs and learning [*ἐφύλαξαν τὰ ἐξ ἀρχῆς ἤθη καὶ τὰ μαθήματα*], even though their sect [*αἵρεσις*] was dwindling, until they vanished in honor."

period—when he ascribes two different philosophical systems to the Pythagorean sects. As I argued in chapter 1, the philosophical division of "so-called" or mathematical Pythagoreans and acousmatic Pythagoreans, which has been established in Aristotle's lost writings on the Pythagoreans, posits a methodological fissure that occurred in the Pythagorean philosophical system and attributes that intellectual revolution to Hippasus of Metapontum, the fifth-century Pythagorean who was said to have published the secrets of the Pythagoreans by making a written demonstration of the sphere.

What survives of the accounts given by Aristotle's associates Dicaearchus of Messana and Aristoxenus of Tarentum does not necessarily suggest any explicit derivation from Aristotle's lost works on the Pythagoreans: they do not describe the split according to the methodological classification of Aristotle but only obliquely refer to the followers of Cylon as overambitious slanderers who retaliated against Pythagoras because they had been rejected by him. The telescoping of Pythagorean history results in the personalizing of what were otherwise political battles between two Pythagorean *ἑταιρίαι*, whereby Aristoxenus and Dicaearchus collapse two anti-Pythagorean uprisings (one around 510 BCE, the other in the mid-450s BCE) into one story and attribute it to personal rivalries between Pythagoras and Cylon. Aristoxenus might have taken his information from his teacher Xenophilus of Chalcidice, who was in Aristoxenus's estimation one of the "last of the Pythagoreans," or perhaps from one of the other Pythagorean exiles in Phlius. This might be thought to explain Aristoxenus's obvious bias against Cylon and his followers, whatever historical information Aristoxenus might have obtained on them. So far as their fragments survive, neither Dicaearchus nor Aristoxenus discuss Hippasus's role in Pythagorean political history. As I will show in chapter 6, when Aristoxenus does indeed speak about Hippasus, he refers to him as a pioneer in mathematical harmonics, whose method of scientific inquiry Aristoxenus nevertheless considers insufficient.

The historical evidence that pertains to the Pythagorean "pretenders" also leads to classification of two groups of Pythagoreans by Timaeus of Tauromenium in his lost *Italian and Sicilian History*: "exoterics" and "esoterics." Timaeus classifies those who are "outside" Pythagoras's curtain as figures who published the secrets of the Pythagoreans, a move that, I suggest, can be associated with an ideology of the democratization of knowledge in the fifth century BCE. Generally, Timaeus exhibits hostility to previous historians' activity of mythologizing charismatic individuals in the biographical tradition. His collection of Pythagorean "exoterics," those who published the Pythagorean secrets, diverges from Aristoxenus's list of the "last" or "most important" Pythagoreans: it includes Cylon and Ninon of Croton, Hippasus of Metapontum, Epicharmus of Syracuse, Theorides and Eurytus of Metapontum, Empedocles of Agrigentum, Diodorus of Aspendus, Cleinias and Philolaus of Heracleia, Archytas of Tarentum, and perhaps even Plato. Especially important for my purposes is the fact that Timaeus refers to the documentary evidence in the registers of the city of Croton from which he derived his historical information concerning the political stasis among the Pythagoreans. There is no reason not to believe Timaeus's account, which presents a story of the split among the Pythagoreans that is marked by neither mythological lore nor character-based innuendo.[131] The exoteric figures described by Timaeus are not said to promote their democratic causes or to publish the secrets of the Pythagoreans for the sake of self-aggrandizement or fame, as some of the other early accounts of Pythagoreans that originate in an Athenian context imply. Instead, there are two goals to these activities in Timaeus's presentation of the Pythagorean exoterics: the denunciation of oligarchical Pythagoreans before the people of Croton and the democratization of arcane Pythagorean knowledge. The latter goal corresponds with Aristotle's description of Hippasus of Metapontum, which suggests that there might be some overlap between the presentation of Aristotle's mathematical Pythagoreans and Timaeus's exoteric Pythagoreans, even if each historian posits different consequences for the publicizing of Pythagorean secrets. Aristotle was interested in the development of Pythagorean natural speculation into scientific demonstration. But Timaeus of Tauromenium shaped Pythagorean exoterism in order to implicate it in broader democratic movements in city-states throughout Western Greece (both Sicily and Southern Italy) from the period of roughly 473 BCE, when the constitution of Tarentum became "democratic," until the expulsion of the oligarchical Pythagoreans and change in the ancestral constitution in Croton during the mid-450s BCE.

131. See Rostagni 1913–14: 394–395.

4

Mathematical Pythagoreanism and Plato's *Cratylus*

The first half of this study has focused on the historiography of Pythagoreanism. I have sought to provide a new account of the Pythagoreans based on analysis of the earliest historical sources concerning the split in the Pythagorean brotherhood in the first half of the fifth century BCE associated with Cylon and Ninon of Croton on the one hand and Hippasus of Metapontum on the other. The primary goal of the first three chapters, then, has been to develop a more robust picture of the split among the Pythagoreans along both philosophical and political lines. Chapter 3 advanced a historical account of the Pythagoreans based on the fragments of the late fourth- to early third-century BCE Western Greek historian Timaeus of Tauromenium, who has generally been underutilized in the most recent scholarly accounts of the history of Pythagoreanism. In particular, I attempted to show there that Timaeus's history, which laid claim to having been derived from civic registers of the city-states of Magna Graecia,[1] documented the same schism in the Pythagorean brotherhood that had been described by Aristotle in his lost works on the Pythagoreans, but in different terms: as argued in chapters 1–2, whereas Aristotle and the other Peripatetics understood the split in the Pythagorean brotherhood between traditional acousmatic and progressive mathematical Pythagoreans to be rooted in diverse objects of knowledge—the fact or the "what" (*τὸ ὅτι*) and the "reason why" (*διότι*), respectively—Timaeus identified the division along political lines. For Timaeus, the traditional Pythagoreans called "acousmatic" by Aristotle were to be understood as "esoterics" or "those inside the curtain" (*ἐσωτερικοί*, *οἱ ἐντὸς σινδόνος*, etc.), whose political philosophy was to be identified with aristocracy, whereas those progressive Pythagoreans identified as "mathematical" or "so-called" Pythagoreans by Aristotle were seen as exoteric, described as "those from the outside" (*οἱ ἔξωθεν*), democratic revolutionaries who sought to encourage fairness and distribution within and even possibly beyond the

1. Iambl. *VP* 262–263, 141.2–19.

confines of the *polis*. At the end of chapter 3, I explored the possibility that both Aristotle and Timaeus were describing the same activity in their illustrations of the mathematical and exoteric Pythagoreans, namely, the democratization of esoteric knowledge through publication and demonstration of the Pythagorean *acusmata*.

In the second half of this study, I build on this revised historical picture of Pythagoreanism in order to use it as a means to reconfigure our understanding of Plato's response to mathematical[2] Pythagorean philosophy. Chapter 4 traces out some of the vibrant philosophical ideas developed by the early mathematical Pythagoreans Epicharmus of Syracuse, Empedocles of Agrigentum, Philolaus of Croton, and Eurytus of Metapontum by analyzing the genuine fragments and *testimonia* concerning their philosophy. This analysis will focus especially on the metaphysical propositions given by these intellectuals, especially their approaches to personal identity and number, priority and preexistence, and epistemology, in response to two influential philosophical systems: Parmenidean monism and Heraclitean fluxism. We will see that each of these figures—regardless of the diversity of their positions—reflects on the problem, central to the mathematical Pythagoreans, of understanding the relationship between the identity and number of an object. Taking my cue from Aristotle and Timaeus, who both saw Plato's philosophy as related to mathematical Pythagoreanism,[3] I seek here to evaluate how Plato engaged with the metaphysical concerns raised by the mathematical Pythagoreans, especially the issues that arise out of Epicharmus's "Growing Argument," in the early sketches of objective grounds for ontology and epistemology found in the *Euthyphro* and *Cratylus*. I will show that Plato actually arranges the mathematical Pythagoreans Philolaus and Epicharmus *against one another* in a bid to respond to the question of how to grasp "the essence of things" (ἡ τῶν πραγμάτων οὐσία) as it is subject to the conditions imposed by language. It thus becomes clear that Plato's characterization of and response to mathematical Pythagoreanism is complicated, in that he seems to associate Philolaus with Parmenidean monism, and Epicharmus and Empedocles with Heraclitean fluxism. In chapters 5 and 6, then, I will further elaborate on the ways Plato critically responds and appropriates mathematical Pythagoreanism, especially the philosophy of Philolaus and Archytas of Tarentum, in the pursuit of various metaphysical accounts given in the *Phaedo*,

2. From this point forward in this book, I will refer to those Pythagoreans called "so-called" or "mathematical" by Aristotle and "exoteric" by Timaeus as "mathematical" for the sake of convenience, unless I am making a specific point about the difference between Aristotle's and Timaeus's treatments of these figures.

3. FGrHist 566 F 14 (= D.L. 8.54) and Arist. *Metaph*. 1.5–1.6, 987a9–31 and 10.2, 1053b11–13; see Thphr. *Metaph*. 11a26–b12. On the Peripatetic *testimonia*, also see Horky: forthcoming.

Republic, and *Philebus*. Plato emerges from the analysis in the second half of this study as a philosopher who appropriates several of the methodological strategies of the mathematical Pythagoreans in order to respond to the central issues raised by their explanations of the primary *acusmata* involving wisdom: (1) "What is second-wisest? What assigned names to things" (*τί τὸ δεύτερον σοφώτατον; τὸ τοῖς πράγμασι τὰ ὀνόματα τιθέμενον*), and (2) "What is the wisest? Number" (*τί τὸ σοφώτατον; ἀριθμός*).[4]

GROWING AND BEING: MATHEMATICAL PYTHAGOREAN PHILOSOPHY BEFORE PLATO

At the end of the previous chapter, I highlighted as an overlooked piece of evidence of the history of mathematical Pythagoreanism, the list of "exoteric" Pythagoreans presented by Timaeus of Tauromenium, but I did not analyze it in detail. Considering the possibility that it is an account of the Pythagoreans that is not based on the information derived from the Phlian Pythagoreans who had fled Magna Graecia, and who seem to have been the source for the historical accounts of Pythagoreanism given by Aristotle, Aristoxenus, and their circles, it is worth quoting a substantial portion of it:

> Sometime later, however, Aresas from Lucania, who had been saved by some guest-friends, led the school; to him came Diodorus of Aspendus, whom he received because of the scarcity of men in the community. . . .[5] In Heracleia, Cleinias and Philolaus devoted themselves to writing about the men; in Metapontum Theorides and Eurytus, and in Tarentum, Archytas. And Epicharmus became one of the disciples outside the school [*τῶν δ' ἔξωθεν ἀκροατῶν*], but he was not from the inner circle of men. When he arrived in Syracuse, he abstained from philosophizing openly because of Hieron's despotism, but he put the thoughts of the Pythagoreans in meter, and under the guise of foolery, published the secret teachings of Pythagoras.
>
> (Iamblichus, *On the Pythagorean Way of Life* 266, 143.2–15; translation after Dillon and Hershbell 1991)

4. Iambl. *VP* 82, 47.17–19, but the former (1) is often given in the masculine (e.g. in the elaborate version given at Iambl. *VP* 56, 30.20–31.1, possibly derived from Timaeus of Tauromenium, and Ael. *NA* 17).

5. I place an ellipsis in the excerpt here for two reasons: first, Iamblichus ceases quoting Apollonius in indirect speech and shifts to direct speech (διέδωκε) in his description of the activities of Diodorus of Aspendus in mainland Greece, and second, the line of successors is replaced by a discussion of "exoteric" Pythagoreans who wrote about the men who preceded

This testimony, which is not without its problems,[6] nevertheless suggests evidence of an exoteric group that passed down knowledge of Pythagoreanism from one to another through textual records: the source, again most likely Timaeus of Tauromenium, suggests that individuals from Heracleia Italica (Cleinias and Philolaus), Metapontum (Theorides and Eurytus), and Tarentum (Archytas) produced written accounts of the Pythagoreans who came before them.[7] The evidence for "writings" concerning the Pythagorean men who came before them is not extant for figures such as Cleinias[8] and

them in the Pythagorean succession. It is possible that Aresas of Lucania, who did not die in the uprising against the acousmatic Pythagoreans (in the 450s BCE), was the last "leader" of the brotherhood that traced itself back to Pythagoras. The Aristoxenian catalogue of Pythagoreans (at Iambl. *VP* 267, 145.11) lists an "Aresandrus" from Lucania, and Stobaeus preserves an extract of *On the Nature of Man* attributed to Aresas of Lucania (1.49.27). Plutarch, too (*de Gen.* 583a–c), whose account is at least partially based on Timaeus's (see chapter 3, section entitled "Pythagorean Exoterics in the Fifth Century BCE? The Historical Evidence of Timaeus of Tauromenium"), refers to Aresas as a Pythagorean who remained in Magna Graecia, specifically in Sicily, and who was in contact with Gorgias of Leontini following the Cylonian conspiracy.

6. In particular, the ellipsis I have inserted appears to mark an interpolation by Iamblichus or by Apollonius, who is likely to be Iamblichus's direct source. There are minor textual problems as well (on which see the supplements adopted by Deubner-Klein). The most problematic phrase in F's manuscript reading is what I (following Dillon and Hershbell 1991) have translated "devoted themselves to writing" (ζηλωτὰς δὲ γράφειν γενέσθαι) about the men who came before. But Burkert (1965: 25, followed by Brisson and Segonds 2011: 142) has suggested that we should emend the text to γράφει, which would make Diodorus of Aspendus the subject and obtain a meaning in which Diodorus wrote about men who were ζηλωταί ("emulators"; see Iambl. *VP* 80, 46.13–17, which occurs right before a summary derived from Timaeus's works; and Polyaen. *Strat.* 5.22.1, probably derived from Timaeus, which narrates how those Parians who were ζηλωταὶ Πυθαγορείων λόγων were dispersed throughout Italy after the arrival of Dionysius II in Italy). For a defense of F's manuscript reading of γράφειν in complementary infinitive, and a return to indirect statement, see Deubner 1935: 676, also accepted by Macris (2004: 128). Either way we read the text, it is clear, I think, that Cleinias, Philolaus, Theorides, Eurytus, Archytas, and Epicharmus are being distinguished from the figures listed before as deviant followers of Pythagoras.

7. None of the recent editors of Iamblichus's work *On the Pythagorean Way of Life* determines the source of *VP* 266, but Burkert (1972: 203, with n. 63), following Delatte and Rostagni, suggests that it could be Timaeus.

8. The Peripatetics were interested in Cleinias and wrote about his approaches to ethics. For example, the Peripatetic historian Chamaeleon of Pontus (at Athen. 14, 624a = DK 54 F 4) speaks of Cleinias as "preeminent in life-practice and ethics," and Aristoxenus (F 131 Wehrli) claims that Cleinias was one of the Pythagoreans who prevented Plato from burning Democritus's works. Moreover, assuming (as DK do) that he is the source for Diodorus Siculus's information concerning Cleinias in book 10 of his *Histories* (10.4.1–2), we see Aristoxenus celebrate Cleinias (this time of Tarentum) for helping out a Pythagorean "friend," Prorus of Cyrene, after Prorus lost all of his fortune due to political upheaval.

Philolaus,[9] nor for Theorides[10] and Eurytus of Metapontum. Still, Eurytus (b. ca. 450 BCE?)[11] presents a special case, for two reasons. First, as Ian Mueller has shown, the testimonies concerning Eurytus's prephilosophical activities in the development of Pythagorean scientific methodology are rich and suggestive.[12] Second, it is clear that Archytas of Tarentum *did* write about Eurytus and his other philosophical predecessors, in one way or another.[13] In particular, we have it on the reliable authority of Theophrastus that Archytas wrote about Eurytus's speculations concerning pebble-arithmetic:

> But one might expect that they would straightaway give an account of things that follow in succession from this principle or these principles and not advance to a certain point and stop. For this is the characteristic of an accomplished and sensible person, to do exactly what Archytas said that Eurytus did by arranging certain pebbles. For he [Archytas] reports that Eurytus would say that this is the number of man, this of horse, and this of something else. As it is, most people advance to a certain point and stop, just like those who make the one and the indefinite dyad principles. For, having generated numbers and surfaces and solids, they pretty nearly leave out the rest.
>
> (THEOPHRASTUS, *Metaphysics* 6a15–27 = Archytas A 13 Text H Huffman)

The report as it stands does not make clear whether Theophrastus is reading Archytas's work himself or looking at one of Aristotle's works on Archytas; of course, if Theophrastus is unclear about documenting his source here, it would not be surprising, since his chief goal in the *Metaphysics* is to raise worries concerning various methodological assumptions of his competitors, including Aristotle and the early Platonists, whose approach to ontological reduction is

9. We only have reliable evidence for Philolaus having composed *On Nature* and perhaps a text called *Bacchae*, and none of the surviving fragments of Philolaus demonstrate a historiographic project on his part. For Philolaus as mathematical Pythagorean, see below.

10. This Theorides is otherwise unknown, but Thesleff (1965: 201) proposes that we should read Thearidas, brother of Dionysius I, who is said to have written a text *On Nature* by Clement (*Strom*. 5.133.1), for which only one probably spurious fragment survives: "The origin of things—since an origin is true—is one; for that [i.e. truth] is both unified and singular in an origin."

11. See Huffman 1993: 4–7. Eurytus was from Metapontum, as Timaeus holds (Iambl. *VP* 266, 143.9), or Croton (Iambl. *VP* 148, 83.24–25, unknown source), or Tarentum (*VP* 267, 144.11, probably from Aristoxenus).

12. See Mueller 1997: 294–298.

13. I discuss the other philosophical predecessors in Archytas's writings in chapter 6.

contrasted against Eurytus's method in this particular *problema*.[14] Aristotle, too, approaches Eurytus's pebble-arithmetic from a methodological point of view, presenting it as an *endoxon* that exemplifies how some of the mathematical Pythagoreans attempted to assimilate natural objects to numbers and to arrange these numbers, which now represent particular natural objects, in ratios.[15] Perhaps Eurytus was approaching meaningful definition of objects through something like the "figurate numbers" that were arrived at through the use of a gnomon, an early Pythagorean arithmological procedure that, as Mueller argues, offers "mathematically interesting results independently of anything resembling stylized Euclidean deduction."[16] It's not necessary to delve here into a technical understanding of how Eurytus "defined" a man, a horse, and so on; it is clear enough that he was worried about how to define natural objects through natural numbers.[17] What Aristotle was interested in—and in fact what is of interest to this study—is the notion that mathematical Pythagoreans such as Eurytus sought to provide definitions of natural objects by means of some sort of assignment of their particular numerical values, which inherently assumes *assimilation of natural objects and numbers*. As I discussed in chapter 1,

14. On Xenocrates's claim that Plato and the Pythagoreans reduced everything to the one and the indefinite dyad, see Horky: forthcoming.

15. Arist. *Metaph.* 14.5, 1092b8–26. The principle seems to be that numbers determine natural objects by acting as points that limit the magnitude of the objects, but it remains impossible to deduce whether Aristotle is manipulating Eurytus's approach or simply reporting it.

16. Mueller 1997: 294–295. Speculation along these lines with pebble-arithmetic is also associated with the mysterious fourth-century BCE Parian Pythagorean Thymaridas (Iambl. *in Nic.* 88, 62.18–63.2), on which see Burkert 1972: 442 n. 92.

17. See Zhmud 1989: 276–277 and Barnes 1982: 391. Raven, too (1948: 103–105), attempted to develop an elaborate reconstruction of Eurytus's theory of pebble-definition based on Ps.-Alexander's account (*in Metaph.* p. 827.13–28 Hayduck = DK 45 F 3). Part of the problem is that we don't know with confidence where Ps.-Alexander got his information on Eurytus, although it is suggestive that he contrasts Eurytus's theory against someone who argued that reason is the "concordance of numbers" (*ἡ συμφωνία τῶν ἀριθμῶν*), which is later elaborated as a reference to the concordant intervals of the fourth, fifth, and the eighth, according to Ps.-Alexander (*in Metaph.* p. 833.18–834.4 Hayduck). Knowledge of these concordant intervals was attributed to Pythagoras by Xenocrates (F 87 IP), but it is more likely that Hippasus of Metapontum was the earliest Pythagorean who knew about them, along with Lasus of Hermione (DK 18 F 12–13, on the reliable authority of Aristoxenus). It is therefore possible that Ps.-Alexander was deriving his information from a text that compared both Eurytus's and Hippasus's, or Pythagoras's, physical theories (Archytas? See Ross 1924, vol. 2: 495). Be that as it may, we can be more confident that Aristotle, and Ps.-Alexander after him, are contrasting Eurytus's theory of number with another Pythagorean theory, probably that of a mathematical Pythagorean. On Hippasus as "first discoverer" of these concordant intervals, see chapter 6, section entitled "The Mathematical Pythagoreans and the Heurematographical Tradition."

a chief characteristic of the "so-called" or mathematical Pythagorean approach to understanding nature was the activity of associating natural objects and even concepts to numbers by means of what Aristotle describes either as imitation (μίμησις) or assimilation (ὁμοίωσις).[18] Elsewhere (*Metaph.* 13.8, 1083b17–19), Aristotle describes the Pythagoreans as "say[ing] that existing things are number" and "apply[ing] mathematical theories [τὰ θεωρήματα] to bodies as if they (i.e. the bodies) consisted of those numbers."[19] Irrespective of whatever way Eurytus undertook to define natural objects numerically, and irrespective of how insufficient *we* might think it was (on little evidence), his *approach* must be considered to represent what "so-called" or mathematical Pythagoreans did, at least in the eyes of Aristotle.[20] Thus, in Eurytus of Metapontum, we can finally identify a specific historical figure who was considered both an "exoteric" Pythagorean by Timaeus of Tauromenium and a "so-called" or "mathematical" Pythagorean by Aristotle.[21]

Importantly, however, Eurytus is not the only mathematical Pythagorean whose practice of (in Aristotle's terms) assimilation of natural objects to numbers characterized his peculiar philosophical activity.[22] The other figure listed in Timaeus's grouping of those who "published" concerning Pythagorean matters, and whom I have not sufficiently discussed hitherto, was Epicharmus of Syracuse (fl. ca. 500 BCE),[23] a controversial figure in Greek intellectual history who—it is important to note—is never *explicitly* associated with Pythagoreanism by

18. Arist. *Metaph.* 1.5, 985b23–986a21, where we hear about the "so-called" Pythagoreans "seeing" (θεωρεῖν) numbers as ὁμοιώματα of natural objects; believing that the nature of objects "is assimilated" (ἀφωμοιῶσθαι) to numbers; and "collecting and harmonizing" (συνάγοντες ἐφήρμοττον) the ὁμολογούμενα of the characteristics and parts of the heavens. Also see Aristotle F 203. See Burkert 1972: 44, with n. 87.

19. In general, I follow Huffman's (1993: 57–61) reading of Aristotle's manipulation of Philolaus's fragments, although I would emphasize the likelihood that Aristotle had access to Pythagorean writings other than those of Philolaus, and that it is possible that he employed (at least) Archytas's writings on the Pythagoreans (such as Eurytus) in his own characterization of Pythagoreanism.

20. Pace Zhmud 1989: 278. See Barnes 1982: 390–391.

21. For a more thorough analysis of the evidence concerning the "so-called" or "mathematical" Pythagoreans, see chapter 1.

22. Primavesi (2012: 252–258) helpfully evaluates the evidence for Philolaus of Croton and "so-called" Pythagoreanism, as designated by Aristotle.

23. Iambl. *VP* 266, 143.10–15; see Iambl. *VP* 166, 93.24–94.12, which associates Empedocles, Parmenides, and Epicharmus with the rise of Magna Graecia. This passage might be ultimately derived from Timaeus of Tauromenium (see Burkert 1972: 216 n. 32) or from Nicomachus of Gerasa, but we cannot be sure.

Aristotle, Aristoxenus, or Dicaearchus.[24] Only the Western Greek tradition associates Epicharmus with Pythagoreanism.[25] Several fragments attributed to Epicharmus are preserved by a certain Alcimus of Sicily (mid-fourth century BCE), who, in the context of his analysis of these fragments, actually charges Plato with stealing the central tenets of his philosophy from Epicharmus.[26] In one of these fragments of Epicharmus, which seems to date to the second quarter of the fifth century BCE,[27] we have evidence of something resembling the pebble-arithmetic ascribed to Eurytus by Archytas of Tarentum:

24. Aristotle does speak of Epicharmus no less than eight times, and he ascribes to him philosophical activity: by describing how things come from other things in a pseudo-syllogistic "climax" (*Rh.* 1.7, 1365a16–19), Epicharmus seemed to speak in particular about the efficient cause (*GA* 724a28–35); Epicharmus criticized Xenophanes in some fashion (*Metaph.* 4.5, 1010a5–7); and Epicharmus employed antitheses, however spurious (*Rh.* 3.10, 1410b4–6). Aristoxenus (F 45 Wehrli = Athen. 14, 648d) refers to *Pseudepicharmeia,* composed by Chrysogonus the flute-player (on which see Cassio 1985: 47–50). Dicaearchus might have purposefully left Epicharmus off his list of the Seven Sages (see White 2001: 204–205), and Eudemus's surviving fragments concerning the history of arithmetic and geometry do not make reference to Epicharmus.

25. The earliest possible association of Epicharmus with Pythagoreanism after Timaeus of Tauromenium (if Iambl. *VP* 266 comes from him) is Ennius, whose longest surviving fragment of the *Epicharmus* (Varr. *LL* 5.64 = Epicharmus F *285 K.-A.) betrays Presocratic natural speculation of a sort that recalls the cosmology of the Derveni Papyrus (e.g. Cols. XVII, XIX, and XXV), Archelaus of Athens (DK 60 A 12 = Aët. 1.7.14; see Betegh 2004: 321–324), and Diogenes of Apollonia (DK 64 B 4–5), in which the lead god (Jupiter/Zeus/Mind/"the god") is associated strongly with air and its attributes. Plutarch (*Numa* 8) calls Epicharmus a disciple of Pythagoras, but we cannot know what his source was (probably fourth or third century BCE: Timaeus? Aristoxenus?; see Humm 2005: 553, with n. 45). It may be significant that Diogenes Laertius places the life of Epicharmus in the eighth book of his *Lives of the Eminent Philosophers* along with the other Pythagoreans, between Empedocles and Archytas.

26. FGrHist 560 F 6. On Alcimus's charges against Plato as a reflection of the political environment of Sicily and South Italy in the mid-fourth century BCE, see Cassio's lucid study (1985: 43–47). The fragments preserved by Alcimus are taken to be possible evidence for early Pythagorean arithmetic by Rostagni (1924: 6–25), Burkert (1972: 289 n. 58), and Zhmud (2006: 223, with n. 41). Mueller's fine study (1997) does not mention Epicharmus.

27. If Rostagni's reconstruction is correct (1924: 18–21). This date would mark Epicharmus's philosophical comedies (1) late in life and (2) roughly contemporaneous with the first generation of mathematical Pythagoreans, such as Hippasus and Lysis (first or second quarter of the fifth century BCE). Generally, on the authenticity problem, see Willi (2008: 119–124), who wavers on whether this fragment is authentic or inauthentic but accepts the ideas discussed in its content as thematically derived from Epicharmus's "Growing Argument" (by reference to Epicharmus F 187 K.-A., describing Pythagoras (?): "thrice again life was life given back [to him]" (τρὶς ἀπεδόθη ζόος)), whether the text of F 276 K.-A. is original to Epicharmus or not. But, in addition to Menn's discussion (see below), see Battezzato 2008, Álvarez Salas 2007, and Cassio 2002, who consider the fragment authentic.

A. But suppose someone chooses to add a pebble to an odd or even number [*πὸτ' ἀριθμόν τις περισσόν . . . πὸτ' ἄρτιον, ποτθέμειν*], whichever you choose, or to take away from those [pebbles?] that preexist [*τᾶν ὑπαρχουσᾶν λαβεῖν*][28]—does it seem to you that it [i.e. the number] would be yet the same [*δοκεῖ . . . ὡὐτὸς εἶμεν*]?

B. Not to me, it doesn't.

A. Nor even if someone were to choose to add another length [*ποτθέμειν . . . τις ἕτερον μᾶκος*) to a cubit-measure, or to cut off from what was there beforehand [*τοῦ πρόσθ' ἐόντος ἀποταμεῖν*]—would that measure still be there [*ἔτι χ' ὑπάρχοι κῆνο τὸ μέτρον*]?

B. Surely not.

A. Now look at humankind this way: for one man grows, while another dwindles [*ὁ μὲν γὰρ αὔξεθ', ὁ δέ γα μὰν φθίνει*], and all men are in a state of exchange throughout all time [*ἐν μεταλλαγᾶι δὲ πάντες ἐντὶ πάντα τὸν χρόνον*]. But whatever naturally is in a process of exchange and never remains the same [*ὃ δὲ μεταλλάσσει κατὰ φύσιν κοὔποκ' ἐν τωὐτῶι μένει*] should be always different [*ἕτερον εἴη*] from what it had been changed from [*τοῦ παρεξεστακότος*]. Even so you and I were different men yesterday, and even now we've turned out to be other [*ἄλλοι τελέθομες*] today, and again we will turn out to be other [tomorrow] and never the same, according to this argument.

(Epicharmus, DK 23 B 2 = Diogenes Laertius 3.9 = Ps.-Epicharmus F 276 K.-A.)

Some scholars have not been inclined to take this fragment very seriously, either as an early testimony to types of mathematical inquiry or as evidence for early Pythagorean number theory.[29] In his earlier investigation into Pythagorean number philosophy, Leonid Zhmud did not discuss Epicharmus, although more recently he has seen in Epicharmus's fragment a parody of a real Pythagorean science.[30] Reviel Netz claims that the evidence of arithmetical speculation

28. It isn't totally clear whether the antecedent would be "pebbles" or "numbers" here.

29. Exceptions among contemporary scholars would be Andreas Willi (2008: 170–175), who, while he avers to use of B 2 in his reconstruction of Epicharmus's "Growing Argument," nevertheless associates the "Growing Argument" as described by Plutarch and the Anonymous Commentator on Plato's *Theaetetus* with Pythagorean rhetorical tricks (by reference to Timaeus of Tauromenium's comment on Heraclitus's description of Pythagoras at FGrHist 566 F 132); and Luigi Battezzato (2008), who reinterprets papyrus for the Anonymous Commentator (71.12–13 Bastianini and Sedley) in order to render this reading: "Epicharmus since he was a pupil of the Pythagoreans, explained well a number of philosophical opinions, and brought to completion the argument about the growing man in a systematic and reliable way."

30. Zhmud 1989 and Zhmud 2012: 153.

presented by Epicharmus is "part of the pre-scientific background for mathematics" and "must be ruled out of court" for a study of how mathematics developed in the Greek world.[31] It may be true that this fragment does not play an obvious role in the development of *axiomatic-deductive* mathematics of the sort found, say, in books 7 and 8 of Euclid's *Elements*, but Euclidean mathematics was not the only sort found in the ancient world. In fact, scholars have had reason to think that Epicharmus had a significant effect on the broader development of ancient philosophy, especially with regard to the problem of numerical identity.[32] This problem was also associated with Epicharmus in antiquity by Chrysippus and the Anonymous Commentator on Plato's *Theaetetus*.[33] Epicharmus's fragment itself represents an early attempt at analogical reasoning, of a sort that assumes numerical relationships between "natural" objects such as pebble-piles, cubit-measures, and human beings.[34] David Sedley and Stephen Menn have variously explored how Epicharmus's fragment anticipates philosophical problems in the writings of Plato and the Stoics. Sedley sees this fragment as the earliest evidence of a Sophistic logic puzzle that especially concerned the Stoics, the so-called Growing Argument, which concerns the problem of "whether material objects and numbers . . . behave alike," or, as Sedley ponders: "Can a material object be individuated by a numerical specification of its ingredients, so that any alteration in these constitutes a change in identity?"[35] According to Menn, who defends the authenticity of the fragment,[36] Plato and the Stoics, in their critical response to the "Growing Argument" as expressed in Epicharmus's Fragment B 2, responded to Epicharmus's concern with how numbers and matter could be confused and how this relates to their peculiar

31. Netz 1999: 272.

32. See Sorabji 2009: 36–39, Barnes 1982: 106–107, and Knorr 1981: 148, who suggests that "the efforts by early mathematicians (say, the fifth-century Pythagoreans) to put in order the body of arithmetic and geometric techniques they were assimilating from the older traditions gave rise to the awareness among philosophers that the same ideas and patterns of reasoning might be applied in their cosmological speculations."

33. Plut. *Comm. Not.* 1083a1–2 and Anon. *in Theaet.* 70.5–26 with 71.12 Bastianini and Sedley.

34. Thus, the assumption here is that "number" is predicated of all natural objects. Epicharmus does not obviously distinguish between discrete and continuous number here, as Aristotle does in his description of quantity as one of the ten categories (*Cat.* 6, 4b20–5a36).

35. Sedley 1982: 255–256. Also see the analysis of Sorabji (2009: 36–43).

36. See Menn 2010: 64–68 (with bibliography). In addition to Battezzato 2008, Álvarez Salas 2007, and Cassio 2002, it is also regarded as genuine by Rodríguez-Noriega Guillén, who is not included in Menn's study, in her edition (1996).

properties of existence.[37] As rich as their studies are, neither Sedley nor Menn has examined the possibility that Epicharmus was a mathematical Pythagorean whose fragments might preserve early Pythagorean speculation concerning identity that operated according to, in the assessment of Aristotle, the related principles of imitation of (μίμησις) and assimilation to (ὁμοίωσις) number.[38]

Epicharmus's version of the "Growing Argument" exhibits, I suggest, the most plausible example of how an early mathematical Pythagorean might have gone about answering one of the chief questions of the type of *acusma* that involves definitions (the τί ἔστι sort): what is a human being?[39] By inducing an analogical relationship between number and natural objects, on the assumption that "number" is an *essential property* of various natural objects, Epicharmus investigates the problem of human identity in the broader context of Presocratic natural philosophy. The assumption seems to be that all those objects that are in a state of "natural exchange" (μεταλλάσσει κατὰ φύσιν),[40] and are measurable (i.e. are subject to "number" or "measure" in some essential way), are subject to comparison. There is a further assumption that comparison between objects that share the attributes of "natural exchange" and measurability according to some standard (i.e. anything that has the ability to grow or diminish, such as a pile of pebbles, a cubit-measure, and a human being) is made possible through shared attributes. For all objects that are in a state of exchange, "number" and things analogous to it (such as "length" in the case of

37. Menn 2010: 43–50. Menn also notes (2010: 45 n. 9) that this problem was taken up by the author of one of the Sophistic *Dissoi Logoi* (DK 90 5.13–15). See below.

38. As asserted explicitly by the Anonymous Commentator on Plato's *Theaetetus*, who must represent a later tradition, since he (70.5–26; also see 71.12 Bastianini and Sedley) refers to the "Growing Argument" as originating with Pythagoras and then being taken on by Plato in the *Symposium* and the Academics under Philo (see Long and Sedley, vol. 2: 170). See Arist. *Metaph.* 1.5, 985b23–986a21 and 1.6, 987b10–15, on which see chapter 1.

39. Another relevant fragment in the Epicharmean corpus that exhibits acousmatic qualities is DK 23 B 10 (Clem. *Strom.* 4.7.45 = F 166 K.-A.), which, in its original formation (retained by Rodríguez-Noriega Guillén) is "nature itself of humans: puffed-up wineskins" (αὕτα φύσις ἀνθρώπων, ἀσκοὶ πεφυσαμένοι). Note that this text closely resembles Philolaus's description of "nature herself" in F 6 Huffman (αὐτὰ μὰν ἁ φύσις). See below.

40. The significance of the principle of exchange (emphasized in the previous line by the phrase ἐν μεταλλαγᾶι) is not wholly unusual among philosophical writers of the first half of the fifth century BCE. Aeschylus describes a *daemon* as capable of being "exchanged" (*Th.* 706), and Empedocles (DK 31 B 115) elaborates further on this *topos* by describing how the *daemon* (soul in the period of strife?) is "naturally born throughout time into all sorts of shapes of mortals, exchanging one difficult path of life for another" (φυόμενον παντοῖα διὰ χρόνου εἴδεα θνητῶν / ἀργαλέας βιότοιο μεταλλάσσοντα κελεύθους). Inwood (2002: 60–63) has plausibly inferred from Empedocles's fragment a reference to metempsychosis. For a more extensive treatment of Empedoclean "exchange," see chapter 5.

the cubit-measure and, we must assume, age in the case of the human being) are necessary conditions for identity. Thus, we have an early speculative inquiry into what it is that makes objects in the natural process of change similar to one another, and whether or not that property helps to guarantee the particular identity of the object. Essentially, what Epicharmus presents us with is an endorsement of a fluxist ontological position, similar to that of Heraclitus, against Parmenidean monism.[41] If Epicharmus indeed is taken to present us with a mathematical Pythagorean position in Fragment B 2, it exemplifies what Aristotle said concerning the Pythagorean matter in a fragment of Aristotle's work *On the Philosophy of Archytas*: the Pythagoreans "called matter 'other,' on the grounds that it was flowing and always becoming other" (*ἄλλο τὴν ὕλην καλεῖν ὡς ῥευστὴν καὶ ἀεὶ ἄλλο γιγνόμενον*).[42]

It is important to note that Fragment B 2 does not present the only example of Epicharmus's inquiry into the related problems of identity and predication of number. We also have evidence elsewhere in the fragments of Epicharmus that exhibits such confusions, apart from a primitive "argumentative" framework that plays on the complex relationship between an object's name, essential identity, and the number of parts that make it up:

> (A.) What's this here? (B.) Tripod, obviously. (A.) Um, why does it have four feet? *It's no tripod*; seems like a tetrapod to me . . .
> (B.) It's *called* "tripod," but it's really got four feet.
> (A.) Well, if it were a "dipod," you'd think it the Riddle of Oedipod!
>
> (A.) *τί δὲ τόδ' ἐστί;* (B.) *δηλαδὴ τρίπους.* (A.) *τί μὰν ἔχει πόδας τέτορας; οὔκ ἔστιν τρίπους, ἀλλ' <ἐστὶν> οἶμαι τετράπους.*
> (B.) *ἔστι δ'ὄνομ' αὐτῷ τρίπους, τέτοράς γα μὰν ἔχει πόδας.*
> (A.) *εἰ δίπους τοίνυν ποκ' ἦς, αἰνιγματ' Οἰ<δίπου> νοεῖς.*[43]
>
> (EPICHARMUS F 147 K.-A. = Athenaeus, *Epitome* 2, 49c)

41. Still, Epicharmus has adopted slightly different language: he gives *μεταλλαγή* instead of Heraclitus's *μεταβάλλον* (DK 22 B 84a), but he retains the language of Parmenides (DK 28 B 8.29: *ταὐτόν τ' ἐν ταὐτῶι τε μένον*). Plato also saw Epicharmus as a fluxist, who joined the ranks of Protagoras, Heraclitus, Empedocles, and Homer against Parmenides (*Theaet.* 152e1–9). On Plato's "wise" antecedents and the problem of flux, see below and Horky 2011. It should be remarked that Epicharmus's fragment does not draw reference to the problem of "irrational" number in discussion of numerical identity, which implies that it may have been composed before "irrationality" of number became a philosophical issue (in the 430s? See Knorr 1975: 22–49).

42. Arist. F 207 Rose = Archytas T A13 Text F Huffman. I adopt Huffman's reading of this fragment as preserving what Aristotle claimed about the Pythagoreans, not Pythagoras himself.

43. Accepting Kassel-Austin's emendation, following Wilamowitz.

This is pretty cheeky humor but also perhaps the sort of humor that might help inform the statement, possibly derived from Timaeus of Tauromenium, that Epicharmus published the secret doctrines of Pythagoras "under the guise of foolery."[44] Indeed, the joke here appears to turn on the "riddle of Oedipus," an oblique reference to the riddle of the Sphinx, for which we have contemporary evidence and which may derive from an original version written in an epic poem. Recall that the earliest versions of that riddle, too, are concerned with the persistence of identity qua number of parts of the body.[45] What is distinctive in the case of Epicharmus, however, is the emphasis on how the numerical prefixes ("tetra-," "tri-," and "di-") of compound names affects the meanings of the words, an early example of the problematization of the naturalness of names. This fragment suggests that Epicharmus's jokes about numbers and naming represent some of the earliest evidence of philosophical speculation concerning what would later become thematized by Plato and brought forth by Aristotle as problems of identity, parts of wholes, alteration, and predication.

With the exception of Eurytus, whom I have discussed briefly earlier, there is no firm evidence of mathematical Pythagoreans immediately after Epicharmus who may have been developing Pythagorean critical responses to the philosophical speculations of the day, especially the debate between pluralist and monist physics.[46] And the evidence for Eurytus is sketchy at best. But there is a great deal more to say about Philolaus of Croton (ca. 470–ca. 385 BCE), allegedly Eurytus's teacher, and Empedocles of Agrigentum (ca. 492–ca. 432 BCE), important figures who have remained suspiciously absent from this book's account of mathematical Pythagoreanism but who were considered in the late fourth century BCE to have been among the first to publish the content of

44. Iambl. *VP* 266, 143.14–15: μετὰ παιδιᾶς κρύφα ἐκφέροντα.

45. Compare, for example, the versions of Aeschylus (at Ath. 10, 456b; see Lloyd-Jones 1990: 332–334), Euripides (F 540a Collard and Cropp = P.Oxy. 2459 F 2), and—perhaps closest to Epicharmus's—Aristophanes (F 545 K.-A.).

46. In this light, however, Zhmud (1989: 277) adduces the interesting case of Ecphantus of Syracuse, not listed among the exoteric Pythagoreans according to Iamblichus at *VP* 266 but said by Aëtius (1.3.19 = DK 51 F 2) to have been the first to have practiced number atomism. Hippolytus (*Ref.* 1.15 = DK 51 F 1) adds that Ecphantus was an epistemological skeptic who posited three indivisible primary bodies that have three "variations" (παραλλαγαί) of these primaries that "already exist" (ὑπάρχειν) and from which are generated sensibles: magnitude, shape, and power. It is difficult to know the source for this information, but it is not likely to have been Theophrastus (see Mansfeld 1992: 37–43). In the absence of Ecphantus's actual fragments, unfortunately, we cannot corroborate this evidence.

Pythagoras's discussions.[47] Like Epicharmus, Philolaus and Empedocles were concerned with the sorts of philosophical worry that Heraclitus and Parmenides had brought to the fore in the first decades of the fifth century BCE: (1) how do we determine what is there in the natural world? (2) how can we talk about things in nature that seem to be related to one another? (3) is something that exists subject to becoming? and (4) what is the relationship between the parts and the whole of an object? Parmenides in particular had set the parameters for debate concerning these questions probably in the 480s BCE, and it was left to pluralists such as Leucippus and Anaxagoras on the one hand and mathematical Pythagoreans such as Epicharmus, Empedocles, and Philolaus on the other to respond to Parmenides in the wake of his arguments concerning the uniqueness of What-Is (*τὸ ἐόν*).[48] Of special significance for both ancient and modern debates about Parmenides's philosophy is one of the many perplexing passages of the longest extant fragment of his poem, Fragment 8:

The same thing is thinking and that because of which there is a thought;
For not without What-Is, depending on which[49] it (thinking) has been expressed,
Will you discover thinking.

τωὐτὸν δ' ἐστὶ νοεῖν τε καὶ οὕνεκέν ἐστι νόημα·
οὐ γὰρ ἄνευ τοῦ ἐόντος, ἐν ᾧ πεφατισμένον ἐστίν,
ἑυρήσεις τὸ νοεῖν·

(PARMENIDES, DK 28 B 8.34–36)

These lines, and what follows them, are notoriously difficult to make sense of, and even more difficult to translate;[50] but Parmenides seems to be saying that the object of thinking, which is also a necessary condition of thinking, is what oral expression of that object depends on, namely, What-Is.[51] What-Is is the

47. By Neanthes of Cyzicus (FGrHist 84 F 26 = D.L. 8.55), on whom see above in chapter 3, section entitled "Pythagorean Exoterics in the Fifth Century BCE? The Historical Evidence of Timaeus of Tauromenium."

48. See Palmer (2009), who treats the responses of Leucippus, Anaxagoras, and Empedocles but does not substantially examine Philolaus and other Pythagoreans (see Huffman 2011: 303).

49. For this interpretation of *ἐν ᾧ πεφατισμένον ἐστίν*, see Palmer 2009: 164 n. 40.

50. See, for example, McKirahan 2008: 202–204, Long 2005: 237, KRS 1983: 252, and Barnes 1982: 206–207.

51. See Palmer 2009: 164–165. "What-Is-Not" cannot, however, be known or spoken of (DK 28 B 2.7–8).

guarantor of thinking, and What-Is also makes it possible to speak about itself along "the Way of Conviction."[52] That may be the case for What-Is qua What-Is. When humans, however, confounded by opinion, "fixed their minds on *naming* the two shapes (*μορφὰς . . . κατέθεντο δύο γνώμας ὀνομάζειν*)" of What-Is, they "distinguished [*ἐκρίνοντο*] opposites in body and bestowed signs [*σήματ' ἔθεντο*] apart from one another."[53] The opposites they seem to distinguish[54] by means of signs[55] or attributes[56] are (1) "fire of flame," which is described as "aetherial," "mild," "immensely light," and "the same to itself in every direction but not the same as the other," and (2) "night," which is characterized as "dark," "opposite in itself," and "a dense and heavy body."[57]

One of the many problems we encounter when we attempt to make sense of Parmenides's poem is that we need to develop an account that can explain the relationship between knowledge of What-Is—which has attached to it various predicates, including "not divisible" (*οὐδὲ διαιρετόν*), "all alike" (*πᾶν ὁμοῖον*), and "continuous" (*ξυνεχές*)—and mortal opinion concerning the same object, which is marked by, among other things, the language of coming-into-being and destruction.[58] It is not my goal to provide a sufficient account of how Parmenides would claim that we could come to understand, and subsequently speak about, What-Is.[59] For my purposes, it is enough to say that these philosophical problems do not appear easy to resolve in Parmenides's fragments, and early critics of Parmenides, such as the Pythagoreans, appear to have developed

52. As it is now referred to, on Palmer's reading (see DK 28 B 8.50).

53. DK 28 B 8.53–56. Note that when the goddess suggests that by giving a name to one of the "shapes," probably "night," they err in doing so (*ἐν ᾧ πεπλανημένοι εἰσίν*), she poetically contrasts that activity with speaking about understanding (*ἐν ᾧ πεφατισμένον ἐστίν*) (8.35).

54. Long (1963: 101–105) suggests that the opposites, shapes described are "what is" and "what is not," and while that remains a possibility, at least in this passage (structurally following after the goddess's differentiation of the "way of conviction" and the "mortal opinions") the most obvious *nominata* are fire and night.

55. Parmenides seems to understand *σήματα* as attributes of objects that we use to identify them. Compare DK 28 B 8.1–6. See Cordero 2004: 167–168, who refers to *σήματα* as "proofs" of What-Is, and McKirahan 1994: 176.

56. Referred to as *δυνάμεις* in DK 28 B 9.1–2.

57. DK 28 B 8.56–59.

58. See, for example, Thanassas's attempt (2005) to differentiate the types of "doxai" (false and appropriate) from truth in Parmenides's poem.

59. For comprehensive recent approaches to this problem, see Long 2005 and Cordero 2004: 168–173.

critical responses to them. I have already shown how Epicharmus argued that whatever is measurable and subject to *natural* "exchange" (ἐν μεταλλαγᾶι; μεταλλάσσει κατὰ φύσιν) cannot ever "stay the same" (κοὔποκ' ἐν τωὐτῶι μένει), an explicit adaptation of Parmenides's description of What-Is (ταὐτόν τ' ἐν ταὐτῷ τε μένον).[60] It is also clear that Empedocles agreed with various aspects of Parmenides's conception of What-Is and how it is described. In particular, Empedocles sought to elaborate further on what seem to be instances of coming-to-be and destruction, or even apparently qualitative changes, in natural objects, and how human beings speak about them.[61] Objects in nature, including the sun, the four elements, and the soul of humans, are understood to undergo change that is limited by a cycle.[62] But such consistency in cyclical changes indicate the unchanging aspect of the one cosmos, that it is all alike in itself. Even so, there was room for disagreement with Parmenides: like Epicharmus, Empedocles sought to investigate further the claim that What-Is is continuous *simpliciter* and cannot be internally divided into parts.[63]

Among the early mathematical Pythagoreans, however, Philolaus offered the most sustained and comprehensive response to the arguments concerning the nature of What-Is as advanced by Parmenides.[64] Scholars have detected a number of challenges to Parmenides in Philolaus's fragments, including (1) a rehabilitation of the concepts of nature (φύσις), order (κόσμος), and harmony (ἁρμονία), all Heraclitean watchwords; (2) use of plurals rather than the singular in terms that Parmenides uses, as an indirect challenge to Parmenides's monism; and (3) employment of the structure of Eleatic "proof by refutation" in order to achieve opposite ends.[65] Philosophically, Philolaus tries to bridge a possible gap between ontology and epistemology by examining the conditions that make it possible both for natural objects to exist and for known objects to be known *by human beings*. The key to understanding the relationship between natural existents and known objects, according to Philolaus, is "number" (ἀριθμός) in its three modalities (even, odd, and even-odd,

60. See above in note 41.

61. DK 31 B 8. On Empedocles's take on the process of alteration, see chapter 5, section entitled "What Is Wisest? Three."

62. DK 31 B 17 (μεταλλάσσον . . . κύκλῳ); DK 31 B 26 (φθίνει . . . καὶ αὔξεται ἐν μέρει); DK 31 B 115 (φυόμενον παντοῖα διὰ χρόνου εἴδεα θνητῶν / ἀργαλέας βιότοιο μεταλλάσσοντα κελεύθους).

63. See Inwood 2002: 24–31.

64. As argued most extensively by Nussbaum (1979) and, with quite different results, Barnes (1982: 385–386) and Huffman (1993: 64–74).

65. See Nussbaum 1979: 83.

which is dependent on the first two types). In order to pursue a broader understanding and application of this claim, it is important to consider three fragments from Philolaus's book, probably entitled *On Nature*.[66] I will start with two fragments that inform about the meaning of "number" in Philolaus's philosophy:

> And, indeed, all the things that are known have number. For it is not possible that anything whatsoever be thought or known without this.
>
> *καὶ πάντα γα μὰν τὰ γιγνωσκόμενα ἀριθμὸν ἔχοντι. οὐ γὰρ ὁτιῶν <οἷον> τε οὐδὲν οὔτε νοηθήμεν οὔτε γνωσθῆμεν ἄνευ τούτω.*
>
> (Philolaus F 4 Huffman = Stobaeus, *Eclogae* 1.21.7b; translation after Huffman 1993)

> Number, indeed, has two proper kinds, odd and even, and a third mixed-together from both, the even-odd. Of each of the two kinds there are many shapes, of which each thing itself gives signs.
>
> *ὅ γα μὰν ἀριθμὸς ἔχει δύο μὲν ἴδια εἴδη, περισσὸν καὶ ἄρτιον, τρίτον δὲ ἀπ' ἀμφοτέρων μιχθέντων ἀρτιοπέριττον. ἑκατέρω δὲ τῶ εἴδεος πολλαὶ μορφαί, ἃς ἕκαστον αὐτὸ σημαίνει.*
>
> (Philolaus F 5 Huffman = Stobaeus, *Eclogae* 1.21.7c; translation after Huffman 1993)

Fragment 4 of Philolaus is practically a direct adaptation of a claim made by Parmenides in Fragment B 8, that thinking (*τὸ νοεῖν*) could not be discovered "without *What-Is*" (*ἄνευ τοῦ ἐόντος*). Philolaus appropriates Parmenides's basic idea by claiming that it is not "without *number*" (*ἄνευ τούτω* [*ἀριθμῶ*]) that anything whatsoever could be either "thought or known" (*οὔτε νοηθήμεν οὔτε γνωσθῆμεν*).[67] In the context of the epistemological fragments, Philolaus adapts Parmenides's language and concepts in at least two ways: (1) he replaces the "What-Is" in Parmenides with "number," and (2) he extends the cognitive activity beyond mere "thinking" to "knowing." By introducing knowledge and its objects into his ontological discussion, Philolaus directs his epistemological statement to an argument that Parmenides had made against knowledge of or ability to express What-Is-Not: "for you could neither know . . . nor speak about

66. D.L. 8.85, from Diogenes of Magnesia.

67. I am convinced by Huffman's arguments (1993: 116–118) against Nussbaum's claims that Philolaus means only grasping or apprehending the identity of an object when he uses the verb *γιγνώσκειν*.

What-Is-Not" (οὔτε γὰρ ἂν γνοίης τό γε μὴ ἐὸν . . . οὔτε φράσαις).[68] As Martha Nussbaum has persuasively argued, Parmenides's argument in Fragment B 2 assumes that "any discourse about the world succeeds by touching or grasping what it is all about" and "all speaking is understood to be like naming."[69] Thus, for Parmenides, speaking about What-Is-Not would be impossible, on the grounds that naming something that does not have an actual *nominatum*, as its object cannot be achieved.[70] Philolaus, for his part, when he appropriates the style of Parmenides's Fragment B 2, shifts the object of philosophical inquiry from What-Is-Not to *things that can be known*. Finally, and perhaps most important, it is *number* that is said to guarantee the knowledge of things that can be known.

What, then, does Philolaus mean by "number"? If I am right that the mathematical Pythagoreans sought to "demonstrate" in some fashion the doctrines of Pythagoras, then it may be relevant to consider whether Philolaus was attempting to explain the *acusma* concerning number: "What is the wisest? Number" (τί τὸ σοφώτατον; ἀριθμός).[71] The evidence from Philolaus goes beyond the Euclidean definition of "number" as a "multitude composed of units" (τὸ ἐκ μονάδων συγκείμενον πλῆθος), since it demonstrates what we might consider a protodialectical concern with classification of number according to "proper" kinds (odd and even) and a derivative kind (odd-even).[72] As Huffman has argued, when Philolaus refers to "number," he implies an ordered plurality *that is counted*.[73] This interpretation of early Pythagorean "number" is corroborated by Epicharmus's usage in Fragment B 2, where "number" is understood to be composite, can be either even or odd, and loses or retains its identity according to whether a pebble is added or subtracted or not.[74] The concerns with identity and analogy exhibited in Epicharmus's Fragment B 2 (as discussed above), as well as with classification and epistemology in Philolaus's Fragments 4 and 5,

68. DK 28 B 2.7–8.

69. Nussbaum 1979: 71.

70. Although the survival of the paraphrases of Gorgias's *On What-Is-Not/On Nature* (DK 82 B 3) testify to continued fascination with the implications of this claim.

71. Iambl. *VP* 82, 47.17. For other relevant sources, see Burkert 1972: 169 n. 22. For Plato's response to this question, see chapters 5–6.

72. Eucl. *Elem.* 7.def.2.

73. Huffman 1993: 174–176. Italics mine. See Burkert (1972: 266), where the principle of "having number," which is central to Philolaus's epistemology, is contextualized with similar usages in Hippocratic corpus, in which this phrase is taken to refer to an object that is ordered and counted.

74. See above.

both exhibit non-Euclidean aspects and may even go beyond the Hippocratic and, more generally, Presocratic practice of conflating an object's existence and its possession of properties.[75] Indeed, when Philolaus refers to known objects *having number*, he seems to be adopting an Eleatic linguistic and conceptual apparatus, but to different ends: in Fragment 5, Philolaus says that particular objects "themselves give as signs" the "shapes" of each "kind" of number, that is, even or odd (ἑκατέρω δὲ τῶ εἴδεος πολλαὶ μορφαί, ἃς ἕκαστον αὐτὸ σημαίνει). This constitutes an adaptation, or rather a reversal of sorts, of Parmenides's argument in Fragment 8, where mortals who make distinctions bestow signs (σήματ' ἔθεντο) and decide to "name shapes" (μορφὰς . . . ὀνομάζειν) of What-Is.[76] For Philolaus, to be sure, it is *each object in and of itself* that signifies (ἕκαστον αὐτὸ σημαίνει) by giving forth many shapes of each kind of number.[77] Thus, in Philolaus's epistemology, the objects in nature, which are knowable, are not subject to the problem of the status of human discourse as formulated by Parmenides and others, on the grounds that they themselves give reliable *signs* or *tokens* of their numerical shape.[78] Indeed, it should be noted that the extant fragments of Philolaus—by contrast with both early natural philosophers and Sophists—never speculate about human language as a potentially "deceitful" medium for communication about what we in fact know. It is possible that Philolaus did not seek to engage in discussing the potentially problematic relationship between the objects of (human) knowledge, our perceptions of them, and our ways of speaking about them.[79] Be that as it may, Philolaus's classification of "kinds" of number in Fragment 5 is abstract enough

75. Pace Burkert 1972: 266. Burkert is probably thinking of the Ionians, along with Heraclitus. Long (2005: 245–246) speaks of the "coextension," for example, of the properties attributed to "thinking" and "being" in Parmenides's poem, although with Parmenides we might detect a differentiation on the truth value of the properties detected.

76. As Palmer argues (1999: 210–212), Parmenides suggests that the predicates that are applied to the universe by the gods and by mortals are different; thus Palmer locates Parmenides within a tradition that had maintained that gods and humans gave different names to things, including Pherecydes and the Hesiodic *Astronomia*.

77. Accepting Huffman's conjecture (following Heeren) of ἕκαστον αὐτὸ σημαίνει from the manuscripts' ἕκαστον αὐτ' αὐτὸ δημαίνει.

78. It is difficult to know for sure what Philolaus means by this "signification," but Mourelatos (2006: 66) provides a possible interpretation: "[Philolaus] is contrasting the mediated relation that obtains between, say, four pebbles and the type picked out by the expression 'even' with the more immediate relation of signification that holds between the tetrad of pebbles and the type picked out by 'four.'"

79. In this way, Philolaus would stand in contrast to figures such as Xenophanes, Heraclitus, Parmenides, and Empedocles, who worried about falsification of the object of knowledge through speech either intentionally or unintentionally, on which see Lesher 1999.

to accommodate Archytas's description of Eurytus's approach to definition of a human being by pebble-arithmetic—even in the rather cryptic explanation given by Ps.-Alexander.[80]

In Fragments 4 and 5 of Philolaus's work *On Nature*, then, we get a rather good sense of how early mathematical Pythagoreans might have responded to Parmenides's declarations concerning human knowledge of objects and the conditions of linguistic representation. Still, these two fragments cannot alone account for how Philolaus advanced on the *ontological* status of natural objects. For that, we need to look at a very interesting fragment of Philolaus, Fragment 6:

> Concerning nature and harmony the situation is this: the being of things, which is eternal, and nature herself admit of divine and not human knowledge—except that it was impossible for any of the things that are and are known by us to have come to be, if the being of the things from which the cosmos came together, both the limiters and the unlimiteds, did not preexist. But since these beginnings preexisted and were neither alike nor even related, it would have been impossible for them to be ordered, if a harmony had not come upon them, in whatever way it came to be. Well then, like things and related things did not require any harmony additionally, but things that are unlike, being neither related nor of equal speed—it is necessary that such things be bonded together by harmony, if they are going to be held in order.
>
> *περὶ δὲ φύσιος καὶ ἁρμονίας ὧδε ἔχει· ἁ μὲν ἐστὼ τῶν πραγμάτων ἀΐδιος ἔσσα καὶ αὐτὰ μὰν ἁ φύσις θείαν τε καὶ οὐκ ἀνθρωπίνην ἐνδέχεται γνῶσιν πλάν γα ἢ*[81] *ὅτι οὐχ οἷόν τ' ἦν οὐδενὶ τῶν ἐόντων καὶ γιγνωσκομένων ὑφ' ἁμῶν γεγενῆσθαι μὴ ὑπαρχούσας τᾶς ἐστοῦς τῶν πραγμάτων, ἐξ ὧν συνέστα ὁ κόσμος, καὶ τῶν περαινόντων καὶ τῶν ἀπείρων. ἐπεὶ δὲ ταὶ ἀρχαὶ ὑπᾶρχον οὐχ ὁμοῖαι οὐδ' ὁμόφυλοι ἔσσαι, ἤδη ἀδύνατον ἦς κα αὐταῖς κοσμηθῆναι, εἰ μὴ ἁρμονία ἐπεγένετο ᾡτινιῶν ἂν τρόπῳ ἐγένετο. τὰ μὲν ὦν ὁμοῖα καὶ ὁμόφυλα*

80. At Ps.-Alexander *in Metaph.* p. 827.13–28 Hayduck = DK 45 F 3. Barnes (1982: 391) usefully speculates about what philosophical concerns underlay this "jejune" process: "Philolaus and Eurytus saw their [i.e. the Neo-Ionian and Atomist] failing, and attempted to meet it: the shapes of things are essential to them (we recognize things by virtue of their shapes); shapes can be expressed arithmetically; and the consequent arithmetical definitions of substances may be expected to function as the foundations of a mathematical physics."

81. If indeed we should adopt the emendation of Badham (followed by KRS and Huffman). Other alternatives found in the manuscripts include *πλέον γα ἢ* (FGVM) and *πλέοντα ἢ* (E).

ἁρμονίας οὐδὲν ἐπεδέοντο, τὰ δὲ ἀνόμοια μηδὲ ὁμόφυλα μηδὲ †ἰσοταχῆ,[82] ἀνάγκα τὰ τοιαῦτα ἁρμονίᾳ συγκεκλεῖσθαι, εἰ μέλλοντα ἐν κόσμῳ κατέχεσθαι.

(PHILOLAUS F 6 HUFFMAN = Stobaeus *Eclogae* 1.21.7d; translation after Huffman 1993)

In Fragment 6, Philolaus clarifies with greater precision how it is that human beings come to know knowable objects. He states that, generally speaking, "the being of things" (ἁ ἐστὼ τῶν πραγμάτων) and "nature herself" (αὐτὰ ἁ φύσις) are not receptive to human attempts to understand them, with the notable exception that objects that exist and are knowable could not have existed (and been knowable?) without the *preexistent* "being of things."[83] What Philolaus means by "nature herself" is not clear, although clues might be obtained through the fact that it is associated with "being."[84] According to Philolaus, "being," which is prior to the two other primary forces (limiters and unlimiteds), is a necessary precondition for the existence of all things in the cosmos, which have been constituted through the "harmonization" of limiters and unlimiteds.[85]

82. This is the manuscript reading. Huffman places a dagger next to this word, but there is no reason to do so if we consider the importance of speed of objects in deducing pitch intervals for early Pythagorean harmonic theorists such as "the associates of Hippasus" (Theon Sm. *Math.* p. 59.4 Hiller = DK 18 13: τῶν κινήσεων τὰ τάχη) and Archytas (F 1A Huffman = Porph. *in Harm.* 1.3: μὴ ἴσῳ δὲ τάχει). On the speed as an essential property of objects in Archytas's philosophy, see chapter 6.

83. Or, as Barnes (1982: 384–385) interprets, "subsistent." I cannot find any reference to "nature itself" among sixth- to fourth-century BCE writers except for Epicharmus DK 23 B 10 (αὕτα φύσις ἀνθρώπων, ἀσκοὶ πεφυσαμένοι, if we accept the manuscript tradition). Also relevant is Philolaus's own formulation (F 5 Huffman), when referring to the signification of knowable objects, of the objects as "each itself" (ἕκαστον αὐτό). On the authenticity of the term ἁ ἐστὼ τῶν πραγμάτων see Huffman 1993: 130–132.

84. Of course, in F 1 Huffman "nature *in the cosmos* is fitted together" out of limiters and unlimiteds, but F 1 fragment makes no strong claims about priority or being. The closest comparison I have found is preserved by Alexander of Aphrodisias, from Aristotle's lost works on the Pythagoreans (F 203 Rose = *in Metaph.* p. 40.11–15 Hayduck), where Aristotle, summarizing F 6 Huffman, claims of the Pythagoreans that "since they thought that numbers were prior to all nature and to things in nature [τοὺς δὲ ἀριθμοὺς ἡγούμενοι πάσης τῆς φύσεως καὶ τῶν φύσει ὄντων]—for it is not possible for anything that exists to exist or be known at all without number, whereas numbers can be known even without those things—they posited that the elements and first principles of numbers are the first principles of all things that exist." Aristotle thus correctly identifies numbers as preexistent in Philolaus's ontology but says nothing here about "number" and "being" (although see Arist. *Metaph.* 5.8, 1017b20–21) and, moreover, muddies the status of nature.

85. See Philolaus F 1 Huffman.

One might speculate that the "being of things" and "harmony," for Philolaus, could be the same thing, or at least strongly related to one another; this would account for the surprising supervenience of "harmony," which is not described as ontologically prior, and seemingly appears out of nowhere in the fragment.[86] In the absence of further corroborating evidence, however, it is not possible to confirm this speculation.[87] Moreover, as Huffman has noted, Philolaus does not accept the Eleatic claim that What-Is and the plurality of things that (for Parmenides and Melissus) *seem* to be its constitutive parts are essentially alike; rather, like Empedocles, Philolaus assumes that the principles of the universe were essentially diverse and, in some sense, did not recognize one another until the supervenience of harmony.[88] If we are to trust Theophrastus's description of Empedocles's epistemology (*On Sense* 11 = DK 31 A 86), understanding belongs to things that are alike, and ignorance to things that are unlike. Empedocles, moreover, appears to have "enumerated" (διαριθμησάμενος) the ways humans "recognize" (γνωρίζομεν) each element according to its similar and concluded this section by claiming that the "fitting together" (ἁρμοσθέντα) of things is a necessary condition for thinking.[89] As Huffman has noted, Philolaus's use of harmony reflects similar usages in cosmology by both Heraclitus and Empedocles—I would add that the epistemic context is relevant here as well—but Philolaus seems to focus on the relationship between harmony and the principles of nature.[90] In some sense, for Philolaus, too, nature likes to hide herself, but she does admit of human knowledge, insofar as humans have access to the signs that natural objects give by way of their classifiable shapes.

By contrast with Empedocles, however, Philolaus provides a rationale for how to associate natural objects with numbers, and he thereby further develops the stimulating, but ultimately rudimentary, response to the concern over numerical identity formulated in Epicharmus's fragments. It is not simply that Philolaus lays down a plurality of elements ("being of things," "limiters," and

86. If this were the case, then the structure of F 6, which proceeds "concerning (a) nature and (b) harmony" and then goes on to discuss the "(b) being of things and (a) nature," would be chiastic.

87. Huffman (1993: 141) doubts the possibility that harmony is an attribute of being: "It remains unclear whether *harmonia* belongs to 'the eternal being of things' in the same sense as limiters and unlimiteds do." It might be relevant, however, that when Theophrastus (F 717 FHS&G = Porph. *in Harm.* 96.21) ascribes to the Pythagoreans the claim that *harmonia* was the concord that was "through all" in a scale, he was implying that *harmonia* was universal.

88. See Huffman 1993: 137–138. On Parmenides's and Melissus's approaches to identity, see the clear analysis of Rapp (2005: 294–304).

89. DK 31 B 107, by reference to DK 31 B 109.

90. See Huffman 1993: 138–140.

"unlimiteds") that are meant to diversify Parmenidean What-Is. By positing three entities that are said to "preexist" (*ὑπᾶρχον*) *all* the objects that make up the universe, Philolaus seems to adapt the language that Epicharmus used to describe the various modalities of numerical existence found in pebble-piles, cubit-measures, and human beings.[91] In Fragment B 2 of Epicharmus, it will be recalled, pebbles (or numbers) are said to "preexist" (*τᾶν ὑπαρχουσᾶν*) before an odd or even number is added to them, as is the "measure" (*ὑπάρχοι . . . τὸ μέτρον*) before a cubit is added or subtracted from it. The activity of adding or subtracting from the preexistent state of a countable or measurable object is then analogized to the growth and deficiency in a human being, whereby Epicharmus's speaker raises the question of the stability of human identity over time. Philolaus and Epicharmus seem to agree that an essential property of all things in existence and understandable (at some level)[92] is that they can be counted. But Philolaus goes further by (1) naming the primary preexistent entities as "being," "limiters," and "unlimiteds," and (2) elaborating on this schema by describing two primary "kinds" of number, "odd" and "even," as well as a derivative kind, "odd-even." The former (1) seems to be the way Philolaus speaks about cosmological entities and the chain of predication. The latter (2) is employed chiefly in epistemological contexts, involving how we can know knowable objects in nature.[93] In the absence of further evidence from Philolaus himself, it is difficult to know to what extent Philolaus implicates number epistemology in ontology.[94] But in the light of the comparative evidence adduced

91. Although it might be possible that, for Philolaus, the meaning of *ὑπάρχειν* is closer to "subsist," in anticipation of Aristotelian ontology. On subsistence in Philolaus, see Barnes 1982: 384–385 and 443.

92. This epistemological aspect is not made explicit in Epicharmus's fragment, but it is suggested by the use of *δοκεῖ τοί*.

93. See Huffman 1993: 181. As Malcolm Schofield suggests to me, however, the links between epistemology and ontology might be stronger than I've implied. He notes that Philolaus certainly does believe that we cannot know much about nature and being themselves, but what we can infer is that whatever we do know would be inexplicable without the being of limiters and unlimiteds. Philolaus F 6, then, might exhibit one of the earliest instances of what Peirce called "abduction," or "inference to the best explanation" (see Josephson 2000: 31–33).

94. Aristotle's account (*Metaph.* 1.5, 987a16–19) explicitly links number and being, with a particular emphasis on predication: "they believed . . . that the unlimited itself and the one itself were the substance of the things of which they are predicated [*οὐσίαν εἶναι τούτων ὧν κατηγοροῦνται*], and hence that number was the substance of all things [*ἀριθμὸν εἶναι τὴν οὐσίαν πάντων*]." Thus, Aristotle's criticism, in which the Pythagoreans' numbers both maintain an independent reality from perceptibles but are not separated from them, seems to be reasonable. See Burnet 1945: 286–287.

for mathematical Pythagoreans Epicharmus and (especially) Empedocles, we can acknowledge that, for Philolaus, the force of "harmony" is a precondition both for the arranged ordering of the limiters and unlimiteds in the cosmos, as well as the "mixing" of odd and even numbers, whatever the relationship of the limiter/unlimited and odd/even pairs itself might actually be.[95]

In this section, then, I have traced the intellectual history of the mathematical Pythagoreans before Plato and Archytas who are named by Timaeus of Tauromenium and whose philosophical writings or thoughts survive: Epicharmus of Syracuse, Empedocles of Agrigentum, Philolaus of Croton, and Eurytus of Metapontum. Even this cursory examination of a few targeted fragments of these intellectuals reveals a sophisticated and philosophically relevant engagement with the contemporary speculations about things in the universe, whether and how they are one or many, as well as the related issues of stability and flux, polarized in the writings of Parmenides and Heraclitus. Furthermore, the fragments of the early mathematical Pythagoreans demonstrate some of the earliest (and richest) evidence for concern over defining number in terms of identity, which concerned not only Plato but also the Stoics after him. The Pythagoreans practiced crude modes of demonstration, even in the form of comic expression, which do not obviously anticipate Euclidean axiomatic-deductive mathematics but obtain philosophically significant results nonetheless.[96] The evidence of early mathematical Pythagoreanism also suggests that Aristotle was not totally wrong in his presentation of "so-called" Pythagorean philosophy as postulating a relationship of natural objects and numbers as mediation by "imitation" or "assimilation," as Cherniss and others have maintained.[97] And the evidence from Philolaus and Epicharmus also corroborates Aristotle's claim that—at least from his own particular vantage point—the Pythagoreans did make advances in speculating about the essences of things, while they nevertheless were too superficial in their definition of things.[98] It is surely the case that Aristotle sought to understand Pythagorean philosophy within his own framework—I

95. Pace Schibli 1996, who presents many compelling arguments for why Aristotle's testimony must remain central to our analysis but does not finally convince me that analogies between these oppositional pairs are original to Philolaus.

96. See Barnes (1982: 391), with regard to Philolaus and Eurytus: "Is all this mere comical arithmology? Or is it the first scrabbling essay towards a quantitative and mathematically-based science? Surely it is both of these things." Also see Sorabji (2009: 39): "The idea that there is no continuous self was introduced early in the fifth century BCE by the playwright Epicharmus, if the text is his, but only as a joke. Philosophers, however, often take jokes seriously."

97. Cherniss 1935: 386 and Kahn 1996: 83.

98. Arist. *Metaph.* 1.5, 987a19–28.

sought especially in chapter 1 to develop a more robust account of Aristotle's classification of the two types of Pythagoreans he distinguished. But even if Aristotle's history of philosophy was tailored to suit of his own project of the development of a new approach to science, the evidence of Philolaus's and Epicharmus's fragments suggests that Aristotle's characterizations of "so-called" Pythagoreanism were not inaccurate.

PLATO AND MATHEMATICAL PYTHAGOREAN "BEING" BEFORE THE *PHAEDO*

If it is true that Timaeus of Tauromenium listed Plato among the "exoteric" Pythagoreans,[99] and that Aristotle' claimed of Plato that his *pragmateia* succeeded that of the mathematical Pythagoreans,[100] it would be appropriate to investigate the possibility that Plato's philosophy exhibits the qualities of what I have been calling "exoteric" or "mathematical" Pythagoreanism. Is there any evidence that Plato, like Epicharmus, Empedocles, and Philolaus before him, sought to "publish" and "demonstrate" the principle doctrines (i.e. the *acusmata*) of Pythagoras? Moreover, do Plato's dialogues exhibit any particular responses to the sorts of approaches to explanation of the *acusmata* preserved in the fragments of the mathematical Pythagoreans Epicharmus and Philolaus? When scholars of the past century sought to find Pythagoreanism in Plato's philosophy, they often gravitated to the *Phaedo* as an imprimatur that indicates both Plato's endorsement of Pythagoreanism and the license to expand on Pythagorean philosophical ideas in his middle dialogues. In diverse ways, Burnet and Taylor saw a Pythagorean theory of the Forms as anticipating Plato's ruminations on the relationship of intelligible to sensible.[101] Bostock was more skeptical that the Pythagorean number theory had anything to do with Plato's theory of the Forms, but he was willing to speculate that "it may be . . . that contact with the Pythagoreans led Plato to pay more attention to the notion of number than he had done before."[102] Chapter 5 will undertake a more detailed investigation of the possibilities that mathematical Pythagoreanism

99. FGrHist 566 F 14 = D.L. 8.54. See chapter 3, section entitled "Pythagorean Exoterics in the Fifth Century BCE? The Historical Evidence of Timaeus of Tauromenium."

100. Huffman (2008: 284–291) has recently challenged this claim, based on a famous passage of Aristotle (*Metaph.* 1.5–1.6, 987a9–31), but see Horky: forthcoming.

101. Burnet 1945: 308–309, without evidence. Taylor (1937: 385–386) followed Proclus (*in Parm.* p. 562 Stallbaum) in speculating that the elusive "friends of the forms" in Plato's *Sophist* (248a4–5) were Pythagoreans.

102. Bostock 1986: 13.

presented Plato with material from which to derive his defining approach to metaphysics in the *Phaedo*. Before this, however, it will be useful to examine two dialogues written prior to the *Phaedo* that offer the opportunity to consider the more advanced theory of the Forms sketched in the *Phaedo* and *Republic* in the light of mathematical Pythagoreanism: the *Euthyphro* and the *Cratylus*.[103] In general, these two works have not been considered loci classici for possible responses to Pythagoreanism on the part of Plato.[104] One possible reason for this could be the influence of the "historical" account of the development of Plato's thought given by Aristotle, reinforced by the early Hellenistic traditions that accuse Plato of stealing his ideas from the Pythagoreans, and corroborated by the *Seventh Letter*, which emphasizes that Plato only came to know about Pythagoreanism after his first trip to Italy and Syracuse, around 388/7 BCE.[105] I have little doubt that Plato's knowledge of Pythagoreanism was deepened by his visits to Sicily and Southern Italy, but we should not assume that Plato could not have had knowledge of Pythagoreanism, at least as it might have been known popularly, before then. When the *Euthyphro* and *Cratylus* are contextualized with the sorts of philosophical concerns raised by Epicharmus and Philolaus, a different picture emerges, in which the concerns over number, identity, and predication hazarded by these mathematical Pythagoreans stimulated innovation to Plato's own approaches to providing explanations for how we are to understand the properties of objects in the light of their essential "being."

It is worth examining a few relevant passages from Plato's early and early-middle dialogues in which he discusses the problems of ontology in the light of the definition and predication of attributes. I start with the *Euthyphro*, on the grounds that (1) it preserves what probably constitute Plato's earliest attempts to formulate a more complex theory of ontological priority than his predecessors, and (2), reasonably or not, has been associated with Pythagoreanism by some scholars.[106]

103. Sedley (2007: 73 n. 13) recants his previous argument (2003: 6–14) and now accepts the *Cratylus* as earlier than the *Phaedo*. But see Ademollo 2011: 20–21, who argues that *Cratylus* should be read after *Phaedo*, and before *Theaetetus*.

104. Notable exceptions to this trend are Burkert (1972: 85, with n. 12, following Boyancé) and Herrmann 2007: 299–307.

105. See Riginos 1976: 169–174. Contra Lloyd 1990: 168–171, who conjectures that the story is more complicated and that the author of the *Seventh Letter* seeks to "block any interpretation of Plato that would assimilate him to the Pythagoreans or worse as positively plagiarising them."

106. E.g. Burnet (1911: 85–86) and Boyancé (1941: 167–175), who has been challenged by Baxter (1992: 110), who does concede, however, that he sees Pythagoreanism behind various etymologizations in the *Cratylus* (see pp. 142–143).

> I'm afraid, Euthyphro, that when you were asked about the pious, what in the world it is [τὸ ὅσιον ὅτι ποτ' ἐστίν], you did not want to make clear to me its essence [τὴν οὐσίαν . . . δηλῶσαι], but you wanted to tell me a certain attribute it has [πάθος δέ τι περὶ αὐτοῦ), that the pious happens to have this attribute:[107] "to be loved by all the gods"; but what it is [ὅτι ὄν] you have not yet told me. So if it's all right by you, please do not keep it hidden from me, but say again, from the beginning, what in the world the pious [is] [τί ποτε ὄν], whether it is loved by the gods or whatever other attribute it happens to have—for we do not disagree about that—but take courage and tell me: what is [τί ἐστιν] the pious and the impious?
>
> (PLATO, *Euthyphro* 11a6–b5)

Plato has Socrates set out to obtain a sufficient definition of "the pious," in pursuit of an objective point of comparison (ἰδέα, εἶδος, παράδειγμα)[108] by which we can better evaluate whether things called "pious" or "impious" are indeed so. That is to say, the definition of "what in the world" the pious is helps us to decide whether "pious" can be predicated of objects, people, actions, and so on. Whether or not Plato is treating Euthyphro as a Pythagorean, as Boyancé believed, need not be of concern here. What is clear from this passage is that Plato, from the initial steps he takes toward a comprehensive metaphysics that would later seek to negotiate Parmenidean stability with Heraclitean flux, formulates the problem of predication as a problem of essential definition: one cannot understand how attributes are predicated of objects without having a clear sense of the "being" (οὐσία) of the object in the first place.[109] As I showed above, the mathematical Pythagoreans Epicharmus and Philolaus responded, in their own unique ways, to concerns over the identity of things in the universe and the ways we might describe their "being." In the *Euthyphro*, as well as in other early-middle dialogues, Plato reformulates these issues as problems of predication and definition of objects. In the *Cratylus*, a dialogue involving a Parmenidean (Hermogenes)[110]

107. I have translated πάθος as "attribute" and πέπονθε τοῦτο as "happens to have this attribute," on the grounds that Socrates here is distinguishing between an essential definition of the pious (which makes it just what it is and nothing else) and what Aristotle would later describe as accidental states or conditions that are applicable to the pious but do not define it universally. For a useful discussion of this problem, see Vlastos 1981: 309–310, with n. 2.

108. Pl. *Euth.* 5d3–5, 6d9–e6. This usage anticipates, but is not necessarily the same thing as, Plato's later Form theories (see Allen 1970: 67–69 and Moravcsik 1992: 60–61).

109. For the variability of Socrates's "What-is-X" question, see Robinson 1953: 53–60.

110. According to D.L. 3.6, an association that might derive from interpretation of the *Cratylus* itself. If Hermogenes had a conventionalist naming theory of the sort described at the beginning of the *Cratylus* (384c10–e2), it would not be impossible to square that theory with

and a Heraclitean (Cratylus) engaged with Socrates in a discussion that takes place in the wake of the *Euthyphro*,[111] there is a stated desire to describe the correctness of names with precision (σαφῶς) before one can advance with a theory of how particular words, both names and the attributes that constitute them, are properly fit to their objects of reference.[112] This concern with a correct understanding of names and their attributes informs what Rachel Barney refers to as the "ontological subplot" of the *Cratylus*, a dialogue especially concerned with "the gradual specification of what both names and objects consist in, and how they are related."[113] Thus these concerns are of a growing importance to Plato's thought, as evidenced by the extensive treatment of them in the *Phaedo*, *Republic*, *Sophist*, and *Philebus*. One way of advancing on such a proper understanding of names that remains uncontroversial for the interlocutors of the dialogue is by identifying a certain essence (οὐσία) appropriate to each object and seeking to describe attributes (sound, shape, color) that are attached accidentally to it.[114] The person who might be said to be capable of ascribing the names to each object that are appropriate to it (i.e. as "imitations" of its essence) would be the expert "name-giver" (ὀνομαστικός), who would understand the essence of each object and is able to "show what it is" (δηλοῖ ἕκαστον ὃ ἔστιν) through letters and syllables.[115] The various "name-givers" described here share many characteristics with the "first-discoverer" figures described in various parts of the middle and later dialogues of Plato, and whose importance to my analysis of the role of Pythagoreanism in Plato's philosophical methodology I will discuss further in chapter 6.

The approval of the expert "name-giver's" place in obtaining correct names for the essences of objects at this part of the *Cratylus*—halfway through the dialogue and just before Socrates turns away from Hermogenes and engages in a dialectic with Cratylus—looks backward to an important passage earlier in the dialogue, where, in what we might see as an argument prompted by Parmenides's

a broader monist view that suggested that all names are applicable to What-Is. It may be relevant that Hermogenes (*Cra*. 421a1–4) refers to the greatest and noblest words as truth (ἀλήθεια), falsehood (ψεῦδος), and What-Is (τὸ ὄν), and name (ὄνομα), all of which are key terms in Eleatic thought. The case for Cratylus as a Heraclitean, broadly speaking, is less problematic: it is explicit by the end of the *Cratylus*. For a good discussion of the biographical evidence for the interlocutors, see Ademollo 2011: 14–19.

111. See Pl. *Cra*. 396d4–397a2.

112. Pl. *Cra*. 383b7–384a4 and 427d4–8.

113. Barney 2001: 88. In fact, it was originally Rachel Barney's suggestion that I examine Philolaus's fragments in the light of the relevant passages of the *Cratylus*.

114. Pl. *Cra*. 423d4–e6. See Kahn 1973: 163–164.

115. Pl. *Cra*. 423e7–424a6. See Schofield 1982: 61–62.

"Doxa" passage,[116] Hermogenes asks Socrates to leave off etymologizing the soul and the body and begs him to discuss the names given *by men* to the gods:

> Well, it's obvious to me that it was people of this sort [lofty thinkers and subtle reasoners] who gave things names, for even if one investigates names foreign to Attic Greek, it is equally easy to discover what they mean. In the case of what we in Attic call "*οὐσία*" [essence], for example, some (1) call it "*ἐσσία*" and others (2) "*ὠσία*." First (1), then, it is reasonable, according to the second of these names ["*ἐσσία*"], to call the essence of things [*ἡ τῶν πραγμάτων οὐσία*] "Hestia." Besides, *we ourselves say* that what partakes of essence [*τὸ τῆς οὐσίας μετέχον*] "is" [*ἔστιν*], so essence is also correctly called "Hestia" for this reason. *We, too*, seem to have called essence "*ἐσσία*" in ancient times. And, if one has sacrifices in mind, one will realize that the name-givers themselves understood matters in this way, for anyone who called the essence of all things "*ἐσσία*" would naturally sacrifice to Hestia before all the other gods. On the other hand, those (2) [who call the essence of things] "*ὠσία*" would think pretty much along the same lines as Heraclitus that the things are all on the go and nothing remains [*τὰ ὄντα ἰέναι τε πάντα καὶ μένειν οὐδέν*]. Hence [they think] that the cause and ruler [*τὸ αἴτιον καὶ τὸ ἀρχηγόν*] of these things is that which pushes [*τὸ ὠθοῦν*], for which reason [they think] it well that it has been named "*ὠσία*."
>
> (PLATO, *Cratylus* 401b11–d7; translation after Reeve in Cooper and Hutchinson 1997)

This passage helps us to contextualize Plato's larger inquiry into the essence and predication of objects by placing it in an ambiguous, but remarkable, dialectic. The "name-givers" whom Socrates identifies as "lofty thinkers and subtle reasoners" (*μετεωρολόγοι καὶ ἀδολέσχαι*) here appear to be cosmological theorists of various sorts.[117] Is it possible to identify these anonymous "name-givers"?[118] The context of Plato's own works points to Anaxagoras and Pericles, who are associated in the *Phaedrus* (269e4–270a8) with *μετεωρολογία* (as an inquiry into the stars) and *ἀδολεσχία* (as a sort of rhetorical knack), respectively; but, as I will show, these are not the only possible, and perhaps not the best, referents. With the first term, *μετεωρολόγος*, one might detect with Ademollo a reference

116. See Barney 2001: 73–80.

117. See Sedley 2003: 100–101. We would add, however, in the light of *Plt.* 299b3–9, that Plato distinguishes the *μετεωρολόγος καὶ ἀδολέσχη*, whom he associates with the Sophist, from the expert doctor and steersman.

118. As Barney notes (2001: 49 n. 1), the possibility that these figures are divine here is explicitly excluded.

to (among others, including Ionian natural scientists) the Pythagoreans: he cites the passage in *Republic* 7 (530d6–531a3) where Socrates recognizes sciences of astronomy and harmonics as "sisters," "as the Pythagoreans say" (ὡς οἵ τε Πυθαγόρειοί φασι).[119] One might object, however, that in the *Republic*, the reference is to ἀστρονομία rather than μετεωρολογία, which would suggest that Plato is not intending to solicit comparisons between the Pythagoreans discussed in the *Republic* and the "name-givers" in the *Cratylus*. Indeed, ἀστρονομία may be distinguished from μετεωρολογία by the fact that μετεωρολογία supplants rigorous scientific method with techniques of persuasion, as is implied in the *Cratylus* passage, and as attested elsewhere in fifth-century BCE intellectual culture.[120] But we also need to keep in mind the fact that Plato does not find the association of the Pythagoreans with Sophists worrisome.[121] Moreover, there is ample evidence that mathematical Pythagoreans in the generation before Plato and Archytas undertook to investigate the heavens in speculative ways. The chief representative among the Pythagoreans for this activity is Philolaus of Croton, whose astronomical system constitutes, in the words of Huffman, "the most impressive example of Presocratic speculative astronomy," in part because it is concerned with "*a priori* notions of order and fitness."[122] I would add to that assessment the broader concerns with developing a cosmogonic model that is also attentive to predication and ontological priority. Consider, for example, Philolaus's description of the cosmic ordering of the universe:

> The first thing fitted together, the one in the center of the sphere, is called ἑστία [hearth].
>
> *τὸ πρᾶτον ἁρμοσθέν, τὸ ἓν ἐν τῷ μέσῳ τᾶς σφαίρας, ἑστία καλεῖται.*
>
> (PHILOLAUS F 7 HUFFMAN = Stobaeus, *Eclogae* 1.21.8; translation after Huffman 1993)

This fragment, when read along with Fragments 1 and 6, helps us to understand how Philolaus conceived of the primordial establishment of natural cosmic

119. Compare the description of what appear to be Pythagoreans at *Cra*. 405d1–3 (ὥς φασιν οἱ κομψοὶ περὶ μουσικὴν καὶ ἀστρονομίαν, ἁρμονίᾳ τινὶ πολεῖ ἅμα πάντα).

120. See Gorg. *EH* 13, where Gorgias uses the examples of epideictic speeches, philosophical debates, and the "speeches of astronomers" (λόγοι τῶν μετεωρολόγων) to show that the soul is susceptible to affection because its "opinion" (δόξα) can be changed through persuasive speech. For an analysis of Gorgias's theory of the affection of the soul, see Horky 2006.

121. See Huffman 2002: 251–270, and see chapter 6.

122. Huffman 1993: 241 and 244. One relevant example is Arist. *Mete*. 1.8, 345a14–17, where some of the "so-called" Pythagoreans claim that the Milky Way is the path of a star that fell from heaven, and others that the motion of the sun burned the path.

order, which is built up out of limiters and unlimiteds.[123] The order that the natural universe comes to take—here imagined as a sphere, in concert with Parmenides and Empedocles[124]—is temporally posterior to the original definition of "the one in the center of the sphere," a definition that obtained thanks to the supervenience of harmony on limiters and unlimiteds.[125] Debate about the precise meaning of this difficult fragment is still raging, but I want to focus on something overlooked in the debate: the way it addresses the problem of naming. Of particular interest is the phrase ἑστία καλεῖται, which in this particular case *qualifies* "the one in the center of the sphere."[126] It is a well-known fact to Aristotle and other later commentators that the Pythagoreans tended to refer to the fire at the center of the universe as "the hearth," and Philolaus himself does not seem to be claiming to innovate when he says that "the one at the center of the universe *is called* [or *calls itself*][127] ἑστία." Given the ubiquitous religious significance of Hestia in Greek and other Indo-European wisdom-traditions, we might be dealing with Philolaus's own explanation of an unrecorded Pythagorean *acusma*.[128] After all, Aristotle preserves several *acusmata* in which the Pythagoreans assigned divine names to astronomical bodies.[129] Be

123. For good discussions of how Philolaus's cosmogony proceeded after the initial "fitting together," see Huffman 1993: 41–43 and 210–214 and, in response, Schibli 1996: 125–126.

124. Parmenides: DK 28 B 8.43. Empedocles: DK 31 B 28–29. See Huffman 1993: 229–230.

125. Aristotle (*Metaph*. 14.3, 1091a13–18) is probably nearly quoting from an actual Pythagorean text when he claims that "they clearly state that when the one had been constituted . . . immediately the nearest part of the infinite began to be drawn in and limited by the limit."

126. Pace Schibli (1996: 121), who claims that it is the "central fire," which "is described as the one." Philolaus has it the other way around.

127. That is, if we take the verb to be in the middle voice. Would this be one of the ways objects themselves give signs (i.e. of their own proper names)?

128. The hearth appears in an *acusma* of the τί πρακτέον type at Iambl. *VP* 48, 27.1–3 and 84, 49.4–6 (one should believe that he has taken his wife like a suppliant from the hearth in the presence of the gods and has led her to his home; see Ps.-Arist. *Oec*. 1.4.1, 10–13). Socrates, in a similar vein, claims (*Cra*. 401b1–2) that "we should begin with Hestia, according to custom" in their listing of the names of the gods. In Socrates's Palinode in the *Phaedrus* (247a1–2), Hestia is the only one of the gods who "remains in the house of the gods alone" (μένει ἐν θεῶν οἴκῳ μόνη), which implies her place at the center of the cosmological order of the universe, as well as her divine stability. A similar model is adopted for the civic design of Magnesia in the *Laws* (745b6–c3), in which Hestia is expected to occupy the acropolis at the center along with Zeus and Athena. But the religious significance of Hestia to cosmic order originates with the Vedic texts, on which see Pinchard 2009: 504–514.

129. Arist. F 196 Rose = Porph. *VP* 41. According to the Pythagoreans, the Bears are the "hands of Rhea"; the Pleiades are "the lyre of the muses"; and the planets are "the dogs of Persephone." See Burkert 1972: 320.

that as it may, we can see that Philolaus, in his account of the primordial generation of the unity that gives order to the objects of nature, has appropriated the concepts and language used by Parmenides to describe What-Is, to diverse ends.

Given Philolaus's attribution of Hestia to "the one in the center of the sphere" and the association of Pythagoreanism with "lofty thinking," we might seek to inquire whether Plato was actually describing Philolaus when he referred to the various ways "name-givers" associated *οὐσία* with *ἑστία* in the *Cratylus*.[130] There is good reason to think this. In particular, Socrates mentions that the name Hestia can be appropriately attributed to "the essence of things" (*ἡ τῶν πραγμάτων οὐσία*), a phrase that never occurs in any formulation elsewhere in the extant fragments of the Presocratics, only one other time in Plato, and twice in Aristotle.[131] In referring to a primary "essence of things" that underlies all other objects, Plato would appear to be adapting the language and concepts of Philolaus, who used the term "the being of things" (*ἁ ἐστὼ τῶν πραγμάτων*) in order to establish the a priori argument that things that are known could not have been known without its preexistence. As I showed, Philolaus's actual use of the term in Fragment 6 seems quite limited: while the Philolaic "being of things" roots all human knowledge and makes it possible for things posterior to exist, no other information concerning it is given among surviving fragments. It remains unclear from his own fragments whether Philolaus actually associated the "being of things" with Hestia, and testimony elsewhere from Aristotle demonstrates a certain confusion of the terms that might be used to refer to the primordial force that gives definition to the universe.[132]

130. As has been argued by Pinchard (2009: 500–504).

131. Also see Anceschi 2007: 61–63. Plato refers to the *οὐσία τῶν πραγμάτων* at *Cra.* 431d3 and, with slight modification, at 393d3–5 (*οὐσία τοῦ πράγματος*). On these passages, see below. I cannot find any formulation of this term attested anywhere else in Archaic, Classical, or Hellenistic philosophical texts before Proclus, except at Arist. *Metaph.* 7.17, 1041b28-29 (*ἐπεὶ δ' ἔνια οὐκ οὐσίαι τῶν πραγμάτων*), where it is not clear that *τῶν πραγμάτων* should be construed with *οὐσίαι*, and at *Metaph.* 1.9 (991b1–3; repeated at *Metaph.* 13.5, 1080a1), where Aristotle attacks Plato's separation of the Forms from particulars (*ὥστε πῶς ἂν αἱ ἰδέαι οὐσίαι τῶν πραγμάτων οὖσιαι χωρὶς εἶεν·*). Aristotle, of course, uses the term in the plural, so as to reject the sort of single unified essence for all things that may have been assumed by Philolaus, Plato, and some of the Platonists.

132. A *testimonium* apparently from Aristotle's lost fragments on the Pythagoreans, and preserved by Alexander of Aphrodisias (Aristotle F 203 Rose = Alex. *in Metaph.* p. 39.16–20 Hayduck), establishes a chain of associations between various primary entities, and possibly with an eye to Philolaus: "And they used to call the One 'mind' and '*οὐσία*.' For he [?] said that soul is mind. And they used to call mind 'unit' and 'one' on the grounds that it is stable, alike everywhere, and authoritative. But they also used to call it [i.e. the One] *οὐσία*, because *οὐσία* is primary. And they used to call the number two 'opinion' on the grounds that it is subject to change in both directions; they also called it 'motion' and 'addition.'" This passage

With Plato's treatment of the naming of the "essence of things," however, comes (I suggest) a paradigmatic case of response to and appropriation of mathematical Pythagoreanism. Plato states the "doctrine" of his anonymous intellectual antecedent ("according to the second of these names [i.e. ἐσσία][133] [they] call the essence of things 'Hestia'"), provides an evaluation of it ("it is reasonable" [ἔχει λόγον] for them to do so), and then appropriates it to his own uses ("*we ourselves say* that what partakes of being 'is' [ἔστιν], so being is also correctly called 'Hestia' for this reason"). He even goes so far as to claim that the ancient Athenians recognized οὐσία as ἐσσία as well, drawing his kin who in times past understood strong connections between essence and the hearth together with the Italian Philolaus. The close connections between this description of the Philolaic "being of things," which is employed by Plato as an etymological justification for Platonic participation of the names of things in their ultimate Form, gives us what is likely to be the very earliest attestation of Plato's inheritance of mathematical Pythagoreanism in the dialogues. Philolaus's claims concerning the "being of things" are approved of, then subordinated, and finally totally appropriated to Plato's own project of describing a mechanism by which imitations can participate in the Form that they imitate. It is precisely this sort of implication of Pythagorean thought into what would become Platonic doctrine that provided the data for Aristotle to develop an account of the relationship between "so-called" Pythagorean and Platonic ontology in the *Metaphysics*.[134] Accordingly, I suggest, it is possible to identify Philolaus as one of the

comes on the heels of Aristotle's description of Philolaus's astronomical system, whereby the ten astral bodies circulate "around the center and hearth" of the sphere, as well as a programmatic discussion of how numbers, planets, and the names of gods are correlated. Now the fact that Aristotle later in this fragment appears to be summarizing Fragment 6 of Philolaus (e.g. at *in Metaph*. p. 40.11–15 Hayduck) might lead us to assume that Aristotle is reinterpreting the content of Philolaus's book. But it is equally possible that Aristotle was referring to Ecphantus in this passage (see note 46 above), who equated mind with soul and may have been the first Pythagorean to introduce monads as units with extension. At any rate, Aristotle himself seems to be unable to get a clear story out of the Pythagoreans regarding the generation of the primary one (see *Metaph*. 13.6, 1080b16–21).

133. There is an issue of concern here, however. Socrates uses the term ἐσσία rather than Philolaus's ἐστώ, which, although they are very similar in formation, could lead one to think that Plato is not referring to Philolaus at all. That must remain a possibility, but we should keep in mind that Plato's goal in this passage is to evaluate his predecessors' etymologization of key *concepts*, not preserve their thought for posterity. He is a bit fast and loose with his etymologization here, too: note that he goes on to claim ἐσσία for the Athenians of old immediately thereafter. Philolaus's use of ἐστώ itself, in fact, is an anomaly, since it is an Ionic word couched in a vividly South Italian Doric, which might suggest that Philolaus borrowed the term from someone else (Aeschylus or Democritus? see Herrmann 2007: 206).

134. Sedley's study of the *Cratylus* (2003: 16–17) admirably reassesses the evidence for Cratylus as an influence over Plato's life and philosophy, but he does not fully examine how the Pythagoreans might have influenced this dialogue (but also see Sedley 2003: 25).

chief targets for the "name-givers," specifically those who fall under the umbrella of those (1) who call the essence of things "Hestia" and whose philosophy is assumed to anticipate Plato's concept of participation in οὐσία. Most important, the philosophical approach to essence described here is understood to represent an approach to stable identity in the face of fluxist physics and language theory, which prompts the philosopher to consider the nature of the entity that gives objects their identities. In chapter 5, I will investigate the further implications of Philolaus's description of "the being of things" in the light of Plato's first complete argument for the existence of the Forms in the *Phaedo*.[135] But before doing so, consideration should be given to the question whether it is possible to identify those "name-givers" (2) who are described by Socrates as being associated with Heracliteanism and who deny the possibility of stable existence.

Now, those people who (2) called the being of things ὠσία on the grounds that it is what "pushes" (τὸ ὠθοῦν) other things, and who followed Heraclitus in denying stability, are very difficult to identify. Nobody, to my knowledge, has proposed any figures behind this particular doctrine (as differentiated from that of (1)).[136] One likely candidate, however, is Democritus, who not only is associated with Pythagoreanism by his contemporary the historian of music Glaucus of Rhegium but also is also linked to Philolaus directly by the historiographer Apollodorus of Cyzicus.[137] Democritus seems to have posited "push" (ὠθοῦνται) as one of the natural forces that draws together similars, including pebbles of oblong shape, into the same place through a sort of vortex (δῖνος).[138] Moreover, as Aristotle attests (*On the Soul* 1.3, 406b15–22 = DK 68 A 104), Democritus believed that spherical atoms by nature never remain the same and tend to draw the body together (συνεφέλκειν), a process that seems analogous to the description of the drawing together of similar pebbles.[139] And later on in the *Cratylus* (439b10–c6), when Socrates makes reference to the (2) "name-givers" who

135. Proleptically at Pl. *Cra*. 440b5–c1.

136. Both Baxter (1992: 159) and Ademollo (2011: 210–215) find in this passage overall a probable reference to Democritus, but they do not carefully differentiate the positions advanced by (1) or (2). Anceschi (2007: 56–69) thinks that both (1) and (2) are Heracliteans, which I find doubtful in the context of the project of the *Cratylus*.

137. D.L. 9.38 = DK 68 A 1.

138. DK 68 B 164: κατὰ τὸν τοῦ κοσκίνου δῖνον . . . αἱ μὲν ἐπιμήκεις ψηφῖδες εἰς τὸν αὐτὸν τόπον ταῖς ἐπιμήκεσιν ὠθοῦνται . . . ὡς ἂν συναγωγόν τι ἐχούσης τῶν πραγμάτων τῆς ἐν τούτοις ὁμοιότητος. See DK 68 B 167 and DK 68 A 128.

139. It is relevant context that the Hippocratic writer of *On Regimen*, who presents a natural science informed by both Heraclitean flux and Pythagorean number theory (see Burkert 1972: 262), suggests (1.6) that when men are sawing a log, the acts of "drawing" (ἕλκειν) and "pushing" (ὠθεῖν) are the same.

believed that all things are always on the go and flowing, he parodies their theory of physics by speculating that they themselves fell into their own vortex and have subsequently dragged in Cratylus and Socrates with them.[140] An objection could be raised here: Democritus wrote in Ionic, and *ὠσία*, which is in Doric, is not attested in his fragments, casting doubt on him as the target of this polemic. Nor do his fragments evince any explicit philosophical conceptualization of *οὐσία*. But we have it on the authority of Aristotle (*PA* 1.1, 642a24–28) that Democritus was the first to touch upon a sufficient definition of *οὐσία* as the natural substance of an object, perhaps in the light of advancements in definition made by the Pythagoreans themselves (see Arist. *Metaph*. 13.4, 1078b19–23). And we will recall that Philolaus never used *ἐσσία*, either, although there is good reason to see him as a target for position (1).[141] Given the likelihood of references to the doctrines of (1) Philolaus and (2) Democritus, we might wish to entertain the possibility that the two dialectical variants of *οὐσία* given by Socrates are not meant to represent actual linguistic usages.

Still, another, possibly better, referent for one of the "name-givers" represented by position (2), who has not been considered by any scholars (to my knowledge), is Epicharmus of Syracuse. There are several reasons to entertain this possibility. First of all, Epicharmus wrote in Syracusan Doric, and although *ὠσία* does not occur in his extant fragments, his extant fragments demonstrate that he was accustomed to contract diphthongs with *ο* to *ω*.[142] Also notable are the number of invented words, including otherwise unattested epithets for the gods and other words derived from Italic languages, attested in his fragments.[143] As I showed earlier, Epicharmus's ontology appears to be broadly fluxist, in that he argued that objects in nature cannot maintain their identity and are "in a process of exchange and never remain the same" (*ὃ δὲ μεταλλάσσει κατὰ φύσιν κοὔποκ' ἐν τωὐτῶι μένει*).[144] We will also recall that he was concerned with inquiring after the identity of objects through their etymologies, especially with regard to the number of parts that make up a

140. *ὥσπερ εἴς τινα δίνην ἐμπεσόντες κυκῶνται καὶ ἡμᾶς ἐφελκόμενοι προσεμβάλλουσιν*. Ademollo (2011: 450 n. 2) identifies the possible referents here as Empedocles, Anaxagoras, and Leucippus and Democritus.

141. See Ademollo 2011: 202 n. 54, who notes that whoever recorded Philolaus F 11 Huffman (certainly spurious) corrected *οὐσία* to *ἐσσία*.

142. For a general discussion of Epicharmus's phonology, see now Willi 2008: 126–127.

143. See Rodríguez-Noriega Guillén 1996: xxiii. Willi (2008: 28–34) provides a good list of loan-words in Sicilian Greek, including words employed by Epicharmus.

144. See above in the section entitled "Growing and Being: Mathematical Pythagorean Philosophy before Plato."

composite being as reflected naturally in the meanings of compound names.[145] This suggests playing around with naturalist theories of naming. One might object that these fluxist positions were only adopted by certain interlocutors in his comedies, and that Epicharmus himself did not espouse them, but there is good evidence that Plato took him for a fluxist of a Heraclitean sort. In the *Theaetetus* (152d7–e9), Socrates names Epicharmus, as chief representative of philosophical comedy, along with Protagoras, Heraclitus, Empedocles, and Homer, as the intellectuals who contradicted Parmenides by claiming that all things are in a state of becoming, on the grounds that they constantly move and mix with one another (ἐκ . . . φορᾶς τε καὶ κινήσεως καὶ κράσεως πρὸς ἄλληλα).[146] It is also notable that Plato pokes fun at Epicharmus's fluxist approach to the problem of identity as expounded in the "Growing Argument" in the *Gorgias* (505d4–e3), where, after Callicles sarcastically requests for Socrates to carry on the debate all by himself, Socrates responds: "In that case Epicharmus's saying applies to me: I prove to be sufficient, being 'one man, for what two men were saying before'" (ἃ πρὸ τοῦ δύ' ἄνδρες ἔλεγον, εἷς).[147] The effect on Plato's dialogue of exploiting Epicharmus's convention, in which two men become one, is profound, even if the purpose might remain somewhat obscure.[148] As a chief representative of philosophical comedy, Epicharmus is also known for his linguistic inventions outside Plato's corpus: Aristotle credits him with discovering the aspirated stops φ and χ, "rather than Palamedes," and appending them to the list of eighteen original letters invented by the Phoenicians and brought to Greece by Cadmus.[149] We will want to recall that Socrates in the *Cratylus* places the activity of distinguishing between various letters in order to craft a name proper to the essence of the object being represented. The activity of an expert "name-giver" involves proceeding dialectically from the division (and classification) of letters, an activity that is paralleled

145. Ibid.

146. See Bárány 2006: 318, who usefully comments concerning the debate concerning *ta onta*: "there is also much room for the pair of opposites 'one-many,' yet it seems to be overridden by the distinction of movement and rest."

147. Trans. Zeyl in Cooper and Hutchinson 1997. Epicharmus's original is given by Athenaeus (7, 308c = DK 23 B 16: ἃ πρὸ τοῦ δύ' ἄνδρες ἔλεγον, εἷς ἐγὼν ἀποχρέω). The scholiast on Plato's *Gorgias* (505e1) claims that Epicharmus first brought on stage two speakers who discussed their various positions, but then, later on, only one was on stage, taking up both positions in the conversation all by himself.

148. See Nightingale 1995: 82–83, with nn. 57–59.

149. Arist. F 501 Rose. These letters would probably fall under the category of mixed "intermediates," as described in the *Philebus* (18b6–d2).

by the division (and classification) of entities.[150] Letters are assumed to be the bearers of various properties, and those attributes should optimally be combined in order to reflect the essential semantic content of the *nominatum* under investigation.[151]

Moreover, we can detect an analysis and refutation of (a modified form of) the Epicharmean "Growing Argument" later on in the *Cratylus*, and by reference to the earlier discussions of the "essence of things," which I have argued evinces a concern with mathematical Pythagorean arguments.[152] At *Cratylus* 431c4–d9, Socrates and Cratylus examine whether the "name-giver," who "imitates the essence of things" (*τὴν οὐσίαν τῶν πραγμάτων ἀπομιμούμενος*) by assigning every letter of syllable to an object that is appropriate to it, by contrast with one who "adds or leaves out a little" (*σμικρὰ ἐλλείπῃ ἢ προστιθῇ*) of what is appropriate, is *good* and does his job *well*. Cratylus hesitates to agree here and, after formulating his thoughts, raises a worry concerning Socrates's prescriptive account of "name-giving" of the sort found in the fragments of Epicharmus:

> CRATYLUS: . . . But you see, Socrates, when we assign "*α*," "*β*," and each of the other letters to names by using the craft of grammar, if we add, subtract, or transpose a letter, we don't simply write the name incorrectly, we don't even write it at all, but straightaway *it is a different name* [*ἀλλ' εὐθὺς ἕτερον ἐστιν*], if any of those things happens.
>
> SOCRATES: That's not a good way for us to look at the matter, Cratylus.
>
> CRATYLUS: Why not?
>
> SOCRATES: What you say may well happen to be the case with things that consist of a certain number, which must consist of that number or not consist of it. Take, for example, 10 itself, or any number you like: if you add to or subtract anything from it, it straightaway becomes a different number [*ἐὰν ἀφέλῃς τι ἢ προσθῇς, ἕτερος εὐθὺς γέγονε*]. *But this isn't the sort of correctness that applies to a certain quality, or to images in general. Indeed, the opposite is true of them—an image cannot remain an image if it presents all the qualities of what it imitates.* See if I'm right.

150. With regard to the divisions of letters according to their properties, however, Ademollo (2011: 283) makes a persuasive case that Plato may be referring to Hippias, Democritus, or even Archinus.

151. See Barney 2001: 92–98.

152. The forthcoming examination, concerning Cratylus's version of the Epicharmean "Growing Argument," has already been primed for discussion by Socrates at the beginning of the dialogue, at *Cra.* 393d3–5: "And *even if a letter is added or subtracted* [to/from a name], that too is of no matter, so long as the essence of the thing remains in force [*ἐγκρατὴς ᾖ ἡ οὐσία τοῦ πράγματος*] and is shown in the name."

> Would there be two things—Cratylus and an image of Cratylus—in the following circumstances? Suppose some god didn't just represent your color and shape the way painters do, but made all the internal aspects like yours, with the same warmth and softness, and put motion, soul, and wisdom like yours into them—in a word, suppose he made a duplicate of everything you have and set it beside you. Would there then be two Cratyluses, or Cratylus and an image of Cratylus?
>
> CRATYLUS: It seems to me, Socrates, that there would be two Cratyluses.
>
> (PLATO, *Cratylus* 431e9–432c6; translation after Reeve in Cooper and Hutchinson 1997)

Cratylus's position represents a Sophistic adaptation of Epicharmus's "Growing Argument" to a theory of language: he asserts that if one adds, subtracts, or transposes one constitutive part of a single name, that is, any letter, the word "is straightaway different." Accordingly, the attempt to write the name with different letters has destabilized the identity of that name, and it has therefore become something else.[153] Because the name as applied to the object is different from what it once was, so Cratylus's argument seems to go, the name has not been written at all. It is not a case of not writing the name well, that is, of misspelling, as Socrates maintains; what has been written is simply a *different name.*

Socrates expresses disapproval of Cratylus's take on the "Growing Argument," and he seeks to refute it by appeal to distinguishing those objects whose essential and unique property is numerical from those objects that have properties beyond number and, consequently, can be imitated. Those objects whose essential property is number, and that apparently do not admit of any other properties, are the things that consist of a certain number (ἔκ τινος ἀριθμοῦ εἶναι), that is, "10 itself" (αὐτὰ τὰ δέκα).[154] So it is in arithmetic alone that the "Growing

153. It should be noted that the application of the Epicharmean "Growing Argument" to naming is not original to Cratylus here: it occurs in *Dissoi Logoi* 5 (11–15), written in the fifth century BCE and in Doric, where the anonymous Sophist argues that "changing the arrangement" [ἁρμονίας διαλλαγείσας] constitutes a different thing (ἀλλοιοῦσθαι). His linguistic examples include (1) change of accent or lengthening of vowels, and (2) addition or transposition of letters. Then, he elicits comparison with adding or subtracting in arithmetic, and then finally with growth of a human. Interestingly, the author concludes by asking, with reference to being or not-being, "is the human the same in some way or in all ways?" (τὶ ἢ τὰ πάντα ἔστιν;) If the respondent says that he is not, he is telling a falsehood, if he says "in all ways."

154. Literally, this phrase means "ten things themselves," which helps to make sense of the idea that they are constituted of a certain number. It is interesting that Plato uses what, in the *Phaedo*, will become Form language to refer to the number 10, while qualifying the term by rendering it numerically plural. Is Socrates referring to 10 as a Form? See chapter 5, section entitled "What Is Wisest? Three."

Argument" works, on Socrates's interpretation. In this way, Socrates provides a reason for denying the "Growing Argument" as it had been formulated by Epicharmus, Cratylus (at least as he is represented in this dialogue), and the anonymous author of *Dissoi Logoi* 5.[155] But Socrates needs to provide an explanation for that argument, and he does so through the "Two Cratyluses" thought experiment. This experiment isn't necessarily innovative. A version had been formulated by Stesichorus and Euripides in generations past, with Helen and her phantom as the paradigm.[156] But by the final quarter of the fifth century BCE, those versions of the story had become synonymous with rhetorical *epideixis* and Sophistic challenges to stable identity, both of which worried Plato.[157] Still, as with his treatment of Helen elsewhere in his oeuvre,[158] Plato exploits the "Growing Argument" in order to develop novel arguments concerning metaphysics and identity. In particular, he refutes the essential reduction of an object to its number in the "Growing Argument" in order to establish the conceptual roots of what becomes in the *Phaedo* and the *Republic* a powerful conceptualization to account for the difference between Forms and the particulars named after them: the principle of deficiency.[159] According to Socrates, if an image of something (e.g. a painting, a name, even a sentence) ends up obtaining all the properties of the thing it imitates, *the image no longer remains an image*. This is because an image, by virtue of its deficiency to the thing it imitates, cannot retain its identity if it obtains the very same properties found in the object it imitates. In that circumstance, it would simply become (or be?)[160] what it imitates and would no longer remain an image at all. So if a name actually ended up obtaining all the properties of the thing it is meant to signify, it would no longer be a name.[161] By introducing the principle that names and other imitations are

155. See n. 153 above.

156. See Sedley 2003: 26 n. 44.

157. For the former, see Gorgias's *Encomium of Helen* and Isocrates's *Helen*. For the later, see Euripides's *Helen* (557–596), where the name Helen is mobile and is said to detach itself from its proper body and attach itself to a phantom image. Even Menelaus's identity is problematized in that play.

158. E.g. at *Phdr*. 243a5–b7 and *R*. 9, 586b7–e2, where Stesichorus's palinode of Helen is used as a model for arriving at the truth.

159. *R*. 10, 596a5–598d6 and *Phd*. 74a2–75a3. See Sedley 2003: 44 n. 46 and chapter 5.

160. It is clear that Socrates is intentional when he modifies Cratylus's εὐθὺς ἕτερον ἐστιν to ἕτερος εὐθὺς γέγονε. See Sedley 2003: 103.

161. Cratylus's acceptance of the premise that a name is an imitation is revealed to be a crucial blunder several times throughout the dialogue (e.g. *Cra*. 439a1–4).

naturally deficient to the objects they imitate, Socrates deftly turns the "Growing Argument" on its head. In fact, what Socrates's response to Cratylus reveals is the weakness in Cratylus's theory of naming: had Cratylus simply not accepted Socrates's ontological assumption that names must *imitate* their *nominata*, he would not necessarily be committed to an absurdity.

What is the upshot of this bizarre episode in the dialogue for my analysis of Plato's critical response to mathematical Pythagoreanism? Something remarkable occurs in the process of Socrates's contortion of Cratylus's "Growing Argument": he states that *numbers alone have as their essential property number*. When we think about it, however, this is unsurprising: it is not possible to attribute, for example, sensible qualities to numbers themselves. Moreover, it appears that Plato would like to deny that numbers can be considered images of anything else at all, on the grounds that they only possess the property of number. It is not clear from the argument whether, by virtue of the fact that they are not images, they do not possess deficiency. Still, from this point of view, numbers might start to look like very good candidates for primary entities of some sort, and while Plato does not go further in evaluating numbers here—the chief goal of the *Cratylus*, after all, is to deal with various types of manufactured things, and not with essences—this passage raises the expectation that numbers will play a crucial role when Plato gets around to discussing the highest and most primary sorts of beings. This will be the subject of the next chapter.

CONCLUSIONS

This chapter, then, has sought to show that the ideas of the mathematical Pythagoreans play a significant role in the broader questions concerning ontology and predication of attributes throughout Plato's metaphysical treatments of *οὐσία* in the dialogues that anticipate the more robust treatments in the *Phaedo* and *Republic*. In particular, two positions are distinguished by Socrates in a rich and complex passage of the *Cratylus* (401b11–d7), where he attempts to describe two approaches to etymologizing the term *οὐσία*: the first (1), which shares the strongest affinities with the fragments of Philolaus of Croton, identifies the "essence of things" (*ἡ τῶν πραγμάτων οὐσία*) with the "hearth" (*ἑστία*) and is praised as an anticipation of Plato's proposition concerning the participation of natural objects (including names) in their own peculiar stable essences. The second (2), which ends up underpinning the discussion of the entire dialogue, posits an etymological relationship between *οὐσία* and *ὠσία* and is criticized as a fluxist position that assumes that the lengthening of the *ο* to *ω* indicates incessant motion. Later in the *Cratylus* (431e9–432c6), the Heraclitean metaphysics associated with position (2) presents Plato with an opportunity to evaluate an offspring of Epicharmus's "Growing Argument," when Cratylus develops a linguistic version of the "Growing Argument" in which the identity of names is challenged when there is addition, subtraction, or transposition of letters. There, Socrates attacks the "Growing Argument" as it has been developed by Cratylus (and others) by arguing that number is only an essential quality for numbers themselves, whereas names, which are sorts of imitations, are subject to many more predicates than number. Unlike numbers, names lose their identity as imitations if they obtain all the properties of the things they imitate. The upshot of Socrates's challenge to Cratylus's "Growing Argument," then, is that numbers cannot be imitations of other things. What class of objects numbers constitute, however, remains undiscussed in the *Cratylus*.

If it is the case that on the one hand Philolaus is extolled for his contribution to Plato's development of the theory of participation of perceptibles in the Forms, as is suggested by (1), and on the other hand Epicharmus is challenged and thrown together with the other Heraclitean fluxists (2), such as Democritus, Protagoras, Empedocles, Orpheus, and Homer, what is the status of mathematical Pythagoreanism in Plato's *Cratylus*? We can conclude that Plato does not classify mathematical Pythagoreans as a single, unified group in the early and early-middle dialogues. He seems less to worry about providing historians of philosophy with doctrines properly assigned to his antecedents than he is concerned to distinguish the doctrines from one another according to whether they identify the proper modes of existence or confuse them. Philolaus of Croton, although he is not named in the *Cratylus*, seems to fall on the side of Parmenides by undergirding his cosmogony with a stable essence that guarantees the existence and knowability of objects. By contrast, other mathematical Pythagoreans, such as Empedocles and Epicharmus, appear to be in a large company of fluxists who deny stable existence entirely and whose theories cannot support claims to knowledge, correct naming, or determination of objects.[162] Concern with the generation of the universe and with the state of change of natural objects can be detected throughout Plato's entire philosophical work, and from this point of view, Plato does not really come off more "Pythagorean" than "Heraclitean" in the *Cratylus*. But even if Plato was not precise in his documentation of the individuals who influenced his ways of thinking about the problems of ontology and predication—he was not much of an autobiographer, after all—and continued to busy himself with the problems of flux and becoming, we can detect one important way the *Cratylus* presents us with a specific critical response to Pythagoreanism. Adapting a suggestion of David Sedley, I propose this possible explanation for the project of that dialogue: Plato's *Cratylus* constitutes the most comprehensive surviving

162. See *Cra.* 439c6–440d7, on which see, inter alia, Horky 2011: 154–155.

analysis of the *τί μάλιστα acusma*, as reported by Aristotle, which responds to the question "What is second-wisest?" (*τί τὸ δεύτερον σοφώτατον;*), to which the answer is "What assigned names to things" (*τὸ τοῖς πράγμασι τὰ ὀνόματα τιθέμενον*).[163] If I am right about this connection, it prompts investigation into whether Plato undertakes to respond to or evaluate other Pythagorean *acusmata*. Of course, this sort of inquiry is a fundamental characteristic of what scholars tend to recognize as the *Socratic* dialogues of Plato: an investigation of the basic definitions of concepts that need to be understood in order to live an ethically good life. Plato's definitional project in the Socratic dialogues, then, would appear to show affinities with the sorts of demonstrations the mathematical Pythagoreans undertook in their attempts to explain the doctrines of Pythagoras. Signs that the Socratic project of definitional inquiry might be related to Pythagoreanism can be detected not only in the Socratic dialogues of Plato but even in Aristophanes's lampooning of his "Presocratic" definition of thunder in the *Clouds*.[164] In the next chapter, I seek to expand our understanding of Plato's approaches to the Pythagoreanism by analyzing the *Phaedo* with regard to the most metaphysically striking *acusma* that survives: "What is wisest? Number" (*τί τὸ σοφώτατον; ἀριθμός*).

163. Sedley 2003: 25. These *acusmata* are reported by Iamblichus (*VP* 82, 47.17–18), probably derived from Aristotle, but also see *VP* 56, 30.20–31.6, which is more elaborate and is linked to information derived from Timaeus of Tauromenium (FGrHist 566 F 17).

164. Aristoph. *Nub.* 403–407. On Socrates and Pythagoreanism in the *Clouds*, see Burkert 1972: 291 n. 73, Bowie 1998: 60–65, and, more recently, Rashed 2009.

5

What Is Wisest?

Mathematical Pythagoreanism and Plato's *Phaedo*

In the previous chapter, I examined the philosophical ideas attested in the extant fragments of and *testimonia* concerning the mathematical Pythagoreans Epicharmus, Empedocles, Philolaus, and Eurytus, with an eye to how these intellectuals appealed to abstract ways of thinking about "number" in their bids to provide explanations for the Pythagorean *acusmata*. I also argued that in the *Cratylus*, the earliest dialogue of Plato that exhibits dedicated critical responses to the philosophical positions of these mathematical Pythagoreans, the ways of thinking about "number" of Philolaus and Epicharmus were taken to represent two competing metaphysical positions. Philolaus, who believed that the chief proposition by which we are to deduce all knowledge of divine things is the claim that the primary entities, from which all knowable things are derived, *must* have preexisted *if* those knowable things are to be knowable, was taken as an antecedent to Plato's argument that ontological stability is a necessary condition for the possibility of knowledge. Epicharmus, by contrast, was taken as a representative of the Heraclitean fluxism employed by Cratylus in his own linguistic version of the "Growing Argument." Thus, I argued, Plato's response to Pythagoreanism in the *Cratylus* does not present the Pythagoreans as a unified group—he is rather more tasked with evaluating their claims individually, and as they accord with his own philosophical commitments. Moreover, I suggested that Plato maintained the anonymity of those Pythagoreans whose philosophical positions he was evaluating in the *Cratylus* by referring to them obliquely as "name-givers."[1]

While the ideas of the mathematical Pythagoreans Epicharmus and Philolaus concerning number, identity, and priority of being played a significant

1. It should be added that in the *Cratylus* (e.g. *Cra.* 388e7–389a3) the "name-givers" are identical with the "law-givers," a term that was apparently used by Aristoxenus to refer to the Pythagoreans who supported democracy against aristocracy, and of which Archytas was a chief proponent (T A 9 Huffman). See Huffman 2005: 317.

role in the "ontological subplot" of the *Cratylus*, the mathematical Pythagoreans themselves do not appear as characters in the dialogue. As I will argue in this chapter, however, Plato did take on mathematical Pythagoreanism less indirectly in the *Phaedo* by developing and challenging philosophical positions resembling those of the mathematical Pythagoreans Epicharmus, Empedocles, and Philolaus. I will try to show that this dialogue also features an "ontological subplot" that, in particular, investigates the essence of numbers. Like the treatment of names in the *Cratylus*, the analysis of numbers in the *Phaedo* is a secondary project that exists for the sake of another, primary objective, which in this case is the development of an argument for the immortality of the soul. As I will show, though, the inquiry into the ontological status of numbers, and the ways they admit of some properties but not others, plays a crucial role in the formalization of Plato's metaphysics and constitutes our best evidence of what Aristotle would go on to describe in *Metaphysics* A as the intermediary "objects of mathematics." By challenging Empedocles's version of the "Growing Argument," in which alternation of the cycles of the universe constitutes an alteration from one to many, and by appropriating Philolaus's terminology and conceptualizations concerning subsistent entities and the necessary conditions for knowledge of objects in the universe, Plato is able to formulate a new theory of number that both accounts for the ways numbers inhere in things that are subject to generation and lays the groundwork for the demonstration of the soul's immortality. Thus, Plato once again pits the philosophical tenets of Philolaus against those of number-stuff theorists such as Epicharmus and (especially) Empedocles. But this contrapositioning plays an important role, I suggest, in elucidating Plato's philosophy more generally and, in particular, reveals the way his metaphysics seeks to supersede mathematical Pythagorean ideas concerning the nature of number.

PLATO'S *PHAEDO* AND MATHEMATICAL PYTHAGOREANISM: SETTING THE STAGE

Before moving onto analysis of Plato's treatment of mathematical Pythagoreanism in the *Phaedo*, there is one more passage worth discussing from Plato's *Cratylus*. The relevant passage seeks to investigate how Zeus's name itself obtains philosophical meaning when etymologized:

> As for myself, Hermogenes, because I persisted at it, I found out about all these things in statements-not-to-be-divulged [ἐν ἀπορρήτοις]—since that through which [δι' ὅ] a thing comes to be is the cause, it is the just and the cause. Indeed, someone told me that it is correct to call this Δία [Zeus]

> for that reason [διὰ ταῦτα]. Even when I'd heard this, however, I persisted in gently asking, "If all this is true, my friend, what in the world then *is* just [τί οὖν ποτ' ἔστιν δίκαιον]?"
>
> (PLATO, *Cratylus* 413a2–8; translated after Reeve in Cooper and Hutchinson 1997)

What is striking about this passage, which occurs in the midst of a discussion about how to define the just (τὸ δίκαιον), is the fact that it exhibits a curious entanglement of various things associated with mathematical Pythagoreanism by Plato's contemporaries and immediate successors: demonstration, the mystic teachings, and speculation about causation.[2] Immediately thereafter, Socrates mentions that the group of people who assume an etymological relationship between Zeus (Δία) and demonstration (δι' ὅ or διὰ) are critical of him because he asks too many questions.[3] Socrates then claims of these people that they try to give several responses to his definitional question and, with what seems to be a notable example of Socratic wit, claims that "they no longer are in concord" (οὐκέτι συμφωνοῦσαν) with one another. Obviously, the play is more effective if Socrates is criticizing some unnamed Pythagorean music theorists, and the only explicit reference to the Pythagoreans in Plato's corpus, in *Republic* 7 (530d7–9, 531a1–3 and b2–c4), explicitly associates their philosophical activity with relative measurement of audible concords.[4] But we are on safer ground if we assume that in the *Cratylus*, Plato is after the aforementioned Heraclitean flux theorists, who, as I argued in chapter 4, included mathematical Pythagoreans such as Empedocles and Epicharmus. These figures, now separated by their

2. Similar classical usages of the term ἀπόρρητα with the normative meaning "not-to-be-spoken of" that can be found in Athenian popular culture and by reference to the mysteries include Herodotus (9.45 and 9.94), Aristophanes (*Eq.* 648, *Th.* 363, *Ra.* 362), Euripides (*IT* 1331, *Hipp.* 293, *Rh.* 943, by reference to Orpheus, who "showed" [ἔδειξεν] the ἀπόρρητα to others; see Aristoph. *Ra.* 1032) and Andocides (2.19), and Plato himself (*Theaet.* 152c8–11, *R.* 2, 377e6–378a6). Closer to the project of the *Phaedo*, however, is the Derveni Papyrus (Col. VII), which interestingly remarks in the context of the encoded philosophical cosmology of Orpheus that "it is not possible to state the meaning/solution [[λύ]σιν] of the words [of Orpheus] even when they have been spoken [ῥηθέντα]." See Tsantsanoglou 1997: 121.

3. Records of the name of the sky god are attested as early as Pherecydes of Syros (DK 7 B 1), where we hear of the names Δίς, Ζήν, Ζάς, Δήν, and Ζής. Hesiod (*WD* 2–3) seems to have played with the meaning of Zeus as διὰ as well. On the names in the cosmogony of Pherecydes, see KRS 56–59. It is relevant that Timaeus of Epizephyrian Locri in the *Timaeus* (41a7) has the demiurge etymologize his own name as the "work done through me" (ἔργων δι' ἐμοῦ); Plato, then, does not seem to reject this strategy of etymologizing the primary god. Later on, the early Stoics would adopt the model of Zeus (Δία) as cause (δι' ὅν), e.g., at D.L. 7.147 (= *SVF* 2.1021).

4. See Barker 2007: 19–26.

various approaches to etymologization, are also distinguished by their positions with regard to identifying "the just" cosmologically with other things. Four possible identifications are listed: (1) the sun, (2) fire, (3) heat, and (4) mind, which might or might not be historical. As Sedley points out, there is no evidence among Presocratics for expressly associating justice with (1) the sun, but (2) fire could easily refer to Heraclitus himself.[5] Still, that won't do if we're after a Heraclitean, that is, someone who may have adopted Heraclitus's fluxism and identified fire with justice. Hippasus of Metapontum, considered the author of mathematical Pythagoreanism by Aristotle, was coupled with Heraclitus by Theophrastus, and seems to have held both that fire was the first principle and that "things are made from fire by condensation and rarefaction, and are resolved into fire again, since this is the single underlying nature."[6] This description could be taken to refer to a type of natural justice, on the model of something like Anaximander's description of things in opposition "giving justice and paying recompense to one another for their injustice," which was associated with generation and destruction of things by Theophrastus.[7] And it was the case that within a generation of Theophrastus, the Stoics took up the cosmic conflagration from Heraclitus or, we might add, possibly other Heracliteans such as Hippasus. Thus, we cannot discount Hippasus as a possible referent for a Heraclitean who associated justice with (2) fire, although there is no firm evidence for it either. The case with (3) heat is equally problematic, but Archelaus comes to mind; and (4) seems relatively straightforward, since Anaxagoras is cited explicitly by Socrates.[8]

Importantly, however, we are dealing with something rather typical in Plato's response to mathematical Pythagoreanism, or so I would argue, in this passage: the association of mathematical Pythagoreans with other Presocratic natural scientists, including Heraclitus and Anaxagoras, who are the easiest to identify. Just as the *Cratylus* took to task those Pythagoreans (among others) whom Plato implicated in the problems raised by Heraclitus's approach to ontology and epistemology, so, too, as we discover, does the *Phaedo* take to task those Pythagoreans whom Plato implicated in the problems raised by Anaxagoras's physics. And for once it seems

5. See Sedley 2003: 116–117. For a very good analysis of Heraclitus's fragments and the Derveni Papyrus, see Betegh 2004: 343–348.

6. DK 18 F 5 = Arist. *Metaph.* 1.3, 984a7–8 and Theophrastus F 225 FHS&G = Simpl. *in Phys.* pp. 23.21–24.12 Diels. See chapter 2 for a fuller analysis of this testimony and the relationship between Hippasus and Heraclitus.

7. DK 12 B 1, A 9 and A 9a. Of course, Anaximander was said to have eschewed the debate concerning which of the four elements was primary, on the grounds that, for him, the boundless was the source from which everything derived.

8. Some unnamed Pythagoreans are said to have equated mind with "the one," whereas they associated justice with the number 4, according to Aristotle (F 203 Rose = Alex. *in Metaph.* p. 38.10–39.18 Hayduck).

that Plato is actually willing to name those Pythagoreans whose philosophical claims and methods he aims to evaluate: Philolaus of Croton, Simmias and Cebes of Thebes, and finally Echecrates of Phlius, to whom the entire dialogue is narrated.[9] Some scholars have expressed skepticism concerning whether these figures are to be identified with Pythagoreanism. For example, Christopher Rowe suggests that, while Simmias and Cebes are said to have associated with (συγγίγνεσθαι) or heard (ἀκούειν) Philolaus, they should not be considered members of the Pythagorean "school."[10] As I have argued in the first half of this book, however, by the time of Plato there was no single, unified Pythagorean "school" but two groups differentiated by on the one hand political ideology and on the other philosophical methodology—by Timaeus of Tauromenium and Aristotle, respectively. The splinter mathematical Pythagoreans were said by Timaeus to have published in writing the oral doctrines of Pythagoras, and, according to Aristotle, this entailed attempts to develop crude demonstrations of esoteric axioms.[11] Simmias and Cebes, along with Echecrates and Philolaus (for whom the classification as mathematical Pythagorean is not controversial), may be thought to represent mathematical Pythagoreanism more broadly conceived in Plato's *Phaedo*.[12]

A closer examination of the dialogue frame of the *Phaedo* partially confirms these suspicions. Simmias and Cebes appear to be mathematical Pythagoreans who play a special role in the dialogue: to exemplify and even evaluate various faulty Pythagorean approaches to "demonstration" by comparison with Socrates's distinctive approach.[13] Cebes complains at the beginning of the dialogue that the arguments he has heard from Philolaus and "some others" were "not clear" (οὐδὲν σαφές), by which he seems to mean that Philolaus's arguments were not *demonstrated precisely* in accordance with a rigorous philosophical methodology.[14] Such a criticism of the Pythagoreans also recurs in Plato's *Republic* 7, when Socrates complains that the Pythagoreans, in their pursuit of

9. On Echecrates, see chapter 3, section entitled "Pythagorean Exoterics in the Fifth Century BCE? The Historical Evidence of Timaeus of Tauromenium."

10. Rowe 1993: 7. References are to *Phd.* 61d6–7.

11. See chapters 1 and 3.

12. See Burkert 1972: 198.

13. See Bluck 1955: 6–7, who refers to Simmias as having "broken away from orthodox Pythagorean doctrine," and more recently Morgan 2010b: 64, who argues that in the *Phaedo*, Plato "draws heavily on imagery of religious revelation and initiation, including many Pythagorean elements . . . [however Plato's philosophy] does not merely appropriate these models but transforms them." Also see Zhmud (2012: 416), who aptly remarks: "it is [Socrates] who explains to them the difference between even and odd as such and specific numbers."

14. This phrase is reiterated for emphasis (*Phd.* 61d8, 61e8–9), and it is associated with written (rather than oral) instruction in the myth of Theuth and Ammon in the *Phaedrus* (275c5–d2), on which see chapter 6. See Huffman 1993: 409.

the numbers that correspond with the audible consonances, "do not ascend to problems" (οὐκ εἰς προβλήματα ἀνίασιν) and fail to obtain the "reason why" (διὰ τί) numbers should be classified as concordant or discordant, which, so Socrates argues, is the proper activity of dialecticians.[15] To return to the *Phaedo*, Socrates playfully elaborates on the problem of Pythagorean demonstration by invoking something approximating the Pythagorean *acusmata*, and by reference to the language of the mysteries, as I showed above with the etymologization of Zeus in the *Cratylus*:

> There is a saying concerning these issues in the things-not-to-be-divulged [ἐν ἀπορρήτοις λεγόμενος περὶ αὐτῶν λόγος], that we men are in a prison [ἔν τινι φρουρᾷ ἔσμεν], and, moreover, that one ought not to free oneself from it nor run away [οὐ δεῖ δὴ ἑαυτὸν ἐκ ταύτης λύειν οὐδ' ἀποδιδράσκειν]; that seems to me to be an impressive saying, and one not easy to comprehend fully [διιδεῖν]."[16]
>
> (PLATO, *Phaedo* 62b2–6)

Socrates's response to Cebes is noteworthy for two reasons. First, it dramaturgically links the staging and subject of the dialogue, Socrates's current situation (arrested and awaiting his death in a prison) and forthcoming exhortation not to fear one's fate (death, which is the soul's "freeing itself" of and "running away" from the body), to explication of the unspoken mysteries (ἐν ἀπορρήτοις). Second, it may provide early evidence—surely distorted for Socrates's own purposes—of two Pythagorean *acusmata*, one involving a definition of the "human" condition (of the τί ἔστι type) and another prescribing what one should not do (of the τί πρακτέον type).[17] It is not clear whether Philolaus

15. Pl. *R.* 7, 531c1–4. Compare also *R.* 6, 511a4–9.

16. For a similar usage of διιδεῖν, see Pl. *Phdr.* 264c1. The appeal to ἀπορρήτα might be thought to distinguish the wisdom-traditions of the Pythagoreans from those ascribed to the Seven Sages, which do not speak enigmatically and are apparently chiefly concerned with sayings of the τί μάλιστα type. See Burkert 1972: 169.

17. There can be no doubt that the saying is Orphic (see Pl. *Cra.* 400b11–c10), but it seems to have Pythagorean connotations as well. See Rowe 1993: 128 and Peterson 2011: 170. We might also note that the attempt to define rational living beings according to a tripartite division (god, human, and Pythagoras) is described by Aristotle (F 192 Rose = Iambl. *VP* 31, 18.12–16) as being ἐν τοῖς πάνυ ἀπορρήτοις. The earliest possible evidence of rationalized Pythagorean *acusmata* that I can identify is associated with Prometheus in the Ps.-Aeschylean *Prometheus Bound* (459) and with Socrates in Aristophanes's *Clouds* (τί ἔστι definition of thunder, which terrifies Strepsiades, at 404–407; compare Arist. *APo.* 2.11, 94b33–34 as against Anaximander DK 12 A 23 and Anaximenes DK 13 A 17), both likely composed in the 420s BCE. This date would correspond with the arrival of Philolaus and the other Pythagoreans expelled from Italy at mainland Greece. On the possible Pythagoreanism of Socrates in the *Clouds*, see especially Rashed 2009.

himself actually laid claim to "the body is a tomb of the soul" or all things are encompassed by God "as if in a prison" (F 14–15 Huffman) as Pythagorean doctrine. But this concern misidentifies the function of the mystically inspired "saying" in the *Phaedo*, which is ensconced in a broader discussion of *explanation through philosophical argumentation*. As Kathryn Morgan writes, Socrates's dissatisfaction with the mystic secrecy associated with (a particular type of) Pythagoreanism stems from his hope to initiate Simmias and Cebes into philosophy, where "the need for open conversation is paramount."[18] Philosophical dialectic is intended to supplant mystic doctrine, and explanation is key to advancing beyond ipse dixit injunctions. It is possible that Philolaus, as a mathematical Pythagorean, employed a Pythagorean *acusma* in order to demonstrate *why* one ought not to commit suicide, but that he failed to do so with precision and clarity, at least in the eyes of Plato.[19]

The normative aspect of Socrates's description of the "saying" preserved in mystical words is directed at Cebes, who is associated with conviction and piety throughout the *Phaedo*. Despite his characteristic argumentativeness concerning claims made by Socrates and Simmias,[20] Cebes appears to be a mathematical Pythagorean, but with some acousmatic leanings, and after being persuaded by the final argument he no longer requires teaching concerning the justice of the afterlife.[21] In this way, he resembles Socrates, who extols piety and reason, which are not bifurcated in Socratic philosophy.[22] Cebes stands in contrast to Simmias, who at the end of the dialogue portion of the *Phaedo* retains Philolaic skepticism about the arguments undertaken as a whole, on the grounds that he still has a low opinion of human understanding.[23] Socrates's exhortation to Simmias at the end of the dialogue, to go back over the arguments of the day (*Phd.* 107b4–9) until he is satisfied that the whole argument becomes "clear" (σαφές), recalls Cebes's description of Philolaus's argument for *why* one ought not to commit suicide at the beginning of the dialogue and consequently marks the difference between the arguments of Socrates and Philolaus.[24] Thus, from the

18. Morgan 2010b: 76.

19. See Morgan 2010b: 72–73. Note that, irrespective of how the content of Philolaus F 15 has been modified (from Athenagoras, *Legatio* 6), the emphasis remains on demonstration (δεικνύει).

20. See Bluck 1955: 35.

21. Pl. *Phd.* 107a1–3.

22. E.g. Pl. *Apol.* 21b1–c8 and, more playfully, *Phd.* 60e4–61b7. See Sedley 1995: 18–20.

23. As acutely noted by Sedley (1995: 20). There may be a verbal echo of Philolaus F 6 (ἀνθρωπίνην γνῶσιν) at *Phd.* 107a8–b3 (ἀνθρωπίνην ἀσθένειαν).

24. Compare also Pl. *Phd.* 101a7–8.

first steps of initial inquiry until the mythical departure to the other world, the *Phaedo* is staged as a critical response to Pythagorean scientific methodology and, as I will show, underlying assumptions they held concerning numerical essence and identity, and predication of numerical properties.

WHAT IS WISEST? THREE

It is impossible to undertake a comprehensive assessment of the *Phaedo*'s concerns with Pythagorean philosophical methodology in this chapter. Such an analysis would require an exhaustive examination of how Simmias and Cebes respond to each and every Socratic argument, whereas it is the goal of this chapter to examine, in particular, how Plato responded to mathematical Pythagorean theories of number. In this light, it will be useful to compare some of Socrates's claims concerning the determination of the "being" (*οὐσία*) of the soul and the mode of demonstration of that determination in Plato's *Phaedo* against the propositions concerning identity of objects advanced by Epicharmus, Empedocles, Philolaus, and Eurytus. Recall from the previous chapter that Plato's engagement with mathematical Pythagoreanism in the *Cratylus* was couched in an evaluation of the etymology of *οὐσία*; in particular, Socrates was concerned there with whether or not the etymologization of *οὐσία*, obtained in the dialectic variants *ἐσσία* and *ὠσία*, faithfully reflected the properties of the "essence of things" (*ἡ τῶν πραγμάτων οὐσία*).[25] I also argued that Socrates praised Philolaus's activity of analogizing *οὐσία* with *ἑστία*, on the grounds that it anticipated Plato's own theory of participation in essence (*τὸ τῆς οὐσίας μετέχον*) and that it exhibited the proper order of sacrifice to the gods.[26] By contrast, I showed that Socrates criticized Cratylus's linguistic version of Epicharmus's "Growing Argument," on the grounds that it only applied to things whose essential and, indeed, only property is number (i.e. numbers) and not to things that imitate the essences of other things, such as names.[27] By refuting Cratylus's version of the "Growing Argument," Plato also stages an attack on Epicharmus's "Growing Argument," implicating Epicharmus's basic claims in Cratylus's linguistic version and effectively seeking to do away with both. But Plato's *Cratylus* did not advance a detailed *positive* account either (1) of the role of Philolaus's ideas in determining an ontological

25. Pl. *Cra.* 401b11–d3.

26. Compare the proper way to obtain piety in the *Laws* (717a6–b2): the Olympian gods are to be honored "superior and contrasting [*ἀντίφωνα*] honors like the Odd," whereas the Chthonic gods are to be allotted "secondary honors as the Even and the Left."

27. Pl. *Cra.* 432a8–c5.

and epistemological theory that could accommodate Plato's metaphysics of the Form, or (2) of the role of numbers in that innovative theory. Development of this positive account, I suggest, was one of the purposes behind the ontological subplot of the *Phaedo*.

Stephen Menn has recently argued that Socrates's criticisms of the natural scientists somehow naturally lead him to the "Growing Argument" of Epicharmus.[28] We might add that the entire discussion of pursuit of the wisdom inherent in natural science is framed narratively by the problems of addition and subtraction,[29] and when Socrates has concluded aporetically about the material causes of becoming and being as well as the "reasons" (διὰ τί)[30] for these, it is the introduction of the principle of participation in "the one" and "the two" that paves the road for development of a method that proceeds from a best hypothesis.[31] And all of this is needed in order to make it possible to advance an argument concerning the immortality of the soul, which stands to refute Cebes's claim that the soul, once it has entered the body, begins to degenerate and finally dies with the body. So, insofar as Plato's criticism of Presocratic and Hippocratic material causes takes the form of a version of the "Growing Argument," as Menn has persuasively argued, it employs what I consider mathematical Pythagorean strategies for explanation—surely criticized as insufficient for the task at hand in the *Phaedo* but important for this book's study of the role of mathematical Pythagoreanism in Plato's metaphysics.

It becomes relatively clear that what Socrates brings in tow for his challenge to material causation is not simply a Sophistical variant of the "Growing Argument" of the sort employed by Cratylus or the writer of *Dissoi Logoi* 5 but a variant embedded in mathematical Pythagorean speculation about the ontological status of "number."[32] Indeed, Socrates's confusion about the problems posed by Anaxagoras's and other natural scientists' theories of causation is

28. Menn 2010.

29. After restating Cebes's position concerning the mortality of the embodied soul, Socrates playfully adds: "This, I think, is what you say, Cebes; I deliberately repeat it often, in order that nothing escape us, and that you may add or subtract something [προσθῇς ἢ ἀφέλῃς] if you wish" (*Phd.* 95e1–4).

30. See Pl. *Phd.* 96a6–9.

31. Pl. *Phd.* 101d1–e2. I do not have space to develop a thorough analysis of the method of hypothesis expressed here.

32. In the *Theaetetus* (154e7–155d4), the "Growing Argument" is presented as the fundamental basis for *self-examination* for Socrates and the young mathematician-philosopher Theaetetus.

ensconced in a broader concern over the status of number, numbers, and their properties:

> "And what do you think now about those things [i.e. the largeness of comparative lengths]?"
>
> "That I am, by Zeus, far from believing that I know the cause of any of those things. I will not even allow myself to say that, when one is added to one, either the one to which it is added becomes two, or the one added and the one to which it is added become two because of the addition of the one to the other. For I would be surprised if, while each of them was separate from the other, each of them was one, and they were not then two, but, when they came near to one another, this should be the cause of their becoming two: the conversion resulting from their being placed near to one another [ἡ σύνοδος τοῦ πλησίον ἀλλήλων τεθῆναι]. Nor can I any longer be persuaded that whenever someone divides one thing, the division is[33] the cause of its becoming two [ἡ σχίσις, τοῦ δύο γεγονέναι], for just now the cause of becoming two was the opposite [i.e. the coming together resulting from being placed near to one another]. In the former instance, it was because they were brought close together and one was added to the other, but now it is because one is taken and separated from the other.
>
> Nor still can I any longer persuade myself that I know why [δι' ὅτι] a unit comes to be, nor anything else either, nor, in a word, why [δι' ὅτι] [anything] comes to be or perishes or exists—by *this* style of procedure, that is [κατὰ τοῦτον τὸν τρόπον τῆς μεθόδου]; I have a confused style of my own, but *this* style just isn't for me. One day I heard someone reading, as he said, from a book of Anaxagoras."
>
> (PLATO, *Phaedo* 96e5–97c1; translation after Grube in Cooper and Hutchinson 1997)

This passage illustrates what I would call a number-stuff theory, in the sense that Socrates characterizes the theories of collection and division of unities in terms of the "stuff" of the universe, especially the "convergence" (ἡ σύνοδος) of stuff-units through approximation, a concept that only reappears in Plato's works in the *Timaeus* (58b4–5; also see 61a3–6), where it is used in reference to the contracting process that occurs when smaller elemental parts of bodies (such as fire) are squeezed into the gaps left by larger bodies

33. Literally, "became" (γέγονεν). Socrates is characterizing this arithmological inquiry into causation in terms of "becoming," probably in line with the fluxist number theorists whose positions he seems to be characterizing.

(such as water).[34] Given the fact that Socrates is associating this notion of elemental "convergence" with the number-stuff theorists' method of explanation for how units become two, it might be worth inquiring as to whether he is referring to theories of mathematical Pythagoreans.[35]

The connection shared by this passage in the *Phaedo* and the description of the "convergence" as a vehicle for alteration of elements in the universe in the *Timaeus* points to the cosmic cycle of Empedocles's "double" tale:[36]

I shall tell a double tale. For at one time [they] grew to be one alone
From many, and at another, again, [they] grew apart to be many from one.
And double is the coming-to-be of mortal things, double is the waning,
For the convergence of all [of them] gives birth to and destroys the one,
While the other, as [they] again grow apart, was nurtured and flew away.
And these things never cease from continually alternating,
At one time all things coming together into one by Love,
And at another time again each being borne apart by Strife's enmity.
<Thus insofar as they have learned to grow as one from many>
And they finish up many as the one again grows apart,
In this respect they come to be and have no constant life;
But insofar as they never cease from continually interchanging,
In this respect they are always unchanged in a cycle.

δίπλ' ἐρέω· τοτὲ μὲν γὰρ ἓν ηὐξήθη μόνον εἶναι
ἐκ πλεόνων, τοτὲ δ' αὖ διέφυ πλέον' ἐξ ἑνὸς εἶναι.
δοιὴ δὲ θνητῶν γένεσις, δοιὴ δ' ἀπόλειψις·
τὴν μὲν γὰρ πάντων σύνοδος τίκτει τ' ὀλέκει τε,
ἡ δὲ πάλιν διαφυομένων θρεφθεῖσα διέπτη.
καὶ ταῦτ' ἀλλάσσοντα διαμπερὲς οὐδαμὰ λήγει,

34. It is also notable that a similar kind of "division" (ἡ *σχίσις*) as such plays a significant role in the demiurge's cosmology in the *Timaeus* (35b3–36c5). After an initial division of the Being-Becoming stuff and filling in portions according to the ratios of the tetrachord—an integrative activity—he slices (*σχίσας*) again the compound along the length, juxtaposes the centers of the new strips like an *X*, and bends them back into circles, in order to create the inner and outer circles of the world-soul.

35. This unusual employment of the term *σύνοδος* for material explanations also occurs in one of the Pythagorean *acusmata*, preserved by Aelian (*VH* 4.17), in addition to the *acusmata* concerning what is wisest and second-wisest: "What is an earthquake? An assembly of the dead [*σύνοδος τῶν τεθνεώτων*]." See Burkert 1972: 185.

36. It is worth noting that Plato (*Phd.* 96b1–2) has just had Socrates refer to Empedocles's theory of the development of living creatures from the putrefaction that results from the conjunction of the hot and cold.

ἄλλοτε μὲν Φιλότητι συνερχόμεν' εἰς ἓν ἅπαντα,
ἄλλοτε δ' αὖ δίχ' ἕκαστα φορεύμενα Νείκεος ἔχθει.
<οὕτως ἧι μὲν ἓν ἐκ πλεόνων μεμάθηκε φύεσθαι>
ἠδὲ πάλιν διαφύντος ἑνὸς πλέον' ἐκτελέθουσι,
τῆι μὲν γίγνονταί τε καὶ οὔ σφισιν ἔμπεδος αἰών·
ἧι δὲ διαλλάσσοντα διαμπερὲς οὐδαμὰ λήγει,
ταύτηι δ' αἰὲν ἔασιν ἀκίνητοι κατὰ κύκλον.

(EMPEDOCLES, DK 31 B 17.1–13; translation after Inwood 2002)

Empedocles's engagement with the now much-discussed issues of the "Growing Argument" takes the form of a cosmological description. The "double tale" distich, which plays the role of something like a formulaic refrain in Empedocles's poem(s), obtains the qualities of a gnomic wisdom statement, in that it seems broadly applicable to many diverse topics treated by Empedocles.[37] Here, it is applied apparently to the life cycle of "mortal things," which is further explained by appeal to a process of endless alternation (*ἀλλάσσοντα διαμπερές*; *διαλλάσσοντα διαμπερές*).[38] As I showed in chapter 4, it is the concept of eternal "natural exchange" (*μεταλλάσσει κατὰ φύσιν*; *ἐν μεταλλαγᾶι δὲ πάντες ἐντὶ πάντα τὸν χρόνον*) that characterizes Epicharmus's version of the "Growing Argument," and there is good reason to see Empedocles's "double tale" in the light of this puzzle.[39] Of course, two aspects of Empedocles's cosmology can be distinguished from Epicharmus's "Growing Argument" here. First, Empedocles appeals to the "convergence" (*σύνοδος*) of all mortal things into a unity, which apparently explains one aspect of the more abstract "growing" (*ηὐξήθη*) found in the gnomic "double tale" distich.[40] Second, unlike Epicharmus and more in the vein of the Eleatics, Empedocles denies alteration to mortal things (*ἀκίνητοι*) at least insofar as they are fixed in the immortal

37. It might be relevant that Aristotle complained about the "numerous repetitions" (*τὸ κύκλῳ πολὺ ὄν*) of Empedocles's style (*Rh.* 3.4, 1407a35–b2 = DK 31 A 25).

38. I take no certain position on whether the "refrain" lines of Fragment 17 (i.e. lines 1–2) were placed at the beginning of the poem(s).

39. DK 23 B 2.

40. The Strasbourg papyrus now also supplies further explanation of one of the modalities of "convergence" of mortal things. Empedocles declares (DK 31 B 17.35/a (ii), lines 21–30) that he will "show" (*δεί]ξω*) before his audience's very eyes "the convergence and unfolding of life" (*ξύνοδον τε διάπτυξιν τ[ε γενέθλης*) by appeal to various mortal creatures, including beasts, humans, and plants. The appeal to demonstration here—he even claims that his myths will supply "true proofs" (*ἀψευδῆ δείγματα*)—is consistent with Aristotle's (and our) understanding of mathematical Pythagoreanism, and it is also worth noting that Empedocles appeals not only to the ears but also to the eyes of his listener, in the demonstration.

cycle.[41] It is also surprising that "growth" is the engine both for coming-to-be more than one (διέφυ) and, perhaps somewhat counterintuitively, for coming-to-be one (ηὐξήθη).[42] Thus, on Empedocles's interpretation, it is "growth" that facilitates alternation from one to many, and vice versa; conversely, the alternation of one and many as a motif in these lines leads to, in Simon Trépanier's words, "a progressive increase in detail with each successive deployment of the motif."[43] In this way, the poem's form thematizes the eternal recurrence of unification and division that it seeks to outline in its content.[44] Growth and expansion are the final consequences of all natural processes.

As has been argued by a number of scholars,[45] Empedocles's cycle of alteration seems to pursue *reconciliation* of being and becoming, an interpretive position that was (to my knowledge)[46] first asserted by the Eleatic Stranger in Plato's *Sophist*:

> Our[47] Eleatic tribe, starting from Xenophanes and even people before him, goes through in detail in their myths how what they call "all things" are really one [ὡς ἑνὸς ὄντος τῶν πάντων καλουμένων]. Later on, some Ionian and Sicilian muses had the idea that the safest way was to weave the two views together. They say that What-Is is both many and one, and is held together by enmity and friendship. According to the more highly strung of these muses, "by being at variance it ever agrees." The less highly strung muses, though, relaxed the assertion that things must always be this way. They say that everything alternates [ἐν μέρει . . . εἶναι φασι τὸ πάν] and that sometimes it's one and friendly under Aphrodite's influence, but at other times it's many and at war with itself because of some kind of

41. See O'Brien 1969: 76–79. Also see Barnes 1980: 309, who argues, by reference to B 21.13, that the four roots "become different in aspect"; but that seems too narrow an interpretation.

42. As noted by Trépanier (2004: 173). Strictly speaking, Empedocles does not deal with the numerical things "one" and "two" in the sense that Plato does.

43. Trépanier 2004: 175.

44. See Graham 1988: 305.

45. See, e.g., the long excursus (with bibliography) of Trépanier (2004: 152–179).

46. Mansfeld (1990: 46–53) tries to argue that Hippias is the local source for Plato's dialectical survey here, but the synthetic element of the Eleatic Stranger's treatment, which focuses on the interweaving (συμπλέκειν) of positions, seems decidedly Platonic in the light of the role that interweaving plays in other parts of his oeuvre (e.g. *Plt.* 309a8–b7 and *Soph.* 259e4–6).

47. Of course, the "our" here is a reference to the Eleatic Stranger's tribe, and does not indicate that the Ionians and Sicilians who are discussed in the next sentence are part of the Eleatic tribe.

strife. It's hard to say whether any one of these thinkers has told us the truth or not.

(PLATO, *Sophist* 242c8–243a3 = DK 31 A 29; translation after White in Cooper and Hutchinson 1997)

The characteristic alteration associated with the "less high-strung" (*αἱ μαλακώτεραι*) muses, that is, Empedocles, is said to interweave Eleatic monism with the more popular fluxism associated elsewhere in Plato's dialogues not only with Heraclitus, Protagoras, and Epicharmus but also with Homer, Hesiod, and Orpheus.[48] What is unique about this report of Empedocles's cosmology is that his invention of cosmic "alternation" (*ἐν μέρει*)[49] is being contrasted to Heraclitus's injunction that everything is both at variance with itself and in agreement with itself *in all circumstances and at all times* (*ἀεί*).[50] Empedocles advances a rather elaborate system for this cycle in Fragment B 26, where the process of alternation is drawn up alongside the modalities of alteration, in a way similar to the "Growing Argument" of Epicharmus:

And in turn [*ἐν μέρει*] they[51] dominate as the cycle rolls around,
And they dwindle and grow into one another in the turn of fate [*ἐν μέρει αἴσης*],
For these things are the same,[52] and running through each other
They become men and the races of other beasts,
At one time coming together by Love into one cosmos,

48. Pl. *Cra.* 402a4–c3 and *Theaet.* 152d2–e9. I am not, of course, claiming that Heraclitus was a pluralist, but that Plato found it difficult to read monism into Heraclitus's philosophy.

49. The term *ἐν μέρει* is used similarly by Empedocles to refer to the fated alternation of cycles of rule under Love and Strife (DK 31 B 17.19; see Arist. *Phys.* 8.1, 252a5–10 = DK 31 A 38).

50. See Heraclitus's own fragment preserved at DK 22 B 10, which, it must be remarked, does not stipulate that the oscillations between opposites occurs eternally.

51. I suspect that the antecedent is the four elements, but many interpretations are possible, on which see Trépanier 2004: 187–190.

52. As I have emended *αὐτὰ γὰρ ἔστιν ταὐτά* (where the manuscripts have *ταῦτα*), following the translation of Trépanier (2004: 187). Compare two similar instances: (1) B 21.13–14, where Empedocles uses the same language in the refrain to claim, of various created objects (among which are men and beasts), "for these things are the same, but, running through one another / they become different [*γίγνεται ἀλλοιωπά*]; to such a great extent does blending change them," and (2) with less ambiguity, B 82: "hair, leaves, thick feathers of birds, and scales on stout limbs become the same" (*ταὐτὰ τρίχες* [etc.] . . . *γίγνονται*). In the former case, the contrast between being the same and becoming different is made more effective through emendation. Otherwise, we are forced to translate into something not possible, given the Greek, like "these are the very things that are" (Graham 2010) or into a banal statement, such as "these very things are" (Inwood 2002).

And at another time again each being borne apart by Strife's enmity,
Until, all of them grown together into one, the universe is subsumed.
Thus insofar as they learned to grow as one from many,
And they finish up many as the one again grows apart,
In this respect they come to be and have no constant life;
But insofar as they never cease from continually interchanging,
In this respect they are always unchanged in a cycle.
(EMPEDOCLES, DK 31 B 26; translation after Inwood 2002)

Not only does Empedocles in Fragment B 26 employ the same language of "growing and dwindling" as Epicharmus (compare Empedocles's καὶ φθίνει εἰς ἄλληλα καὶ αὔξεται in line 2 with Epicharmus's ὁ μὲν γὰρ αὔξεθ', ὁ δέ γα μὰν φθίνει), but he also seeks to formulate a way things in the process of becoming can nevertheless retain their unique identities while constituting a single universe.[53] Note that through the cosmic alternation owed to fate (ἐν μέρει αἴσης), which is apparently responsible for regulating the "growing and dwindling" of things, the universe is finally subsumed once the things have all "grown together into one" (ἓν συμφύντα τὸ πᾶν).[54] If we are unclear about what explanatory work "fate" (αἶσα) achieves here, or "necessity" (ἀνάγκη) as elsewhere in Empedocles's fragments,[55] we are not alone.[56] Aristotle, who, in his close reading of Empedocles's work, detected the link between necessity and alternation of one and many, nevertheless complained that Empedocles employed his causes inconsistently.[57] And, as I discussed earlier, Plato also expressed confusion about whether "convergence" can rightly be considered a cause of two things becoming one or whether the "division" causes one thing to become two.

In the light of Empedocles's fragments concerning cosmic alternation, we can now return to Socrates's criticism of the number-stuff theorists in the *Phaedo*. It

53. DK 23 B 2.

54. See Trépanier 2004: 246 n. 15. Compare Diogenes of Apollonia (DK 64 B 2): "But, since they all are altered from the same thing, they become different things at different times and return to the same thing."

55. DK 31 B 115. Contrast Philolaus's appeal to necessity (ἀνάγκα) as a *logical* injunction that all things must be limiting, or unlimited, or both (F 2 Huffman), and that harmony must have bonded different things together, if there is to be an ordered universe (F 6 Huffman).

56. See Inwood 2002: 74.

57. Arist. *Metaph*. 1.4, 985a21–985b2 (= DK 31 A 37): ὅταν δὲ πάλιν ὑπὸ τῆς Φιλίας συνίωσιν εἰς τὸ ἕν, ἀναγκαῖον ἐξ ἑκάστου τὰ μόρια διακρίνεσθαι πάλιν. Also see Arist. *Cael*. 3.2, 301a18–20 (= DK 31 A 42).

is relatively clear that Socrates directs his criticism of those who espouse the number-stuff theorist "style of procedure" against figures such as Empedocles, but there are some important interventions on the part of Plato that constitute divergences from Empedocles's own cosmology. First of all, Socrates shifts the problem away from "one and many" to "one and two," which heightens the emphasis on arithmetical number in the more general examination of unity and plurality. Second, we see that the concept of "convergence" (σύνοδος), which was used as an analogue to Love by Empedocles, is further qualified by the explanatory supplement of "approximation" (πλησίον ἀλλήλων) of one to one, which is not used in any of Empedocles's fragments, although it may have played a more general role in his epistemology.[58] How unity achieved through "approximation" might be thought to work is difficult to know precisely, and that is not Socrates's point here anyway.[59] What is clear is that we are dealing with a sort of theoretical science that cannot be demonstrated simply by means of the *concrete objects* employed in pebble-arithmetic. Even if the Empedoclean number-stuff theory employed in *Phaedo* 96e–97c to explain the coming-to-be of a unit and a plurality is to be rejected, the "one" and "two" have a role to play in Socrates's more robust theory of causation. Along the way, I suggest, Plato also takes advantage of the criticisms he leveled against the Pythagoreans in the *Cratylus* in order to develop a new way of thinking about number as informed by teleological causation. After hypothesizing the Forms (the Beautiful itself by itself [καλὸν αὐτὸ καθ' αὑτό], etc.) as causes, Socrates returns to the alternative theory and reevaluates it in accordance with this new hypothesis:

> Then you would not avoid saying that, when one is added to one, it is the addition and, when it is divided, it is the division that is the cause of the two? And you would loudly exclaim that you do not know how else each things can come to be except by sharing in the particular being in which it shares [μετασχὸν τῆς ἰδίας οὐσίας ἑκάστου οὗ ἂν μετάσχῃ], and in these cases you do not know of any other cause of becoming two except by the sharing in the two [τοῦ δύο γενέσθαι ἀλλ' ἢ τὴν τῆς δυάδος μετάσχεσιν], and

58. That is, if Theophrastus (*Sens.* 1.11 and 1.19 = DK 31 A 86) preserves anything like Empedocles's theory of the mixture of elements in human physiology.

59. The best *comparandum* here would be Democritus's description of how objects might *seek* to fit together, but Democritus vehemently denies that the blending of two objects together could catalyze their becoming one (see DK 68 A 37 = Arist. F 208 Rose and DK 68 A 64 = Alex. *de Mixt.* 2). Democritus thus stands against Empedocles and others who might have thought that two could become one and vice versa (see Barnes 1982: 444–445). One might also think that Zeno's arguments against plurality are relevant (especially DK 29 A 21), but Hackforth (1955: 131) is probably right in seeing Plato as exercising authorial license.

> that the things that are to be two must share in this, as that which is to be one must share in the unit [μονάδος], and you would dismiss these additions and divisions and other such subtleties, and leave them to those "wiser" than yourself to answer.
>
> (Plato, *Phaedo* 101b10–c9; translation after Grube in Cooper and Hutchinson 1997)

Here, Socrates effectively replaces the calculative operations of "adding" (πρόσθεσις) and "dividing" (σχίσις) with the ontological "sharing" (μετάσχεσις). The cause of becoming one or two is the sharing in the unit and the two, respectively.[60] One might doubt whether Socrates actually accepts this claim in the dialogue, and it is difficult to know precisely how to interpret his remarks about hypothesis in the lines that follow. But there can be no doubt that Plato represents his Pythagorean characters as willing to hypothesize the Forms and teleological causation: not only do Simmias and Cebes both emphatically agree with Socrates, but even Echecrates interrupts Phaedo's story in order to express his approval of Socrates's demonstration in terms that implicitly contrast the obscurity of Philolaus's arguments noted earlier.[61]

So we cannot be sure at this point whether Socrates's apparent admission of the Forms of the One and the Two reliably constitutes an adaptation of mathematical Pythagorean metaphysics, on the grounds that the best comparison for his analysis of the number-stuff theory in this part of the *Phaedo*, the fragments of Empedocles, do not admit the persistent identity of numbers themselves. Another way to put it is this: Plato seems to be willing to accept that the metaphysical operation of participation, which he subtly links to Philolaus in the *Cratylus*, can effectively explain how things can *become* one or two, but this does not help us to explain, for example, whether the number 2 admits of only one property (i.e. "twoness") or of others. Aristotle also saw the problem in similar terms, and he is the earliest witness to associate a puzzle involving numbers and predication of numerical properties with Plato and the Pythagoreans:

> But while the Pythagoreans have declared in the same way that there are two principles, they made this addition, which is peculiar to them, namely that they thought that the limited and the unlimited were not uniquely

60. It would appear that Socrates and his interlocutors accept the possibility that they are talking about Forms of the One and the Two, but the language is not totally forthcoming yet on that issue.

61. See Pl. *Phd.* 102a2–6, where Echecrates praises Socrates's argument for its vividness (ὡς ἐναργῶς).

different substances,[62] such as fire and earth and anything else of this sort, but that the unlimited itself [αὐτὸ τὸ ἄπειρον] and the one itself [αὐτὸ τὸ ἕν] were the substance of the things of which they are predicated, and hence [διό] that the substance of all things was number [ἀριθμὸν εἶναι τὴν οὐσίαν πάντων]. On these issues, then, they made claims on these lines. And concerning essence[63] [περὶ τοῦ τί ἐστιν], they began to make arguments and definitions [λέγειν καὶ ὁρίζεσθαι], but their treatment was too simple [λίαν ἁπλῶς ἐπραγματεύθησαν]. For they both defined superficially and thought that the substance of the thing [ἡ οὐσία τοῦ πράγματος] was that to which a stated term would first be predicable, for example as if someone were to believe that "double" and "two" were the same because "two" is the first thing of which "double" is predicable. But surely to be "double" and to be "two" are not the same things. If that were to be the case, one thing would be many—a consequence that they actually drew. So much, then, can be grasped from the earlier thinkers and from the rest.

The *pragmateia* of Plato followed on the aforementioned philosophies, according with them in many ways, but possessing certain aspects distinct from [the *pragmateia*] of the Italians. . .[64] Following [Socrates], Plato assumed that this [i.e. pursuit of the universal through definition] is concerned not with perceptibles but with other things for this reason: it is not possible for a general definition to come from any of the perceptible things, which are always in a state of change. Well, then, he named these other sorts of entities "Ideas" [ἰδέας προσαγόρευσε], and he [said that] perceptibles are all called after them and in accordance with them. For the many things that bear the same name as the forms exist by virtue of participation [κατὰ μέθεξιν] in them.[65] With regard to participation, he changed the name only: for whereas the Pythagoreans claim that entities exist by means of imitation of numbers [μιμήσει τῶν ἀριθμῶν], Plato says by means of participation [μεθέξει], modifying the name. As to what participation or imitation is, however, they left it to us to seek it out together.

Furthermore, Plato claims that besides perceptibles and Forms is a middle type of entity, the objects of mathematics [τὰ μαθηματικὰ τῶν

62. See chapter 1, note 82.

63. Or, more literally, "concerning the 'what it is.'" This appears to be an explicit commentary on the failure of the mathematical Pythagoreans' approach to the first class of *acusmata*, those that respond to the question "What is it [τί ἐστι]?"

64. For the sake of convenience, I excise the discussion of Heraclitus and Socrates here.

65. This is a notoriously difficult passage. I have adopted the text of Ross.

> πραγμάτων], which differ from perceptibles in being eternal and immutable, and from Forms in that many [objects of mathematics] are similar, whereas each Form itself is one.
>
> (Aristotle, *Metaphysics* 1.5–1.6, 987a13–b18)

Aristotle assumes that Plato's metaphysics operates by virtue of the same ontological vehicle as the Pythagoreans', and he claims that Plato's only real innovation in this sphere was modifying the name from "imitation" to "participation." From Theophrastus forward, scholarly attention on the problem of making sense of how "participation" and "imitation" can be brought to bear on one another has obscured a more interesting point that we can draw out of Aristotle's classification:[66] Aristotle implicitly links the mathematical Pythagorean *pragmateia* to that of Plato, on the grounds that the mathematical Pythagoreans assumed that what is predicated, not what it is predicated of, is the substance.[67] In *Metaphysics* B (3.4, 1001a9–12), Aristotle is more explicit in drawing these inferences: "Plato and the Pythagoreans hold that neither being [τὸ ὄν] nor the one [τὸ ἕν] are anything other [ἕτερόν τι] [than one another], and their nature is this [i.e. to be nothing other than one another], as if the substance [of things] were to be the same thing as being unified and existing [ὡς οὔσης τῆς οὐσίας αὐτοῦ τοῦ ἑνὶ εἶναι καὶ ὄν]."[68] We can be relatively confident that Philolaus, in Fragment 6, held that "the being of things" is predicated of all other things ontologically. He says as much when he claims that things in the cosmos could not have come to be if their being did not preexist (μὴ ὑπαρχούσας τᾶς ἐστοῦς τῶν πραγμάτων). The case for "number" is more controversial: nowhere in the extant fragments does Philolaus explicitly say that the "one" or any number is ontologically prior to all things. He says, instead, that the "limiters" and "unlimiteds," as principles, preexisted (ὑπᾶρχον), but, given the fact that they were fundamentally different, it was necessary for harmony to supervene in order for them to achieve a proper ordering. Still, in Fragment 4, Philolaus does seem to say that number is a necessary condition for knowledge of all knowable objects, by which he seems to mean objects that have definition.[69] And in Fragment 5, he breaks number down into three classes: the "even" (ἄρτιον) and the "odd" (περισσόν), which are "proper kinds" (ἴδια εἴδη), and the derivative kind "mixed up from these" (ἀπ' ἀμφοτέρων μιχθέντων), the so-called even-odd (ἀρτιοπέριττον), which appears to refer to the "one." Aristotle, too, knew this

66. See Horky: forthcoming.

67. See Schofield 2012: 163–164.

68. Accepting Cavini's (2009: 180–181) interpretation and text, following Thomas Aquinas.

69. See Nussbaum 1979: 93 and Huffman 1993: 177.

fragment, and he is explicit in associating the one with the "even-odd" class and stating that it is derived from the "even" and the "odd."[70] Scholars debate the generation of the "one" in Pythagorean philosophy, and it is difficult to know for sure whether it is Aristotle who is confusing something otherwise clear and consistent in Philolaus's philosophy or the other way around.[71] Be that as it may, Aristotle's testimony concerning mathematical Pythagorean ontology might help us to make sense of Plato's own reception of Pythagorean thought. Or, to be more specific, Aristotle's concern over the Platonic and mathematical Pythagorean confusion of predicate and object of predication might give us insight into the sorts of philosophical concepts Plato might have adopted from the Pythagoreans.

Indeed, it is in the context of Empedocles's cosmological fragments, Aristotle's testimony on Platonic and Pythagorean predication of unity and being, and Philolaus's fragments concerning number that Plato's appropriation of mathematical Pythagoreanism in the *Phaedo* takes shape. Once again, I believe that the underlying problem has to do with responding to the puzzles originally raised by the "Growing Argument" of Epicharmus, around which Plato has structured his recurrent critique of mathematical Pythagorean philosophy. After Phaedo, Echecrates, Simmias, and Cebes have all been shown to accept the crucial assumptions that (1) each Form exists, and (2) things derived from them acquired their name by having a share in them—two assumptions that were also accepted throughout the inquiry into the correctness of names in the *Cratylus*[72]—Socrates momentarily shifts the discussion away from numbers and arithmetic to the classes of number and the properties that attend those classes. All speakers accept the proposition that "an opposite will never be opposite to itself," and Socrates seeks, in the second section of the final argument, to prove the immortality of the soul through, in part, a metaphysical analysis of the two Philolaic "proper kinds" (ἴδια εἴδη), the Forms of the Even and the Odd:[73]

70. Arist. F 199 Rose = Theon Sm. *Math.* p. 22.5–9 Hiller.

71. For the various positions taken and an up-to-date bibliography, see Zhmud 2012: 400–401 with n. 50 as well as 444–445.

72. See chapter 4.

73. That Plato thought that other mathematicians were hypothesizing intelligible classes of the even and odd, as well as other mathematical objects "themselves" (e.g. the "square itself" and "the diagonal itself") is confirmed in the third segment of the Divided Line (*R.* 6, 510c2–511a2), which corresponds with διάνοια (thought). The reference to "those who concern themselves with geometry, calculation, and such sorts of things" is obscure, and it has been thought to refer to Philolaus (see Lloyd 1991: 66). The language might help to clarify the situation: Socrates draws attention to those other hypothesized things that are "akin" (ἀδελφά) to even, odd, shapes, and the "three angles." This is the scientific language generally of "the Pythagoreans" in Plato's *Republic* (7, 530d6–9), and specifically of Archytas (F 1 Huffman).

"It is the case, then, concerning some of these sorts of things [i.e. cases in which a thing appears to share in the property that is opposite to it], not only that the Form itself [αὐτὸ τὸ εἶδος] is entitled to its own name for all time, but also that there is something else that is not it [i.e. the Form] but which consistently has its aspect [ἔχει τὴν ἐκείνου μορφὴν ἀεί], whenever it exists [ὅτανπερ ᾖ]. Perhaps what I'm saying will be still clearer through this example: the Odd should, I suppose, always obtain the very same name we are now saying. Or is that not the case?"

"Certainly."

"Now is this the only thing among things that exist—this is my question—or is there something else, not the same as the Odd, that exists, and that should nevertheless always be called this too [i.e. "odd"] along with its own name, because its nature is such that it is never to be separated from the Odd? What I'm talking about is the sort of thing that characterizes the number 3 [οἷον καὶ ἡ τριὰς πέπονθε], as well as many others.[74] Take the case of the number 3:[75] doesn't it seem to you that it should always be called by its own name and by the name of the Odd [i.e. "odd"], although [the Odd] is not the same thing as the number 3? But still the number 3 and 5 and half of the entire class of number have such nature as to be odd, although each is never the Odd. And, in turn, the number 2 and the number 4 and the entirety of the other array[76] of number, although [each of them] is not the Even, nevertheless it is always even. Do you agree or not?"

"How couldn't I?"

"Pay attention, now; what I want to make clear [δηλῶσαι] is this: that it appears not only that those opposites [i.e. opposite Forms] do not admit of one another, but also that *these* things, even if they are not opposite to

74. As discussed in chapter 4, Socrates in the *Euthyphro* (11a6–b1) distinguishes between essential property (οὐσία) and accidental attribute (πάθος; πέπονθε). It is also notable in that dialogue (12c6–8) that Socrates speaks of oddness as an accident of number, on the grounds that the odd is part of number.

75. It is not yet clear that Socrates is speaking about "the three" as a Form, but he will make that inference later. White (1989: 197–198) helpfully tries to distinguish the various "threes" here as "the triad" (ἡ τριάς), "triple" (τὰ τρία), and "the Form of threeness" (ἡ τῶν τριῶν ἰδέα), whereas I have tried to preserve some of the ambiguity in my translation.

76. Plato uses the term στίχος, which only appears one other time in his work (*Leg.* 959a1), where it clearly refers to a monostich (single line of a poem). Just as here in the *Phaedo*, it tends to appear in military contexts in the fifth and early fourth centuries BCE (e.g. A. *Th.* 924; Eur. *Supp.* 669; X. *Lac.* 11.5.4). Implicit in that usage is that it is (ideally) properly arranged, countable, and limited. It is somewhat surprising not to find a term that might refer to "class" here, such as εἶδος or γένος. One possibility is that Plato is using the term to refer to a "row" or progressive sequence of numbers, as Nicomachus seems to take it (e.g. *Ar.* 2.12.8).

> one another, always possess their opposites[77]—not even *these* things resemble things that admit that Form/character [ἰδέα][78] that typically is what is opposite to what [i.e. Form/character] is in them, but whenever it advances [on them], they either perish or yield [to it]. Shall we not say that [a group of] three things [τὰ τρία] would first perish and suffer anything whatsoever, before, still standing its ground and being three things, it were to become even [ἄρτια γενέσθαι]?
>
> "Definitely."
>
> "And yet the number 2 [δυάς] is not the opposite of the number 3 [τριάς]."
>
> "Surely not."
>
> "Not only, then, do Forms yield their ground when their opposites advance on them, but also certain other things yield their ground when their opposites advance."
>
> (Plato, *Phaedo* 103e2–104c9)

It is initially unclear what Socrates is referring to when he speaks about the "certain other things" that refuse to maintain their essential properties and thus cannot maintain identity when their opposites advance. Given the intensity of the rhetorical flair and buildup of the elenchus here, we would expect Socrates to be marking out a novel argument concerning numbers. Indeed, Socrates goes on to define those "certain other things" that "yield their ground when their opposites advance" as, I suggest, Forms of Numbers.[79] A Form-Number, as described by Aristotle (*Metaph.* 13.6, 1080a28–35), would be an intelligible number that exists in sequence (ἐφεξῆς) and is countable (ἀριθμεῖται) but is "incomparable" (ἀσύμβλητος) because it does not contain anything that might make it similar to other Form-Numbers.[80] The Form of 2 cannot engage

77. It is difficult to know precisely what Socrates is saying here. Initially, it seems like he is referring to numbers, but that cannot be right, since numbers, while they do not have opposites (i.e. 2 is not opposite to 3), nevertheless cannot be both even and odd (pace Rowe 1993: 255). The likely meaning is that "these" things are what follows, namely, groups of things (such as a group of three pebbles) that can "have" (in the sense of possess) their opposites in them (in the same way that a group of three pebbles "has" two pebbles in it) while maintaining essential properties (e.g. to be odd).

78. This ambiguity should be maintained here, since it does not get resolved until *Phd.* 104e5.

79. See Nehamas (1973: 488–489), who evidently feels inclined to use the term "Form" here but hesitates so as "to avoid complications" (n. 32).

80. As Ross (1924, vol. 2: 427) notes, strictly speaking this term means that they cannot "be expressed as a fraction of each other, or at least as greater or less than or equal to each other," but that the context here suggests that they cannot be "capable of entering into arithmetical relations with one another—of being added and subtracted, multiplied and divided." For further analysis, see Burnyeat 1987: 237.

in any relationship with the Form of 3 because they possess contrary predicates, Aristotle seems to be saying.[81] Even if Aristotle does not *explicitly* link Form-Numbers to Plato, there is good reason to see the similarities between Aristotle's account and Socrates's description of the Form of 3 in the *Phaedo*.[82] Socrates explains that "the Form of the [group of] the three things" (ἡ τῶν τριῶν ἰδέα) has no opposite itself, but, while it is the bearer of more than one essential property ("threeness" and "oddness"),[83] it still refuses to admit of the properties of those Forms that are, in a way that remains not fully explained in the *Phaedo*, opposite to the Form from which it derives an essential characteristic (the Odd).[84] And the ontological status of these Form-Numbers, which is not totally clear here in the *Phaedo*, has prompted critics from Aristotle and Xenocrates to today to speculate about whether Plato was hypothesizing intermediaries between complete Forms and incomplete intelligibles, which would be the objects of discursive thought (διάνοια).[85] Be that as it may, the conclusion that follows from this part of the "final argument" appears to be an essential definition of the Form of 3 obtained through the explanation: 3 is uneven (ἀνάρτιος ἄρα ἡ τριάς).[86]

81. Literally, Aristotle speaks of numbers such as two or three as "different and separate from" (ἕτερα ἄνευ) the other numbers.

82. See Mueller 1987: 257.

83. From this point of view, both "threeness" and "oddness" are necessary and sufficient conditions for the Form of 3 and, presumably, for anything that falls under it. As Hackforth (1955: 158), Taylor (1969: 49), and Nehamas (1973: 489 n. 33) have pointed out, however, in the "more sophisticated" (κομψοτέραν) explanation that comes afterward, "oddness" is explicitly rejected, and what makes a number "become odd" (ἐγγένηται περιττός) is said to be the presence of a "unit" (μονάς) left over. Adapting the analysis of Rowe (1993: 258–260), I see the "more sophisticated" explanation as reflecting exactly to what a sort of number-stuff theorist would argue (someone like Epicharmus? Perhaps Ecphantus?) and to what is being rejected by Plato.

84. It is somewhat interesting that Socrates here does not discuss whether or not "number" (ἀριθμός) is predicated of the Form of 3, since, for Philolaus (F 5), all knowable things have number. Certainly, for Plato, stable identity and Forms are necessary conditions for knowable objects (see *Cra*. 439e7–440b1). But number seems only necessary for sensibles that are countable (see Pl. *Theaet*. 198c1–2).

85. Plato has Glaucon distinguish between numbers as instanced in sensibles and numbers "only grasped in thought" ([ἀριθμοὶ] ὧν διανοηθῆναι μόνον) at *R*. 7, 525d5–526a7. The former seems to refer to numbers as compared with one another, and the latter to numbers themselves. See Klein 1968: 76–78, Nehamas 1973: 467–468, and Denyer 2007: 302–306.

86. That this argument has been for the sake of producing a new definition of "three" is emphasized by the phrase ἐπὶ τὸ τοιοῦτον δή at 104e10. It never appears anywhere else in reference to definitions (or even descriptions) of "the odd" in Greek mathematics or philosophy. Contrast the essence of "three" according to Aristotle (*APo* 2.13, 95a35–38) as "number, odd, prime, and prime in this sense [i.e. not composed of number]."

What I would like to draw out of this extensive and difficult passage is how Plato evidently pits the natural consequences of one mathematical Pythagorean position against another. Number-stuff theorists such as Empedocles and Epicharmus, who developed various formulations of the "Growing Argument," seem to have rejected persistent identity. They both allowed for the possibility that objects in nature could be unified and the same as themselves at a particular moment in time, but they also denied that such objects could remain in this state. In doing so, Empedocles and Epicharmus took a middle position between on the one hand early pluralist cosmologists like Pherecydes of Syros and Archelaus of Athens and on the other Eleatic monists like Xenophanes and Parmenides.[87] Where this leaves Philolaus the Crotonian, the silenced teacher of Simmias and Cebes in the *Phaedo*, is difficult to know. I would conjecture that Plato would have classed him with the rest of the number-stuff theorists, although perhaps in a qualified way.[88] Whether or not Philolaus would have understood number as applying not only to epistemology but also to metaphysics, Plato certainly did so, and with an eye to Philolaus as a forerunner. In this way, Plato obtained from Philolaus the vocabulary and rudimentary conceptualization of subsistent entities, including "being," as instantiated in the natural objects humans perceive with their senses.[89] Plato also seems to have adapted from Philolaus the basic concept of classification of essential properties, not only because Philolaus had recognized that the necessary condition for grasping anything that can be understood is that it possess number as a property but also because Philolaus began to think about the basic classes (*ἴδια*) of numbers according to whether they were essentially odd, even, or mixed.[90] As Aristotle argued, Plato sought to separate numbers as instantiated in perceptible objects from numbers themselves, and as a consequence Plato developed Form-Numbers, which he may have considered the "intermediary" group of mathematical objects. In *Metaphysics* M (13.6, 1080a28–29) Aristotle refers to each of these objects as a "number itself," for example "3 itself" (*ἡ τριὰς αὐτή*) or "2 itself" (*ἡ δυὰς αὐτή*). That Plato himself conceptualized Form-Numbers as "numbers themselves" can be inferred from a few passages in his dialogues:

87. If, indeed, those are the figures to whom we should refer the metaphysics of the "three" and the "two" beings at *Soph.* 242c9–d4.

88. That is, qualified because, as Barnes notes (1980: 389), Philolaus "recognizes stuffs, but he insists equally on shapes."

89. See Herrmann 2007: 204–206 and 214–217.

90. It is beyond the scope of this study to examine in more detail the significance of the concept of "mixture" of Forms or classes for Plato's philosophy, especially the metaphysics outlined in the *Sophist* and the *Philebus*.

we hear of "numbers themselves" in *Republic* 7 (525d6), a passage that is reformulated in the Early Platonist *Epinomis* (990c6), as well as "10 itself" (αὐτὰ τὰ δέκα) in the *Cratylus* (432a10).[91] In the eyes of many critics, Plato's reaction to Philolaus in the *Phaedo* is thought to be found somewhere in the discussion of the constitution of the soul, whether the soul is to be considered a harmony or not. But perhaps we might see that this has all been a cover for the ways Plato sought to evaluate the ideas of the mathematical Pythagoreans Epicharmus, Empedocles, and Philolaus concerning *number*. Given that our knowledge of mathematical Pythagorean theories of the soul is far more deficient than our understanding (and the ancient understanding) of their theories of number, we might not need to seek in the *Phaedo* to discover a mathematical Pythagorean psychology.[92] Instead, what we do see is Plato's ultimate rejection of the mathematical Pythagoreans' methods of demonstration as insufficient for the goal of understanding, with clarity and precision, *why* a group of three things can never, under any circumstances, be even. For in that explanation is to be found the ultimate rationale for why the soul will never, under any circumstances, be dead.[93] And if we admit an immortal soul that, under no circumstances, could become anything other than what it is, then we will have stumbled upon the stable identity that is required to refute, once and for all, the problem of alteration in Epicharmus's "Growing Argument."

There is one final question concerning the Form of 3 and mathematical Pythagoreanism: why does Plato seek to prove the existence of the Form of 3 and not 5 or 7, or even 10, the Decad itself?[94] Was there any special significance attached to the number 3 by Pythagoreans or other Presocratics? Anaximander believed 3 to be the root number for the ratio of distances between the cosmic bodies, and Pherecydes of Syros, considered by some in antiquity to have been Pythagoras's teacher, ascribed to three principles

91. On the *Epinomis*, see below. There is also the notorious problem of the "equals themselves" in the *Phaedo* (74c1), which, I think, are likely to be intelligible mathematical objects (like angles) but are not numbers (see Hackforth 1955: 69 n. 2). Generally, on this problem, see Sedley 2007: 82–84.

92. It is possible that Xenocrates himself was responsible for attributing to Pythagoras the theory that the soul is "number moving itself" (F 169 IP = Aët. 4.2.1–3).

93. See Rowe 1993: 260–261.

94. Zhmud (2012: 408–409) effectively dismisses the importance of the Decad for Pythagoreanism, on the grounds that it can only be traced back as far as the writings of Speusippus. We can actually see its significance for Plato in the *Laws* (737e1–738b3), however, where the product of all the numbers 1 through 10 multiplied together is 5,040, the number of households to be established in Magnesia. Also see Pl. *Criti.* 113e8, where 10 is a central number in the design and history of Atlantis.

(Zeus, Chronos/Time, and Chthonia).[95] With the early Pythagoreans, the number 3 took on a new significance, in that it was understood to be totalizing. Epicharmus of Syracuse seems to have Pythagoras in his sights when a character of his on stage says "thrice again was life given back [to him]" (τρὶς ἀπεδόθη ζόος).[96] And Ion of Chios (ca. 490–ca. 420 BCE), who praised Pythagoras and (apparently) Pherecydes alongside one another (DK 36 B 4), attributed in his work *Triad* (Τριαγμός) the poems of Orpheus to Pythagoras, according to Diogenes Laertius (8.8 = DK 36 B 2).[97] The beginning of that work survives:

> [This is] the beginning of my account: all things are three, and [there is] nothing more or less than these three. The excellence of each one [person? thing?] is a triad: intelligence, power, fortune.
>
> *ἀρχὴ δέ μοι τοῦ λόγου· πάντα τρία καὶ οὐδὲν πλέον ἢ ἔλασσον τούτων τῶν τριῶν. ἑνὸς ἑκάστου ἀρετὴ τριάς· σύνεσις καὶ κράτος καὶ τύχη.*
>
> (ION OF CHIOS, DK 36 B 1)

The claim that "all things" are only and exactly "three," which is then further explained as "intelligence, power, fortune," exhibits the same sorts of intellectual habits as those of Ionian natural scientists like Diogenes of Apollonia, whose treatise began with a statement of origin (DK 64 B 1), followed by universal statement (DK 64 B 2): "In general, it seems to me [ἐμοὶ δὲ δοκεῖ τὸ μὲν ξύμπαν] that all things in existence are altered from the same thing and are the same as it."[98] But the focus on the number 3 is distinctive and marks out Ion's account from those of other Ionian natural scientists and historiographers.[99] As

95. For Anaximander, see DK 12 A 10–11 and 21–22. Pherecydes (DK 7 B 1) posits the three primary entities, but he is also said by Damascius, whose source is Eudemus of Rhodes (DK 7 A 8), to have described the generation of three elements (fire, breath, and water), followed by another generation of a class of gods under the title "five-nook" (πεντέμυχον). On this see KRS 1983: 56–57. Speculation concerning the cosmic significance of odd numbers might therefore have played a role in Pherecydes's cosmogony.

96. F 187 K.-A. See Willi 2008: 174.

97. For Ion, Pherecydes, and Pythagoras, see Riedweg 2005: 52–53.

98. See Baltussen 2007: 299–300. Also compare Hecataeus of Miletus FGrHist 1 F 1: "Hecataeus of Miletus speaks in this way: I write these things as they seem to me to be true" (τάδε γράφω, ὥς μοι δοκεῖ ἀληθέα εἶναι).

99. If Philoponus is to be trusted (*in GC* p. 207.16–20 Vitelli = DK 36 A 6), Ion also postulated three elements (fire, air, and earth).

Han Baltussen has argued, Ion's introduction to his work *Triad* exhibits the same sort of strategies for understanding number that are ascribed by Aristotle to the Pythagoreans:

> It is just as the Pythagoreans say, the universe and all things in it are defined by the number 3 [*τὸ πᾶν καὶ τά πάντα τοῖς τρισὶν ὥρισται*]; for end, middle, and beginning comprise the number of the all, and they possess the number of the triad [*τελευτὴ γὰρ καὶ μέσον καὶ ἀρχὴ τὸν ἀριθμὸν ἔχει τὸν τοῦ παντός, ταῦτα δὲ τὸν τῆς τριάδος*]. Therefore we have obtained this number from nature, as if it were one of her laws, and make use of it even for the worship of the gods.
>
> (ARISTOTLE, *On the Heavens* 1.1, 268a10–15)

Aristotle does not tell us his source for this information, and the further elaboration he evinces—on the way the number 3 represents the properties of magnitudes—represents problematic evidence for Pythagorean theories of space and mathematical dimensions, if it constitutes historical evidence at all.[100] What is certain is that Aristotle believed that the number 3 was quite important to the Pythagoreans' ways of defining things: in his lost works on the Pythagoreans, he also identified a Pythagorean division of rational beings into three: gods, humans, and beings "like Pythagoras."[101] And in his work *On the Heavens*, Aristotle presents the Pythagoreans as believing with regard to number, the universe, and the objects in it that what is predicated is the same as what it has been predicated of. "Beginning, middle, and end," which are subjects, are also said to "comprise" (*ἔχει*) the number of the entire universe and its objects, that is, to "possess" (*ἔχει*) the number 3 (presumably as a property: threeness).[102] It is somehow supposed to follow that the "number 3" thus *defines* the universe and all the objects in it. Whether or not this represents Aristotle's version of a Pythagorean argument cannot be easily

100. Pace Baltussen 2007: 302–303.

101. Arist. F 192 Rose = Iambl. *VP* 31, 18.12–16. Note, too, that Aristoxenus (at Iambl. *VP* 182–183, 101.20–102.14) describes the Pythagorean value system in triads.

102. The conceptual antiquity of this statement can be confirmed by the Derveni Papyrus, which partially quotes a line from the poems of Orpheus (Col. XVII.12) that states "Zeus is the head, Zeus the middle, and all things are fashioned out of Zeus." Aeschylus (F 70) similarly claims: "Zeus is the aether, Zeus is earth, Zeus is sky. Zeus is all things, and what is above them." See Pl. *Leg.* 715e7–716a2: "Indeed, just as the story of old [*ὁ παλαιὸς λόγος*] says, god possesses/is [*ἔχων*] the beginning and end and middle of all things in existence, and he advances straight in his natural rotation."

inferred; Aristotle does not critique the argument here one way or another.[103] But it is relatively clear that once again we can see the conflation of predicate and object of predication in Pythagorean metaphysics and logic, at least as it had been reformulated by Aristotle, with regard to the number 3 and the properties it is meant to instantiate.

When Socrates in the *Phaedo* provided the mathematical Pythagoreans Simmias and Cebes with an analysis of addition and subtraction, of the even and odd types, and of the number 3, then, he probably appealed to concepts and strategies familiar to them. This is clear from the fact that the philosophy of number employed by mathematical Pythagoreans such as Epicharmus, Empedocles, and Philolaus concerned itself with these approaches to understanding, in particular the generation of things in the universe. Although Plato seems to have accepted the strategies for enquiry into the nature of generation and persistent being expressed in the fragments of the Pythagoreans, he sought to improve on them in at least two ways. First, just as he did with natural scientists like Anaxagoras and Diogenes of Apollonia, Plato invoked teleological causation in order to reveal weaknesses in the Pythagoreans' arguments and methodology. Second, teleological causation would have the effect of forcing mathematical Pythagoreans such as Simmias and Cebes to hypothesize the Forms. The possibility that Pythagoreans may have admitted Forms to their metaphysics— the old problem of figuring out what exactly entailed the Philolaic "proper kinds" (ἴδια εἴδη) of even and odd number comes to mind—need not trouble us, since the debate itself misses the point: whether or not the Pythagoreans did actually postulate Forms (Aristotle, our earliest critic, didn't think so), Plato took the step to develop a robust and rigorous set of explanations for *how* Forms could improve on the insufficient or tautological arguments of the mathematical Pythagoreans concerning number. The natural consequence was the introduction of Form-Numbers in the *Phaedo*, intelligibles that on the one hand do not necessarily inhere in perceptibles but on the other hand are still subject to the properties of those Forms not predicable in the opposite direction. (The Form of 3 admits of its own predicate, "threeness," and of the predicate "oddness," but the Form of the Odd, while it always admits of "oddness," does *not* always admit of "threeness.") If Plato chose the number 3 to develop a new, more rigorous ontology and epistemology than that of the mathematical Pythagoreans, he did so probably in full awareness of its deep significance for mathematical Pythagoreanism.

103. In the *Posterior Analytics* (2.13, 96a24–b14), Aristotle did, however, follow Plato's lead in using the number 3 as the basic example suited for inquiry into the nature of permanent attributes (τὰ ὑπαρχόντα ἀεί), which are predicated in the definition of an object.

But it is also important to recognize that Plato's appropriation of the number 3 for his own metaphysics provided a new definition for it that superseded its Pythagorean heritage and became the stuff of Platonism for generations to come. In *Republic* 6 (510b4–511d5), Plato associated the realm of mathematicals, especially geometry of the sort practiced by Archytas and probably Philolaus, with the third section of the Divided Line, thought (διάνοια). But Plato also went to extreme lengths there to show that mathematics was not sufficient for completing the understanding of objects in the universe, instead supplying a fourth section corresponding with the Socratic intervention, dialectic, which was the realm where the philosopher could contemplate things totally as such. Indeed, the number 4 comes to obtain a particular significance for Early Platonists, probably as a consequence of its heightened importance in Plato's own dialogues. At the beginning of the *Timaeus*, Socrates initiates the dialogue by counting out a triad (by reference to the speakers there: Timaeus, Critias, Hermocrates) and adding: "Where's number 4, Timaeus?" To which Timaeus responds: "he came down with something or other" (ἀσθένειά τις αὐτῷ συνέπεσεν). This obscure introduction was associated by Iamblichus to the Divided Line in an allegorical manner, and he believed that the fourth man was absent "because he was suited for another subject of contemplation, namely the intelligibles."[104] The Middle Platonist Dercyllides saw in this passage an implied authorial signature of Plato himself, who was famously unable to attend Socrates's death in the *Phaedo* on account of illness (Πλάτων δὲ οἶμαι ἠσθένει).[105] If the implied fourth person were to have been Plato, we would see how he had positioned himself as the final, completing member of the *tetraktys* of speakers.

The number 4, then, comes to obtain a significance in the dialogues of Plato that cannot be detected in the extant writings of the Pythagoreans, and, in some sense, it is their "3" that is subsumed under the fourth dimension, that of the simple intelligibles. We also see this in effect in the description of the mathematical studies in the dialogue attributed to Plato and called *Epinomis*,[106] also known in antiquity under the subtitle *Philosopher*, and probably written

104. Iambl. *in Tim.* F 3 Dillon.

105. Procl. *in Tim.* 1.19.32–20.14. A certain Aristocles thought it was Theaetetus, who was said in the *Theaetetus* (142b1–4) to have been injured and sick from dysentery. If this is right, the clever response by Socrates takes on new valence: "it's for you and your friends to fill in for him in his absence." On the commentary tradition concerning the beginning of the *Timaeus*, see Dillon 2006: 21–22.

106. On the question of the authorship of the *Epinomis*, see now Brisson 2005.

by Plato's amanuensis Philip of Opus.[107] At the end of the dialogue, the Athenian Stranger describes four types of mathematical objects, adapting the educational curriculum for the philosopher-kings laid out in the *Republic.* He associates the mathematical sciences that inform the educational curriculum of the young men who will be members of the Nocturnal Council with a progression of mathematical objects in terms of increase in three dimensions (i.e. triple-growth: τρὶς ηὐξημένους) and upward thrust. First, they must study "numbers themselves, as opposed to numbers that possess bodies," which he elaborates to mean "the entire nature and properties of odd and even—all that number contributes to the nature of existing things." From there, the curriculum proceeds to the combined study of planes (geometry) and geometric shapes in motion (stereometry) and then the sequence based on both the arithmetic and harmonic means, which is harmonics.[108] From these sciences, the students will be ready to analyze the highest of the generated mathematical objects, the heavenly bodies, which will compel the budding philosopher to realize that the saying "all things are full of gods" (τὸ θεῶν πάντα πλέα) has been said excellently and sufficiently.[109] These studies are, in the view of the Athenian Stranger, all undertaken for the sake of comprehension of the unity of things in the universe:

> To the person who learns in the right way it will be revealed that every diagram and complex system of numbers, and every structure of harmony and the uniform pattern of the revolution of the stars are a single thing in concord with all these phenomena [τὴν ὁμολογίαν οὖσαν μίαν ἁπάντων]. And it will be revealed to anyone who learns correctly, as we say, fixing his eye on unity. To one who studies these subjects in this way, there will be revealed a single natural bond that links them all [δεσμὸς πεφυκὼς πάντων τούτων εἷς].
>
> ([Plato] *Epinomis*, 991d8–992a1; translation after McKirahan in Cooper and Hutchinson 1997)

In this passage of the *Epinomis*, we see, once again, that mathematics is finally subsumed under the synoptic inquiry into unity, and the student of mathematics

107. D.L. 3.60, by reference to Thrasyllus's ordering. This title was also found on the oldest manuscripts. Of course, the *Philosopher* was one of the dialogues projected to be the conclusion to the tetralogy of *Theaetetus-Sophist-Statesman-*[*Philosopher*] (Pl. *Soph.* 217a4), but it was never written. See Brisson 2005: 13.

108. Where the line fits into this schema is difficult to know, since the author of the *Epinomis* does not discuss it as such.

109. Or so it appears to my eye, but the passage already assumes the educational curriculum in mathematics from the *Republic* and focuses on exegesis of it, namely the unity of the sciences.

is expected to discover, finally, that all things are one, and that there is a "single natural bond that links them all."[110] The mathematical theories developed in rudimentary and sometimes ambiguous ways by Plato became central to the accounts of generation, unity, and being of the intellectuals associated with the Early Academy: in the *Epinomis*, for example, primordial cosmological growth progresses from point[111] to line, line to plane, plane to solid, a theory that appears late in Plato's dialogues and, I think, in response to the old problem of the "Growing Argument" (*Laws* 893e1–894b1).[112] From the moment when the *Epinomis* was written, though, the predication of universal agreement (ὁμολογία) was located alongside the more familiar predications of unity and being ascribed by Aristotle to Plato and the Pythagoreans. Plato's adaptation of the educational quadrivium as developed by the mathematical Pythagoreans and discussed by Archytas of Tarentum[113] led to more vivid proposals concerning the ontological operations of number and of the objects of mathematics in the Early Academy under both Speusippus and Xenocrates, as well as among those Platonists who participated in the *hetaireia* that based itself in the groves near the river Cephisus.[114] Explanations for what were taken to be the core elements of Plato's philosophy began to proliferate after the death of Plato, with each intellectual vying to characterize the "authentic" version of the sage's philosophical dogma either in order to lend

110. This statement echoes Socrates's confession at *Phd.* 99c5–8 that he has not been able to discover in the writings of the mathematical Pythagoreans, Anaxagoras, and other materialists what "really binds and holds together" all things.

111. The status of this "beginning" (ἀρχή) is a source for dispute among the Platonists. See Burkert 1972: 23–27.

112. One of the more creative adaptations of the progression of mathematical objects comes in the Zoroastrian cosmology given by Hermodorus of Syracuse (at Plut. *Is. et Osir.* 46–47, 369d5–370c4), who described Ahura Mazda as increasing himself threefold (τρὶς ἑαυτὸν αὐξήσας) before he adorns the heavens with stars and creates the twenty-four gods. See Horky 2009: 79–85.

113. On which see Huffman 2005: 388–389 and Barker 2007: 311–318.

114. Among the metaphysicians of the Early Academy, Speusippus (F 28 Tarán = Procl. *in Parm.* 7, pp. 38.32–40.7 Klibansky), who wrote a treatise *On Pythagorean Numbers*, seems to have claimed that the Pythagoreans held the One to be higher than being, thus predicating "oneness" of all things, but not being (see Dillon 2003: 56–57). Hermodorus of Syracuse (F 7 IP = Simpl. *in Phys.* p. 247.30–36 Diels), too, seems to have believed that oneness is predicated of all things, but he argued in contrast that "greater and lesser" is predicated of all things *except* for the One. Xenocrates (F 87 IP = Porph. *in Harm.* 30.1–10) apparently followed Plato (*Phlb.* 17c11–d7) in linking the study of harmonics and sound to the study of objects in motion (and attributed the discovery of the inherence of number in music to Pythagoras), and he clearly believed in Form-Numbers (see Horky: forthcoming).

legitimacy to his own version, as in the case of Xenocrates or Hermodorus, or to use it as a point of attack, as in the case of Aristotle. The principles of philosophical explanation developed by Socrates in the *Phaedo* and the *Republic* left their mark on every intellectual who associated with Plato or his friends in the Academy, and it was only natural that they would apply similar methods in their own explanations of the sometimes ambiguous or unclear definitional statements preserved in the dialogues of Plato. Let us not forget that in the *Phaedo* (61a2–3), Socrates had claimed somewhat enigmatically, and by reference to the dreams that came to him, that "philosophy is the highest of the musical arts" (*φιλοσοφίας οὔσης μεγίστης μουσικῆς*).[115] In its own way, then, the *Epinomis* presents an early account from the next generation after Plato, those figures in the Academy who sought to give explanations of what they thought were the basic tenets of *their* master's philosophy as preserved in, among others, the deep and wealthy mines of the *Phaedo*.

CONCLUSIONS

In this chapter, I have extended my analysis of the ways Plato responded to the puzzles concerning numerical and personal identity raised by the "Growing Argument" of Epicharmus by considering his inquiry into number in the *Phaedo*. I sought here to extend my evaluation of the status of number in Plato's metaphysics in the light of the speculations concerning number, becoming, and being in the fragments of two mathematical Pythagoreans, Empedocles and Philolaus. Plato's critique of Empedocles's one/many metaphysics took the shape of a broader challenge to number-stuff theorists to account sufficiently for the causes of unity and division in the light of the causative and predicative functions of the Forms, which Socrates hypothesized in the *Phaedo* apparently without dissent from his mathematical Pythagorean interlocutors Simmias and Cebes (as well as Echecrates and Phaedo).[116] By contrast, Plato more favorably appropriates and extends Philolaus's formulations of "number," "being," and the "proper kinds" of number ("odd" and "even"), which are brought to bear on the axiomatic presentation of the Forms. By appeal to Philolaic number theory, Plato was able to develop an account that could explain with greater clarity and precision[117] not only *why* a number (e.g. 3) *must* possess the

115. By which Socrates, of course, means the "arts of the muses." See Pl. *R.* 8, 548b7–c2 and *Phdr.* 259d5–7. On Socrates's philosophical music, see Morgan 2010b: 71–77.

116. See Herrmann 2007: 216.

117. Although, it must be admitted, scholars from antiquity have continued to debate whether or not the proof is both valid and sound.

specific property of the Form that is over and above it, from which it derives its name and which has no opposite (e.g. the Form of 3), but also why a number refuses to admit of the properties of Forms that are opposite to a Form that is necessarily over and above its Form (e.g. the Odd). From this point of view, Plato sought to improve on (and not simply to reject) Philolaus's "unclear" (οὐδὲν σαφές) demonstrations of the nature of the universe, of its coming-to-be, and of the things that exist in it.[118] By doing so, Plato was able to develop a new methodological apparatus for inquiry into being and becoming that roughly complements, and reflects the developments in, his approaches to education and knowledge in the *Republic*, which recognized the exalted place of Pythagorean mathematics in philosophical education but subordinated it to the science of dialectic.

What is not treated in the *Phaedo* is Plato's evaluation of Philolaus's description of the cosmological mechanisms, the "limiters" and the "unlimiteds," which do not appear to play any role in Plato's attempts to demonstrate the immortality of the soul. If ontology plays an important but ultimately ancillary role in the *Phaedo* as a subplot to the larger inquiry into the nature of the soul—as it seems to have done in the *Cratylus*, where the primary goal was to arrive at a sufficient theory of naming—then it will not be surprising that Plato's treatment of Philolaic "limiters" and "unlimiteds" occurs in the dialogue that sets itself the goal of discovering "what in the world is good" (ὅτι ποτ' ἐστὶν ἀγαθόν)[119] and elaborates a methodology sufficient for solving the question of its identity: *Philebus*. I turn to this dialogue in the final chapter.

118. Pl. *Phd.* 61d8.

119. Pl. *Phlb.* 13e5–7.

6

The Method of the Gods

Mathematical Pythagoreanism and Discovery

This chapter represents an extension of the investigation undertaken in chapters 4–5 into the ways Plato appropriated and then advanced beyond mathematical Pythagoreanism. My purpose here is to try to understand how Plato strategically employs the "first-discoverer" myths of Prometheus, Palamedes, and Theuth in order to explore what the methods of inquiry practiced by the mathematical Pythagoreans might offer for his philosophy in the later dialogues. This project will require a comprehensive study of "first-discoverer" myths in pre-Platonic literature (especially among the Athenian tragedians Aeschylus, Sophocles, and Euripides and the Sophist Gorgias) in order to establish a foundation from which Plato would develop his own "first-discoverer" treatments. Overall, Plato's employment of the "first-discoverer" myth allows him to attack the positions of his contemporary intellectual competitors without naming them; but, as I argue, his treatment does not allow for easy determination of the objects of his attack because of the literary convention known as "active double-voiced discourse," which resists simple equivalence between literary figure evoked and target of polemic. This chapter will proceed by distinguishing two approaches to the "first-discoverer" myths in Plato's writings: those in the earlier and middle dialogues (*Protagoras*, *Republic*, and *Phaedrus*) and those in the later dialogues (*Statesman*, *Philebus*, and *Timaeus*). Treatments of the "first-discoverer" myth in Plato's earlier dialogues will be shown to respond chiefly to the problems that mathematics and writing, embodied in the sciences of "number" and "letters," as relevant to the pursuance of the Good. These earlier treatments are thus characterized chiefly by polemic against Plato's Sophistic and Pythagorean competitors who were thought to be noteworthy for making discoveries in these subjects. That polemic takes the form of a criticism of the ontological and epistemological status of both the objects of their intellectual pursuits and the medium through which they communicated their discoveries, which Plato considers derivative of the true intelligible reality. But in his treatment of the "first-discoverer" in the later dialogues, which is more positive, Plato demonstrates a reevaluation of what empirical science—especially that employed by the mathematical Pythagoreans in their approaches to harmonic theory—could offer to his own approaches to cosmogony, metaphysics, and dialectic. The kinds of inquiry into harmonic intervals that Plato criticizes for their incapacity to provide a sufficient means to grasp the intelligibles in the *Republic* become fundamental to the metaphysical procedure described as the "gift of the gods" and passed down by a "certain Prometheus" and carried forward by the "forefathers" in Plato's *Philebus* (16c5–10). Speculation since late antiquity on who the "forefathers" might be suggests that the mathematical Pythagoreans, especially Philolaus of Croton and Archytas of Tarentum, are intended referents. Moreover, Plato implicitly ties discoveries in harmonics associated with Archytas and the so-called progenitor of the mathematical Pythagorean method Hippasus of Metapontum to the Demiurge's construction of the world-soul in the *Timaeus*. An overall picture results, in which—even despite the possible multiplication of referents in Plato's myths that result from a surplus of meaning—the late dialogues of Plato appear to associate the "gift of the gods" with mathematical Pythagoreans in particular. As I suggest, the "forefathers" elicit comparisons with Philolaus and Archytas, and the "certain Prometheus" seems to refer to Hippasus of Metapontum, the so-called progenitor of mathematical Pythagoreanism. This chapter thus argues that examination of the "first-discoverer" *topos* throughout Plato's dialogues presents us with what is perhaps the best strategy for this study's investigation into Plato's developmental responses to mathematical Pythagoreanism and its philosophical tenets.

THE PARADIGMATIC FIRST-DISCOVERER IN THE AESCHYLEAN *PROMETHEUS BOUND*

Since Burkert's *Lore and Science in Ancient Pythagoreanism* (1972), the standard way to interpret early Pythagoreanism has been to employ as explanatory poles "religion" and "science." Scholars tend to fall on one side or another on this issue, with some emphasizing the importance of religion and ethics in early Pythagoreanism (e.g. Kahn 2001 and Kingsley 1996), and others the significance of modes of scientific inquiry in Pythagorean philosophy (e.g. Huffman 1993 and 2005, Müller 1997, Barker 2007, and Zhmud 2012). Is this supposed dichotomy simply an anachronism displaced from modern value systems, or is there evidence of the dichotomy between "religion" and "science" in the fifth century BCE? In this chapter, I suggest that such a polarity can be detected in the cultural practices of Greeks in the classical period and, in particular, is invested in the common trope of the "first-discoverer." Concern with human apprehension of the "arts" (*τέχναι*) as an act of impiety against the gods was a canonical *topos* already in the archaic Greek world by the time of Hesiod (*Theogony* 521–616 and *Works and Days* 42–89). The "religion versus science" discourse is invested, in the study of Classical Greek culture in general and of Pythagoreanism in particular, in what Leonid Zhmud usefully calls "heurematography," that is, the surviving written treatments of various "elements of culture as discoveries (*εὑρήματα*)" made by certain "first discoverers (*πρῶτοι εὑρεταί*)," whether divine or human.[1] Again, as I showed in chapter 3, the case of Isocrates's *Busiris* evinces a double-discovery of philosophy, whereby an Egyptian king is said to have discovered it first and Pythagoras to have introduced it again to the Greeks.[2] In the fifth century BCE, a similar *topos* had been developed by the author of *Prometheus Bound*, where the gift of Prometheus is associated with the arts of astronomy, number, and memory:

> *ἀλλ' ἄτερ γνώμης τὸ πᾶν*
> *ἔπρασσον, ἔστε δή σφιν ἀντολὰς ἐγὼ*
> *ἄστρων ἔδειξα τάς τε δυσκρίτους δύσεις.*
> *καὶ μὴν ἀριθμόν, ἔξοχον σοφισμάτων,*
> *ἐξηῦρον αὐτοῖς, γραμμάτων τε συνθέσεις,*
> *μνήμην ἁπάντων, μουσομήτορ' ἐργάνην·*

> But everything they did was
> Without understanding, until I showed them the risings and
> Settings of the stars, a challenge to discern;
> And furthermore, number, eminent among instruments,
> I introduced to them, as well as combinations of letters,
> Memory of everything, industrious mother of the muses.
>
> (AESCHYLUS OR EUPHORION, *Prometheus Bound* 456–461)

1. Zhmud 2006: 12.

2. See chapter 3, section entitled "Pythagoras among the Athenian Philosophers in the Fourth Century BCE."

The author[3] of *Prometheus Bound* has Prometheus pass down his gift of τέχναι to human beings through use of fire,[4] which appears to be paradigmatic for all other τέχναι. But Prometheus here emphasizes the role of the mathematical skills of number and astronomy as antidotes to ignorance. The list of skills introduced here is then completed by the "combinations of letters," a euphemism for writing. Prometheus first fixes on the "risings and settings of the stars," which are called "hard to discern" (δυσκρίτους),[5] followed by number, which is described as "eminent among instruments" (ἔξοχον σοφισμάτων), a euphemism that appears to recall the Pythagorean *acusma* of the "what is to the greatest degree" (τί μάλιστα) sort, namely "What is the wisest? Number" (τί τὸ σοφώτατον; ἀριθμός).[6] The τέχναι bestowed by Prometheus also go on to include carpentry, architecture, animal husbandry, navigation, metallurgy, prophecy, and medicine. The revelation of such divine secrets to humankind by Prometheus is seen as an extreme act of philanthropy in the face of the authoritarian rule of Zeus. Importantly, several of these gifts are described as *arts of discernment*, which include astronomy, mathematics, and grammar; notably, in the context of *Prometheus Bound*, these arts of discernment, which are reflected

3. It should be noted that these lines have given scholars reason to assume that the author of this *Prometheus Bound* is a later poet, perhaps Euphorion in the 430s, on the grounds that a fragment attributed to Aeschylus, slightly different from the one in *PV*, is the original that the author of *PV* modified. The fragment under discussion (Aeschylus F 181a = Stob. *Ecl.* 1 Prologue) attributes to Palamedes (not Prometheus) the claim that "ἔπειτα πάσης Ἑλλάδος καὶ ξυμμάχων / βίον διῴκησ' ὄντα πρὶν πεφυρμένον / θηρσίν θ' ὅμοιον· πρῶτα μὲν τὸν πάνσοφον / ἀριθμὸν ηὕρηκ', ἔξοχον σοφισμάτων": "Then I organized the life of all the Greeks and their allies, which previously had been as chaotic as that of beasts. To begin with, I invented the ingenious art of number, supreme among all techniques!" (translated by Sommerstein). On this thorny issue, see Sommerstein 2000: 121–122, with bibliography. It should also be noted that these lines are the ones known to Plato, who has Socrates (*R.* 7, 522c–d) refer to Aeschylus as the authority behind the claim that the discovery of number (ἀριθμὸν εὑρῶν) should be associated with the general Palamedes in the context of criticizing Archytas's philosophy. See below.

4. Ps.-A. *PV* 253–254 and 110–111. See Thomas 2006: 221–226.

5. The same adjective is applied to "chance utterings" (κληδόνας) in line 486 and dreams (κἄκρινα πρῶτος ἐξ ὀνειράτων), all part and parcel of the mantic art.

6. Cited twice, by Aelian (*VH* 4.17) and Iamblichus (*VP* 82, 47.17). See chapter 5. Griffith (1983: 169) protests that Pythagoras cannot be an intended referent here, on the grounds that medicine receives a similar treatment (*PV* 477–483); but of course Iamblichus's list of Pythagorean *acusmata* also includes "what is the wisest of things *among us*? Medicine." It is also worth mentioning that if the *Prometheus Bound* was composed in the last quarter of the fifth century BCE, it might be thought to correspond with the publication and explanation (ἐξήγησις) of the *acusmata* by Anaximander the Younger (on which see Burkert 1972: 166 and FGrHist 9 T 1), but this can only be conjecture.

in the use of the verb κρινεῖν and its cognates, are chiefly directed toward making sense of what the gods know and what humans could not previously understand. Divine types of judgment, in a sense, become accessible to humankind, with the consequence that the oracles of Zeus, once "indecipherably without meaning" (ἀσήμους δυσκρίτως), can subsequently be understood by mortals.[7] Because he bestowed the arts of discernment on humankind, however, Prometheus is forced to pay the penalty, a relatively common *topos* in the Greek world during the mid-fifth century BCE, exemplified in various defenses of Palamedes such as that of the Sophist Gorgias of Leontini (in Sicily) or those of the Athenian tragedians Sophocles and Euripides.[8]

It is remarkable in *Prometheus Bound* that we do not find among Prometheus's gifts to humankind the capacity to live with one another in a political organization. Politics is not, strictly speaking, associated with those arts of discernment that semantically refer to some type of intellectual "judgment" (κρίσις). Rather, *Prometheus Bound* emphasizes how Prometheus is a first-discoverer of those arts that can be employed in order to understand the natural universe, especially through "discernment" of things.[9] It is not clear who wrote *Prometheus Bound*, but there are still good reasons to see it in the light of Presocratic thought, especially inquiry into natural science.[10] Before the middle

7. Ps.-A. *PV* 662.

8. Aeschylus, Sophocles, and Euripides each wrote plays called *Palamedes*, and the latter's surviving fragments are of significance here. Euripides (F 578 Nauck = Stob. *Ecl.* 2.4.8) presents Palamedes as a philanthropic first-discoverer in a way similar to the author of *Prometheus Bound*: "I alone articulated [ὀρθώσας] the remedies for forgetfulness, both those that are voiced and unvoiced, by setting up syllables; I invented writing for men to know, so that a man, if he is absent and over the ocean's plain, might have good knowledge of all things back at home, and a dying man might thereby record the measure [μέτρον γράψαντα λείπειν] of his wealth, and the inheritor know it. And the evils that befall men in a state of quarreling—these a written tablet sunders [διαιρεῖ], and it prevents the telling of lies." Gorgias's Palamedes (*Palamedes* 30) makes reference to the *topos* in a metanarrative fashion, implying the audience's familiarity with it: "I might say, and saying it I would not lie nor would I be refuted, that I am not only blameless but also a great benefactor of you and the Greeks and all mankind . . . for who else would have made human life accessible out of intractable [πόριμον ἐξ ἀπόρου] and ordered out of unordered [κεκοσμημένον ἐξ ἀκόσμου], by inventing military equipment of the greatest advantage and written laws, the guardians of justice [νόμους τε γραπτοὺς, φύλακας τοῦ δικαίου], and letters, the tool of memory [γράμματά τε μνήμης ὄργανον], and measures and weights, the convenient standards of commercial exchange, and number, the guardian of items [ἀριθμόν τε χρημάτων φύλακα], and the very powerful beacons and very swift messengers, and draughts, the harmless game of leisure? *Why do I remind you of these*?" (translation after Freeman; italics mine).

9. See Irby-Massie 2008: 139–140.

10. Generally, see Irby-Massie 2008.

of the fifth century BCE, such inquiries were undertaken chiefly in two areas of the Greek world: in Ionia, on the coast of Asia Minor, and in Western Greece, especially in Sicily and Southern Italy. Miletus, in Ionia, had been a trading hub that linked Persia to Greece at least since the seventh century BCE, and we can infer strong relationships between the extensive mercantile trade and exchange of ideas there.[11] The same goes for Ephesus, which is likely to have profited commercially following the sack of Miletus in 494 BCE, as it became the western end of the King's Road.[12] In Miletus, Thales, Anaximander, and Anaximenes each employed various approaches to "measuring" things in the universe, which led to what Stephen White has aptly deemed "the first scientific revolution."[13] Measurement also played a significant role in the slightly more abstruse philosophy of Heraclitus of Ephesus, for whom change—a form of measured exchange between oppositional forces—reflects the intelligible order of the universe.[14]

While the author of *Prometheus Bound* draws broadly from Ionian natural philosophy, the philosophy of Heraclitus in particular provides a useful point of context for the gift of Prometheus. This is unsurprising: the author of *Prometheus Bound* demonstrates awareness of Ionian natural philosophy in a variety of ways, but the primacy of fire in the myth naturally solicits comparisons with pyrocentric philosophical models among the Presocratics. In the *Prometheus Bound*, the protagonist himself claims that fire is the means[15] by which humans will eventually come to discover technology.[16] For Heraclitus, similarly, the world is understood to be an eternal process of the measured changes of fire.[17] Knowledge of the γνώμη, how all things are steered, is the mark of

11. See McKirahan 1994: 20–22.

12. See Burkert 2004: 107.

13. White 2008: 122. Of course, it would be more correct to call it the "first *Greek* scientific revolution." See Burkert's contribution (2008) in the same volume.

14. See White 2008: 121 and Long 2009, who summarizes neatly (107): "[Heraclitus] intuited the unifying power of structure, measure and proportion in the world's physical processes; took these to be instantiated in the operation of divine intelligence; and, in his greatest and most far-reaching innovation, posited human capacity to think and speak commensurately—i.e. in accordance with nature, and therefore rationally."

15. Specifically, fire is the "path" or "means" (πόρος) that makes possible the τέχναι (*PV* 477). In Homer, πόρος refers to a path, or to a natural or artificial means to cross something (see *LSJ* 1 and 3), and in Herodotus (2.2) it is explicitly related to intellectual discovery through technology. Also see Eur. *Med.* 1418.

16. Ps.-A. *PV* 253–254.

17. DK 22 B 30.

wisdom.[18] Fire, which appears to be the same thing as soul,[19] is a ruling element in the universe,[20] but apparently it cannot overstep its measures, lest it be punished by the Erinyes.[21] It is difficult to synthesize these obscure statements into a unified theory of the epistemological function of fire, given their gnomic formulation and presentation in isolation from one another in Heraclitus's corpus; but it remains plausible that, for Heraclitus, knowledge of fire and of its chief attributes (e.g. that it is measured and eternally undergoes change within those measures) reflects a more universal understanding of the modalities of the universe.[22]

Just as we see in *Prometheus Bound*, fire and discriminatory understanding of the universe are associated in the fragments of Heraclitus. For Heraclitus, "fire, when it advances upon them, will judge and overtake all things" (*πάντα τὸ πῦρ ἐπελθὸν κρινεῖ καὶ καταλήψεται*).[23] When Hippolytus quotes this line, he relates it to the principle that fire would be that which would judge the entire universe, probably with an eye to the periodic *ἐκπύρωσις* of Stoic cosmology.[24] Be that as it may, it is still likely that *κρινεῖ* in Heraclitus's fragment extends the language and semantics of justice beyond the simple juridical context and into epistemology.[25] Heraclitus elaborates on the basic principle of "judgment" by coupling it[26] with the verb *καταλήψεται*, which, in another

18. DK 22 B 41. On this relationship, see especially Long 2009: 104–105. I will not go into the complexities involved in making sense of Heraclitus's text, for which see KRS 202 n. 1. Also note that *γνώμη* is an operative term in Democritus's epistemology, which can be specified further into two types (DK 68 B 11): "bastard" (*σκοτίη*), which deals with sensibles, and "genuine" (*γνησίη*), the object of which is what is too small to be seen. See Taylor 1999: 218–219.

19. DK 22 B 31 and 36.

20. DK 22 B 64.

21. DK 22 B 94. That is, on the assumption that the sun and fire are equivalent, or that the sun possesses the same attributes as fire.

22. For one attempt to fix these relationships, see Long 2009: 99–102.

23. DK 22 B 66.

24. Obviously, the implications of this contextualization are contested.

25. See Kahn 1979: 274–275 and 337 n. 46. One might contextualize this fragment with another of Parmenides (DK 28 B 8.53–57), who describes how mortals, with regard to Light and Night, "made up their minds to name two forms" (*μορφὰς γὰρ κατέθεντο δύο γνώμας ὀνομάζειν*) and "distinguished them as opposites in outer appearance" (*τἀντία δ' ἐκρίναντο δέμας*). Of course, Anaximander (DK 12 B 1), in some way, understood the principles of order in the universe to be consequent of recompense paid for injustices.

26. Is the *καὶ* epexegetical? Without further evidence, we cannot be sure.

fragment of Heraclitus, denotes an action leveled against those who purport to be wise.[27] When Clement of Alexandria cites this other fragment, he does so in the context of epistemology:[28]

> What the most famous man knows and guards are but opinions. . . . Dike [Justice] will judge those who fabricate lies as well as their witnesses.
>
> *δοκέοντα γὰρ ὁ δοκιμώτατος γινώσκει, φύλασσει· [καὶ μέντοι καὶ] Δίκη καταλήψεται ψευδῶν τέκτονας καὶ μάρτυρας.*
>
> (HERACLITUS DK 22 B 28 = Clement of Alexandria, *Stromata* 2.331.20)

Justice, in Heraclitus's philosophy, seems to be an activity coordinate (and possibly coextensive) with discriminatory thinking. As with Prometheus, discriminatory thinking is somehow coupled with both *fire itself*, as a discriminating agent that makes things in the universe change, and *the understanding of fire*, that it rises and falls in measured parts and is thereby representative of the compensatory activity of Justice herself. Importantly, moreover, Heraclitus holds that Dike will judge the "most famous man" (δοκιμώτατος) who is responsible for fabricating lies, a description that no doubt recalls Heraclitus's reference to Pythagoras as *ἀρχηγὸς κοπίδων* elsewhere (DK 22 B 81). The associations between fire-judgment and Pythagoreanism are early, although we cannot infer from this evidence that Heraclitus considered Pythagoras to have been a Promethean figure *simpliciter*, although Heraclitus clearly disapproved of Pythagoras's intellectual contributions, whatever they might have been.

It is also important to emphasize the differences between Prometheus's gift of fire-*τέχναι* in *Prometheus Bound* and Heraclitus's description of fire as a discriminating or judging agent. We might recall that Prometheus sees himself as a consummate philanthropist whose gift is to the benefit of all humankind; Heraclitus does not discuss the "discovery" or "gift" of fire to precivilized humans, and it is also remarkable that Heraclitus's fragments do not evince any particular concern with *τέχναι*, except, as I will show, in negative terms. Of relevance to my investigation into mathematical Pythagoreanism, Heraclitus not only emphasizes the difference between divine and human knowledge, but he appears to distinguish between *various types* of human knowledge, namely, those that are true and properly discriminated and those lies that are based

27. See DK 22 B 40 and 129.

28. Still, it is probable that Clement has collocated two otherwise unconnected fragments, which is why I have bracketed *καὶ μέντοι καὶ*. See Marcovich 1967: 435–436.

solely on opinion and are unreliable.[29] In the fragment cited above, Heraclitus seems to distinguish between those who have real knowledge of the universe and those who only produce ungrounded human opinions, perhaps for the sake of popularity. Again, the likely referent here is Pythagoras, whose special claim to wisdom Heraclitus elsewhere rejected:

> Pythagoras, son of Mnesarchus, practiced inquiry to the greatest extent of all men, and by making a selection of these writings, he contrived a wisdom of his own: much learning, base trickery.
>
> *Πυθαγόρης Μνησάρχου ἱστορίην ἤσκησεν ἀνθρώπων μάλιστα πάντων καὶ ἐκλεξάμενος ταύτας τὰς συγγραφὰς ἐποιήσατο ἑαυτοῦ σοφίην, πολυμαθίην, κακοτεχνίην.*
>
> (Heraclitus DK 22 B 129 = Diogenes Laertius 8.6; translation after Marcovich 1967)

Heraclitus thus characterizes Pythagoras as a swindler who, by selecting from some unknown writings, "contrived a wisdom of his own" (*ἐποιήσατο ἑαυτοῦ σοφίην*), which Heraclitus characterizes as "much learning" (*πολυμαθίη*) and "base trickery" (*κακοτεχνίη*).[30] The extant fragments of Heraclitus do not seem to endorse the "arts" as such, and it is plausible that the "base trickery" ascribed to Pythagoras here is to be associated with the sort of spurious argumentation, or deficient understanding, that "inquiry" (*ἱστορίη*) into a wide variety of subjects may be thought to produce. It is difficult to know with more precision what the actual philosophical activities of Pythagoras were,[31]

29. See DK 22 B 1, where Heraclitus declares that he makes his explanations by "differentiat[ing] each thing according to its nature" (*κατὰ φύσιν διαιρέων ἕκαστον*). This distinction between reliable and unreliable knowledge is found in Parmenides's poem and suggested by the fragments of Xenophanes, on which see Kahn 1979: 210–211. But the binary of "truth/falsehood" in cosmic terms is strongly associated with Persian religion in both Greek and Persian sources, on which see Horky 2009: 51–66. "Truth" and "opinion" are surprisingly not differentiated as such by other early Ionian historiographers, such as Hecataeus of Miletus (FGrHist 1 F 1).

30. See DK 22 B 81, where Heraclitus apparently called Pythagoras the "prince of lies" (*κοπίδων ἀρχηγός*), derived from Timaeus of Tauromenium (FGrHist 566 F 132), who explicitly denied that Pythagoras was the "discoverer" (*εὑρετής*) of such clever tricks. My treatment of this Pythagoras and the Pythagoreans in the writings of Timaeus of Tauromenium is in chapter 3, section entitled "Pythagorean Exoterics in the Fifth Century BCE? The Historical Evidence of Timaeus of Tauromenium."

31. Scholars' responses to this question have been various, and I will not weigh in on how this evidence can tell us about the actual activities of Pythagoras. For useful recent discussions, see Zhmud 2012: 32–35, Riedweg 2005: 49–52, and Kahn 2001: 14–16.

but we are on sure ground to conclude that Heraclitus found Pythagoras's "wisdom" (σοφίη), whatever it entailed, to be wanting, possibly because Pythagoras was thought to lack the proper discriminatory capacities to distinguish between what Heraclitus would describe as true "understanding" and false "opinion."[32]

If we accept the positive correspondences between fire and the activity of discrimination underscored by both Heraclitus and the author of *Prometheus Bound*, we may admit the possibility that the author of *Prometheus Bound* appeals broadly to a Heraclitean philosophical tradition in his staging of the fall of the philanthropic giver of τέχναι, Prometheus. It can be concluded that the heurematographical *topos* of the "first-discoverer," who gave fire and the arts of discrimination to humans, is embedded more broadly in literary, dramatic, and philosophical discourses concerning epistemology, especially divine versus human knowledge. This *topos* appears in Athens at least as early as Aeschylus (first half of the fifth century BCE), who either described the "first discoverer" of number as Prometheus (if the *Prometheus Bound* is indeed by him) or Palamedes (if the *Prometheus Bound* is a later composition that imitates Aeschylus's *Palamedes*, perhaps written by another Athenian tragedian influenced by Sophistic intellectual culture). The *topos* was probably codified and given classifications by the Sophist Hippias of Elis in his *Anthology*, a work that, as Jaap Mansfeld has argued, may have influenced Plato greatly.[33] It is important for us to keep in mind this heurematographical tradition that ascribes such discovery of the arts of distinction by measurement to Prometheus and/or Palamedes, in great part because, as I will now show, this tradition held a great deal of influence over Plato in his writings about the development of the philosophical and political arts and methods, as described in five dialogues, the *Protagoras*, *Republic*, *Phaedrus*, *Statesman*, and *Philebus*.

PLATO'S EARLIER TREATMENTS OF FIRST-DISCOVERERS: PALAMEDES, PROMETHEUS, AND THEUTH IN THE *PROTAGORAS*, *REPUBLIC*, AND *PHAEDRUS*

In the light of the literary traditions that correlate the discovery of number and an art of discrimination that trace back to the Aeschylean tradition in Athens, and the rich developments that followed in the writings of the other

32. Xenophanes might have found a similar fault with Pythagoras (DK 21 B 7). His criticism of Pythagoras could be formulated as his failure to distinguish properly between a human being and a dog (see Lesher 1992: 80 and, more recently, Schäfer 2009: 52–53).

33. Mansfeld 1990: 84–96.

tragedians and Sophists, it should be unsurprising to detect similar heurematographical tendencies in Plato's writings. An important example of the description of Palamedes's art occurs in the midst of Plato's attack on Archytas of Tarentum and mathematical Pythagorean empiricism in *Republic* 7. After describing "number and calculation" (*ἀριθμός τε καὶ λογισμός*) as what every art and science, of necessity, has a share of (*πᾶσα τέχνη τε καὶ ἐπιστήμη ἀναγκάζεται . . . μέτοχος γίγνεσθαι*) at 522c6–8, Socrates argues that the military art also involves "calculation" by way of a clever reference to Palamedes:

> In the tragedies, at any rate, Palamedes is always showing up Agamemnon as a totally absurd general. Haven't you noticed? He says that, by inventing number [*ἀριθμὸν εὑρών*], he established how many troops there were in the Trojan army and counted their ships and everything else—implying that they were uncounted before and that Agamemnon (if indeed he didn't know how to count) didn't even know how many feet he had. What kind of general do you think that made him?
>
> A very strange one, if that's true.
>
> Then won't we set down this subject as compulsory for a warrior, so that he is able to count and calculate [*λογίζεσθαί τε καὶ ἀριθμεῖν δύνασθαι*]?
>
> More compulsory than anything. If, that is, he's to have any understanding about setting his troops in order, or if he's even to be properly human.
>
> Then do you notice the same thing about this subject that I do?
>
> What's that?
>
> That it turns out to be one of the subjects we were looking for that naturally leads to understanding [*πρὸς τὴν νόησιν ἀγόντων φύσει*]. *But no one uses it correctly, that is, as something that is really fitted in every way to draw one toward being* [*ἑλκτικῷ ὄντι παντάπασι πρὸς οὐσίαν*].
>
> (PLATO, *Republic* 522d1–523a3; translation after Grube and Reeve in Cooper and Hutchinson 1997; italics mine)

This description of Palamedes's discovery of "number" (*ἀριθμὸς εὑρών*) and employment of "calculation" (*λογίζεσθαί*) recalls the gnomic lines of Archytas's book *On Sciences* (F 3 Huffman): "once calculation was discovered, it stopped discord and increased concord" (*στάσιν μὲν ἔπαυσεν, ὁμόνοιαν δὲ αὔξησεν λογισμὸς εὑρεθείς*). Archytas does not tell us who he thinks first discovered "calculation," but he does suggest, in the lines that precede these in Iamblichus's and Stobaeus's quotations, that people are able to attain knowledge by means of either learning or self-discovery.[34] Discovery through one's self thus plays a significant role in Archytas's epistemology—at least once calculation has been discovered[35]—and firmly fixes Archytas in the heurematographical tradition that is most especially associated with the Sophists and tragedians before Plato.[36] The claim that the arts of intellection were discovered not by an Olympian god or Titan but by humankind *without* divine intervention is also found in Sophocles's work, and it is likely that it was a subject of Sophistic debate whether humankind received the τέχναι from the Olympian gods, the titan Prometheus, or the human Palamedes.[37]

In the context of Plato's argument in the *Republic*, Socrates goes on to describe how calculation, which has generally been considered useful for military operations, can be employed for the sake of pursuing subsistent "being." Why would Plato refer to the Aeschylean story of Palamedes and Agamemnon at this very point in his exposition on the educational curriculum of the philosopher-kings, couched in his larger criticism of the mathematical Pythagorean method of employing empirical evidence derived from sensible objects in order to make sense of reality? Plutarch, who is perhaps deriving his source material from Eratosthenes's *Platonicus* (third century BCE),[38] may offer a valuable point of context:

> Eudoxus and Archytas and their followers began to set in motion this prized and famous science of mechanics, by embellishing geometry with its subtlety, and, in the case of problems which did not admit of logical and geometrical demonstration, by using sensible and mechanical models as supports. Thus, they both employed mechanical constructions for the

34. "For it is necessary to come to know those things which you did not know, either by learning from another or by discovering yourself [ἢ μαθόντα παρ᾽ ἄλλω ἢ αὐτὸν ἐξευρόντα]. Learning is from another and belongs to another, while discovery is through oneself and belongs to oneself. Discovery, while not seeking, is difficult and infrequent, but, while seeking, easy and frequent, but if one does not know <how to calculate>, it is impossible to seek" (trans. Huffman). In this light, Aristoxenus (F 23 Wehrli = Stob. *Ecl.* 1. Prooem. 6) argued that Pythagoras "seems to have advanced it [i.e. the *pragmateia* concerning numbers] by withdrawing it from the use of merchants and likening all things to numbers."

35. Following Huffman's reading of the text (2005: 189).

36. See Zhmud 2006: 64–66.

37. E.g. the first stasimon of Sophocles's *Antigone* (332–383). Sophocles's stance on the discovery of the "wise thing that contrives the arts" (σοφόν τι τὸ μαχανόεν τέχνας) is ambivalent, however, since he claims that it can be used for good or ill effects.

38. For an excellent discussion of the interpretive problems involved in making sense of this testimony, see Huffman 2005: 370–392.

> problem of the two mean proportionals, which is a necessary element in many geometrical figures, adapting to their purposes certain mean lines from bent lines and sections. But, when Plato was upset and maintained against them that they were destroying and ruining the value of geometry, since it had fled from the incorporeal and intelligible to the sensible, using again physical objects which required much common handicraft, the science of mechanics was driven out and separated from geometry, and being disregarded for a long time by philosophy, became one of the military arts.
> (PLUTARCH, *Marcellus* 14.5–6 = Archytas A15b Huffman; translation by Huffman 2005)

This passage illustrates a methodological dispute between Archytas and Plato on the use of sensible objects in the demonstration of geometrical problems such as the doubling of two mean proportionals. In Plutarch's account, the reference to Plato's disagreement with Archytas and Eudoxus is meant to evince the importance of applying the abstractions characteristic of geometry pragmatically to military technology. This passage may be folkloric, or perhaps derived from a dialogue that staged a debate about the use of mechanical props in solving geometric problems. Whatever its origins, it represents a Hellenistic account that confirms how substantial parts of *Republic* 7, in particular, are leveled against Archytas's philosophical tenets. We cannot be absolutely sure what the "supports" said to have been employed by Archytas and Eudoxus were; if Plato knew about such things, to be sure, his criticisms in *Republic* 7 would apply well to this use of mechanics in geometry.[39] It would not be out of character, moreover, for Plato to employ figures from mythology by proxy in order to criticize his contemporaries. Plato was a sophisticated comic writer, whose jokes we cannot always fully comprehend, sometimes for lack of context. Was Archytas, the famous seven-time general of what was likely the most powerful military in Western Greece in the first half of the fourth century BCE[40]—as well as the philosopher-mathematician who discovered the two mean proportionals through the use of semicylinders—the target of a playful literary reference to Palamedes's discovery of "number" in *Republic* 7?

It is difficult to determine the object of literary reference in Plato's writings. Even in the case of the mythical "first-discoverer" Palamedes, there could be more than one implicit object of Plato's playful attack; for example, the "Eleatic Palamedes" that we find in Plato's *Phaedrus* (261b7–d8) is a rather certain reference to

39. It is worth noting that another Pythagorean from Tarentum to whom inventions in military mechanics are attributed is Zopyrus, who is plausibly associated with Archytas by Zhmud (2012: 129, with n. 111)

40. Str. 6.3.4. See Huffman 2005: 11, with n. 4.

Zeno of Elea, but this passage of *Republic* 7 could not easily be associated with Zeno.[41] Often, it seems, by employing mythological references, Plato aims at several targets at once.[42] As Andrea Nightingale argues, Plato's use of the "first-discoverer" myth reinforces a recurrent theme: the destructive power of the written word for the "all-wise" man, who, *because* he employs the art of writing, meets with injustice and catastrophic suffering.[43] This is not a simple case of Platonic moralizing. It is also built into the fabric of Plato's ideas about "being" and its relationship to literary performance. As Nightingale notes:

> These retrospective assessments of the subtext . . . invite us to reevaluate the story of Palamedes. In particular, we are asked to see Palamedes' death as due to his ignorance about the nature of his invention; if he had understood the true power of writing, Plato suggests, he could have avoided his fate. Instead of the story of a wise and good man who comes to a tragic end, we are presented with an alternative tale in which a proud and self-deceived man is hoist with his own petard. But this, of course, is not a tragedy. This scene, then, offers an excellent example of parody or "active double-voiced discourse." For Plato foists his own interpretation of the tragic tale and, in doing so, rejects what he takes to be the most distinctive "semantic intention" of the genre of tragedy: the claim that a good man can be reduced to wretchedness. By appropriating the tragedy of Palamedes, Plato transforms the alien voice into his own.[44]

41. See Pl. *Parm.* 127d6–128a1.

42. In addition to Archytas as the object of Plato's ridicule in the reference to Palamedes in *Republic* 7, we might also consider the Sophist Hippias of Elis, who (DK 86 B 12 and A 11) claimed to have been proficient in geometry and may have discovered how to trisect a rectilinear angle by means of a quadratix (B 21), as well as Antiphon the Sophist, who is associated with Hippocrates of Chios in trying to tackle the problem of squaring the circle (F 13a–b Pendrick).

43. Nightingale 1995: 149–153, also citing Pl. *Apol.* 41a–b as well as X. *Apol.* 26 and *Mem.* 4.2.33 as *comparanda* for Socrates.

44. Nightingale 1995: 153–154. I would add to her claims that with parody, especially the kind of parody that denies semantic autonomy for the target of the parody, the degree of dialogic that is predicated on difference between assailant and target is reduced, with at least two results: the division between assailant and target is rendered more porous, and consequently the potential for confusion of assailant and target is higher than in a parody that encourages semantic autonomy and authority for the target. There is a further consequence: because semantic autonomy and authority are denied to the target through various tricks of intertextuality, the identity of the target becomes more elusive, and it becomes more difficult to identify the target of the parody at any given point, to say "here Plato is attacking X." Platonic parody of this sort resists the identification of the target X, since it multiplies the possible referents by confusing the voices of the implied target with other targets and, moreover, with the assailant.

Because of Plato's unique employment of what Nightingale calls "active double-voiced discourse" when drawing references to mythological and historical predecessors, we are both stimulated to seek the identity of these figures in such mythological presentations and frustrated by our failure to determine it.[45] It is, I suggest, a common strategy of Plato to both appropriate the philosophical content of his personal intellectual competitors (Heraclitus, Empedocles, Epicharmus, Anaxagoras, Cratylus, Socrates, Parmenides, Philolaus, Isocrates, Archelaus, Archytas, etc.) and appraise its value for his philosophical project at the same time. Usually, that criticism is related to what we take to be a fundamental novelty in Plato's philosophy, for example, the postulation of the Forms, the advancement of dialectic, the establishment of the rules of predication, or the hypothesizing of teleological causation.[46] But Plato tends to do this in a way that is extremely vexing for historians of thought, since he tends not to name the living figures whose thought he wishes to acknowledge some debt to *in propria persona* and instead recasts them in the garb of mythological figures.

There is an added challenge to interpreting the role the "first-discoverers" played in Plato's philosophy. Because he eschews any first-person authority in his dialogues, we are forced to read his "first-discoverers" through the dramatic lenses of the speakers who bring them up. A classic example of this is the famous "Great Speech" of Protagoras, found in the *Protagoras*, in which (320c2–7) Protagoras gives an epideictic showpiece in anticipation of his demonstration of how virtue is teachable. This showpiece takes the form of a heurematographical myth involving the allotment of powers to all creatures by Epimetheus and Prometheus. At the allotted time, creatures are generated from the mixture of earth and fire and their damp compounds, a statement that recalls especially Heraclitus's cosmology.[47] Initially, so Protagoras's story goes, Zeus orders Prometheus and his brother Epimetheus to "organize and distribute the powers" (*κοσμῆσαί τε καὶ νεῖμαι δυνάμεις*) of the animals carefully. The distribution takes the form of a complex quantitative and qualitative leveling out (*ἐπανισμῶν ἔνεμεν*) of "powers," in which, for example, Epimetheus compensates for the small size of birds by adding wings. Such a distribution is described

45. It should be noted that I consider Nightingale's "active double-voiced discourse" to be applicable not only in parody per se, but more generally in Plato's treatment of his predecessors' thought, especially in mythological contexts. Of course, because Plato is required to restate his predecessors' beliefs (and sometimes does so in humorous contexts), the relationship between Plato's restatement and the original thought is broadly paralogical (i.e. it approximates the original meaning and language, and sometimes adopts it explicitly, but always recontextualizes it in new frameworks).

46. See chapter 5.

47. DK 22 B 30 and 90. See Denyer 2008: 101.

as being what is "appropriate" (ὡς πρέπει) for each creature. There is a strong emphasis on proper distribution according to qualitative and quantitative criteria, especially given the fact that it is Zeus, the paradigmatic distributor in traditional Greek culture, who commands Prometheus and Epimetheus to create animals.[48]

But, Protagoras continues, poor Epimetheus, "After-thought" embodied and therefore "not terribly wise" (οὐ πάνυ τι σοφὸς ὤν), forgets about human beings! At the last minute, Prometheus, or "Fore-thought," who is assumed to be wise, surveys the distribution and recognizes that human beings have been allotted no "powers" to prevent their extinction by other animals or the hostile environment.[49] It is under these circumstances that Prometheus bequeaths humans with the divine gifts:

> It was then that Prometheus, desperate to discover [εὕροι] some means of survival for the human race, stole from Hephaestus and Athena wisdom in the practical arts together with fire [ἔντεχνον σοφίαν σὺν πυρί]—without which this kind of wisdom is effectively useless [ἀμήχανον]—and gave them outright to the human race. The wisdom it acquired was for staying alive; wisdom for living together in society, political wisdom, it did not acquire, because that was in the keeping of Zeus. Prometheus no longer had free access to the high citadel that is the house of Zeus, and besides this, the guards there were terrifying. But he did sneak into the building that Athena and Hephaestus shared to practice their arts, and gave them to the human race. And it is from this origin that the resources human beings needed to stay alive came into being. Later, the story goes, Prometheus was charged with theft, all on account of Epimetheus.
>
> (Plato, *Protagoras* 321c7–322a2; translation after Lombardo and Bell in Cooper and Hutchinson 1997)

Following the distribution of "wisdom in the practical arts together with fire," human beings start to build (ἱδρύεσθαι) altars and images of the gods, to articulate (διηρθρώσατο) oral speech and written words, and to discover (ηὕρετο) houses, clothing, shoes, blankets, and means to grow fruit from the earth, all on

48. Denyer (2008: 102) notes that various forms of νεμ- appear seven times in this passage.

49. Compare the end of the myth of the *Statesman*, where (274b1–d8) the Eleatic Stranger claims that humans, with the care of the gods wanting, were under threat of extinction and had to take care of themselves using the gifts of the gods, which are fire from Prometheus, metalwork from Hephaestus, weaving from Athena, and agriculture from other gods, "along with an indispensable requirement for teaching and education" (μετ' ἀναγκαίας διδαχῆς καὶ παιδεύσεως).

their own. Thus Protagoras posits a story in which the imparting of the *ἔντεχνον σοφίαν σὺν πυρί* leads to the development of another art that makes their survival possible, which Protagoras describes as "the demiurgic art" (*ἡ δημιουργικὴ τέχνη*). Subsumed under "the demiurgic art" in Protagoras's myth are the sister arts of building, language, weaving, cobbling, and agriculture. Each of the various arts that fall under the "demiurgic art" is distributed to individuals who will become specialists, and who will use their own respective skills to the benefit of many other people.[50] It is not immediately clear how fire is a necessary component for wisdom in the practical arts (*ἔντεχνον σοφίαν σὺν πυρί*) to escape being "effectively useless" (*ἀμήχανον*). But if we consider that fire is needed for the creation of mechanisms by which such productive arts can be achieved, that is, through smelting of metal objects that create tools, it becomes clear (with the notable exception of language) that fire is indeed requisite for the fruits of those technologies that would be considered "demiurgic."

Protagoras develops the myth of Prometheus beyond what I have shown so far in the *Prometheus Bound* by explicitly establishing an aetiology for the political art (*ἡ πολιτικὴ τέχνη*).[51] The "demiurgic" art alone suffices to render humans capable of survival on their own, at least until they encounter wild animals or other humans that are capable of overcoming them, according to Protagoras. But without the "political" art, of which the "military" (*πολεμική*) art is a part, human beings are incapable of surviving in the long run.[52] Zeus, concerned about the dissolution of humankind, once again makes a command—this time to Hermes[53]—to bring justice (*δίκη*) and shame (*αἰδῶς*) to humankind, in order to promote order and establish friendships between humans. Hermes asks Zeus about the nature of the distribution:

> "Should I distribute them as the other arts were? This is how the others were distributed: one person practicing the art of medicine suffices for many ordinary people; and so forth with the other craftsmen. Should I establish justice and shame among humans in this way, or distribute it to

50. See Pl. *Prt.* 322c5–d2.

51. Betegh (2010: 222) argues that the appropriate context for a *μῦθος*—whether in Plato's writings or elsewhere—is aetiology.

52. Pl. *Prt.* 322b1–c1.

53. Note that it is, once again for Plato, the Egyptian Theuth (who was the same figure, for many Greeks, as Hermes) who is paired with the "first-discoverer" Prometheus and provides a context for Palamedes. Rowe (1998: 208–209) acutely notes the false etymology between *Theu*th and Prome*theus*. For a useful discussion of Plato's use of these figures, see Nightingale 1995: 149–151. Also see below.

> all?" "To all," said Zeus, "and let all have a share [μετεχόντων]. For cities would never come to be if only a few have a share of [μετέχοιεν] these, as is the case with the other arts. And establish this law as coming from me: Death to him who cannot partake of [μετέχειν] shame and justice, for he is a pestilence to the city."
>
> (PLATO, *Protagoras* 322c5–d5; translation after Lombardo and Bell in Cooper and Hutchinson 1997)

As Fritz-Gregor Herrmann has argued, Zeus's employment of the term μετέχειν here refers to a kind of universal distribution that is not expressly quantitative. At least so far as Protagoras's myth goes, there is no sense that some future citizens, for example, will "share *more* of" shame and justice than others.[54] Later on, in the demonstration portion of his speech, Protagoras will differentiate various degrees of excellence (ἀρετή) in various skills such as flute-playing (e.g. 327a4–7, 328a6–b1), but we should be careful not to confuse the types of skill here: excellence of these sorts is the purview of the "demiurgic art," not the "political art," whose attributes (shame and justice) are available to all future citizens of the *polis*.

In Protagoras's "Great Speech" in the *Protagoras*, then, Plato's treatment of the two arts of culture, the "demiurgic" and the "political," assumes and appropriates a traditional discourse by ascribing the "gifts" to Prometheus and to Hermes, respectively, via the order of Zeus. Likewise, in the *Phaedrus*, Plato effectively synthesizes the philanthropic Titan and the messenger-god, where Socrates tells an Egyptian tale—heard from "those who came before" (οἱ πρότεροι)—in which the *daemon*[55] Theuth, "first discoverer (πρῶτον εὑρεῖν) of number and calculation, as well as geometry and astronomy, and draughts and dice, and what is more, even letters," conducts a discussion with Thamus, the divine king of all Egypt.[56] Theuth visits Thamus and, after "showing off the arts" (τὰς τέχνας ἐπέδειξεν), suggests that they should be passed onto the rest of the Egyptians. Thamus seeks to understand the "benefit" (ὠφελία) each art presents, so they undertake a debate in which Theuth presents each art in order and Thamus gives reasons for or against each of them. Socrates passes over the debates concerning number, calculation, geometry, and astronomy, as well as games, but pauses to elaborate further on Thamus's evaluation of letters, which

54. Herrmann 2007: 37–41.

55. It may be significant that Theuth is described as a *daemon*, like Eros in Plato's *Symposium*, who functions as an intermediary between the gods and humans.

56. It should be noted that Ammon was associated with Zeus in Herodotus (2.42), an association that may have been made as early as Pindar. See Lloyd 1976: 195–198.

Theuth had promised would make the Egyptians "wiser" (σοφώτεροι) and "improve their memory" (μνημονικώτεροι):

> Most crafty [τεχνικώτατε] Theuth, one man has the ability to beget [τεκεῖν δυνατὸς] the elements of an art, but another has the ability to distinguish [κρῖναι] what portion of harm or benefit it holds for those who are intending to make use of it. So now you, since you are father of letters, have been led by your affection for them to confuse the capacity [δύναται] of their art with its opposite. For your invention will produce forgetfulness in the souls of your students through a lack of practice at using their memory, since, through their trust in writing, they are reminded by imprints foreign [to the soul] from the outside, and not from the inside, themselves by themselves [ἀυτοὺς ὑφ' αὑτῶν]; it's a medicine not *for memory* but *for reminding* that you've discovered. To your students you've given an appearance [δόξα] of wisdom, not the reality [ἀλήθεια] of it.
>
> (PLATO, *Phaedrus* 274e7–275a8; translation after Rowe 1999)

As commentators have frequently noted, this is a rather pessimistic view concerning the role the arts play in human success. Theuth's invention of letters, which is intended for the sake of making human beings more able to recall what they have heard, turns out to be no better than a mnemonic device for reminding them of what they don't *actually* know.[57] In order to make sense of this passage in the context of heurematography, it is important to note how Plato distinguishes the two "capacities" associated with, respectively, Theuth and Thamus: the capacities to "give birth" (τεκεῖν) and to "distinguish" (κρῖναι) between what is beneficial or not beneficial to human beings.[58] Theuth thus plays the part of a mad scientist who is unable to predict the ill effects his invention will wreak among its users. Because the art of letters is not developed to produce soul-imprints that are "themselves by themselves" (ἀυτοὺς ὑφ' αὑτῶν)—note the familiar language of the Forms here—it will threaten to

57. Useful treatments of recollection in the *Phaedrus* include Morgan 2010a: 56–63 and 2000: 217–222, as well as Griswold 1986: 204–209. On soul-imprint in Plato's writings, see Horky 2006.

58. It is worth comparing this description with the midwife passage of Plato's *Theaetetus* (150a8–b4), where Socrates compares his own art with those of the midwife who, after the mother has given birth (τίκτειν), is expected to "distinguish" (κρίνειν; διαγνῶναι) whether the child is a "true" (ἀληθινά) offspring or a mere "image" (εἴδωλα) of one. Both accounts thus posit two separate activities in the legitimation of something under consideration, namely its discovery through "giving birth" and its evaluation through "judgment." Later on (150c8–d2), Socrates claims: "In fact, I myself am not notably wise, nor can I claim as the child of my own soul any discovery [εὕρημα] worth the name of wisdom."

function as an impediment toward attainment of the "truth" (ἀλήθεια).[59] Written discourse and the letters that form its elements are alien to what the soul really understands and is capable of recollecting. Even if, however, the art of letters should be understood itself as one of the arts of "judgment" or "discernment" that follow from the arts of number and/or calculation, we should be careful not to assume that *all* the arts listed by Theuth should be considered subject to Thamus's criticism. After all, the "judgment" (κρῖναι) of what is beneficial or not beneficial—that is, the recognition of the good or bad effects of artistic creation—is understood to be something fundamental to the coordinate ontological and ethical evaluations of truth (ἀλήθεια) in Thamus's reply. In the larger context of Plato's desire to promote philosophical dialectic, it may be significant that the activities of "discovery" or "giving birth" to ideas and the "judgment" of what is beneficial or not beneficial in them are practiced by two interlocutors, exemplified in the characters of Theuth and Thamus.

How has Plato modified his apparently earlier accounts of Prometheus's gifts in the *Protagoras* and Palamedes's inventions in the *Republic*? In particular, the "art of letters" plays a role hitherto unstated in Plato's treatment of the first-discoverer *topos*, and never emphasized in his previous dialogues. Theuth's celebrated description of writing as the "medicine for memory and wisdom" (μνήμης καὶ σοφίας φάρμακον) hearkens back to the tragedies of *Palamedes* by Sophocles and Euripides and appropriates their versions to Plato's own purposes in the *Phaedrus*.[60] It also recalls the *Defense of Palamedes* by the Sophist Gorgias of Leontini, who highlights the benefits that Palamedes's inventions of "written laws, the guardians of justice [νόμους τε γραπτοὺς, φύλακας τοῦ δικαίου], and letters, the tool of memory [γράμματά τε μνήμης ὄργανον] . . . and number, the guardian of items [ἀριθμόν τε χρημάτων φύλακα]" have presented to humans.[61] Plato thus engages in Sophistic treatments of the subject of the "first-discoverer" (implicitly, via Sophocles and Euripides, perhaps back to Protagoras's lost writings; and explicitly, in his appropriation of Gorgias's *Palamedes*) and criticizes the misuse of the arts of writing by essentializing γράμματα as lower-order objects that belong in the realm of opinion (δόξα).[62] This obviously has had a significant effect on scholars' readings of the *Phaedrus*, as well as the status of the entire

59. Contrast this passage with that a similar one in the *Theaetetus* (206a5–8), where learning the letters "each by itself" (αὐτὸ καθ' αὐτό) "whether they are spoken or written" leads to a basic understanding of simples and their complexes. See Menn 1998: 300–303.

60. Soph. F 479 Radt and Eur. F 578 Nauck.

61. DK 82 B 11a. 30.

62. Pace Vasunia 2001: 150–151.

Platonic corpus.[63] But, in the context of the heurematographical tradition, it has the effect of discrediting the "wisdom" (σοφία) characteristically attributed to the first-discoverer.[64] This challenge to those who claim to transfer "wisdom," of course, plays into the polemic that Plato held against the Sophists. As I mentioned earlier, however, it would be unwarranted to exclude Pythagoreans—especially mathematical Pythagoreans, who practiced, in contradistinction to the acousmatic Pythagoreans, various modes of written "demonstration" (like Sophists) and preserved their philosophical ideas *in writing*—from the objects of Plato's playful banter.[65] The introduction of letters in Plato's *Phaedrus* thus puts the status of the heurematographical *topos* in Plato's writing on ambivalent footing: given Theuth's incapacity to gauge the *effects* of his invention of letters—whether they will benefit or harm their users—we are forced to reconsider the "forethought" associated with Prometheus's name.[66] The *Theuth* of Plato's *Phaedrus*, then, often compared etymologically with Prom*etheus*, threatens to collapse into Epim*etheus*.

In sum, we can detect a relatively consistent evaluation of the heurematographical tradition found in the writings of the Sophists and tragedians in Plato's early and middle dialogues *Protagoras*, *Republic*, and *Phaedrus*. To the "first-discoverer," whether Prometheus, Palamedes, or Theuth, is attributed the invention of the arts

63. See Vasunia 2001: 151 and 155–159; Nightingale 1995: 150–153; Ferrari 1987: 206–222; Griswold 1986: 202–226.

64. We might, with Nightingale (1995: 153), detect a profound irony in Thamus's address to Theuth as "most crafty" (τεχνικώτατε). Socrates in the *Cratylus* (436a1–439b8) at least provisionally accepts the supposition that the figure (*daemon* or god) who first gave names to things could have done so *correctly*, but that in our inquiries into the true nature of things, we ought to focus on discovering things "in themselves" (ἐξ αὐτῶν) rather than "from their names" (ἐξ ὀνομάτων) because names are images of the real things. Socrates is not clear in the *Cratylus* about whether he accepts or rejects the premise that whoever first set down names did so correctly, but he is rather more explicit in arguing that however names came to be, they ought to be explained by appeal to an account (e.g. at 426a1–7). See Barney 2001: 84–85.

65. From this vantage, it would be another of the deep ironies that Plato (*Phdr*. 278c4–d5) would attempt to distinguish those who consider themselves "wise" (i.e. the Sophists) from those who "pursue wisdom" (i.e. philosophers) by extending his chief criticism of Sophists, i.e. that they deal in fabrications and imitations of true reality, to the Pythagoreans, who, following their master, may have been the first people to call themselves "philosophers" (on which see the balanced treatment of Riedweg 2005: 90–97). On Archytas and the Sophists, see Huffman 2002.

66. It should be noted that technical expertise—an understanding of what something "is" and whether something has been done "correctly"—remains subordinated to ethical judgment with reference to music in Plato's *Laws* (668b9–669b3).

of number, which include calculation, geometry, astronomy, games, and letters, in a rather straightforward way that approximates the "first-discoverer" *topos* in the surviving fifth-century BCE writings of Aeschylus/Euphorion, Euripides, Sophocles, and Gorgias. Nothing is particularly new here. But Plato innovates in the tradition by elevating the significance of the criticism of the efforts of these "first-discoverers" in a way that both points out their failure to evaluate (*κρινεῖν*) the ethical fruits of their inventions and recognizes the importance of an external check on their "demiurgic" discoveries, as they are described by Protagoras in the "Great Speech." That external check, associated with the cosmic lawgiver extraordinaire Zeus, belongs to the realm of the political and deals with universal questions of personal ethics and well-being, which are also writ large in Plato's metaphysical and epistemological propositions concerning the Good, especially in the *Republic* and *Phaedrus*. At the root of the treatment in the *Phaedrus* is the criticism of writing, which is ontologically and epistemologically inferior to the recollection of what is true, that is, the Forms, in Plato's midcareer metaphysics. The complication of heurematography with politics, ethics, metaphysics, and epistemology is also found in the fragments of the mathematical Pythagorean Archytas of Tarentum, especially Fragment 3, which seems to be a major target of Plato's attack in *Republic* 7. It is also possibly the object of Plato's ridicule in the reference to Palamedes's discovery of "number" that is embedded in the larger polemic against the mathematical Pythagoreans in that book. Overall, I suggest, Plato's treatment of the "first-discoverer" might indicate a larger challenge to the mathematical Pythagoreans, a strategy that would fall in line with and could be seen as an extension of similar heurematographical treatments of Pythagoras among his contemporaries Antisthenes and Isocrates, who emphasized Pythagoras's role in invention in rhetoric and philosophy, respectively.[67] As I will argue in the next two sections of this chapter, however, Plato's treatment of the heurematographical *topos* in the late dialogue *Philebus* evinces a revision of his earlier thoughts on the discovery of the arts of number, as Plato came to reconsider the role that mathematical Pythagoreanism, in particular, played in the epistemology and metaphysics of his philosophy.

THE MATHEMATICAL PYTHAGOREANS AND THE HEUREMATOGRAPHICAL TRADITION

The "first-discoverer" myth as illustrated in Plato's *Protagoras* and *Phaedrus*, like its antecedents in the writings of the Sophists and Athenian tragedians, is inscribed within the more general theme of religious piety in the Greek world.

67. See chapter 3, section entitled "Pythagoras among the Athenian Philosophers in the Fourth Century BCE." As I discussed in that chapter, however, we cannot be absolutely sure that Antisthenes is the author of the entire account involving Pythagoras's skills in rhetoric.

In *Protagoras* and *Phaedrus*, especially, it is understood that the inventions of Prometheus and Theuth are contrary to the will of the divine king, whether Zeus or Egyptian Thamus (who were the same god in the minds of the Greeks).[68] This contravention of the divine king occurs in different circumstances and at various grades. As I argued above, Prometheus in the *Protagoras* and in the *Prometheus Bound* receives judgment for the transgression, whereas in the *Phaedrus* a far gentler Thamus gently reproaches Theuth in the manner of a philosopher interlocutor, just like Socrates when he offers dialectical challenges to any Sophist who lays claim to making discoveries in Plato's dialogues. That reproach does take on the overall valence of a "religion versus science" paradigm, since Socrates goes on (*Phdr.* 275b5–276a7) to suggest that one should attend to the authority behind certain words, whether that authority is divine or human.[69] Gods are said to speak prophetic (μαντεία; μαντικούς) truth, and humans approximate said truth in a mimetic "image" (εἴδωλον) of it *by means of* τέχνη.[70] Thus, religion and science are understood to fall into two categories, with the objects of science understood to be a *derivative imitation* of the objects of religion. Overall, this "religion versus science" paradigm reflects a larger concern, on the part of Plato, with issues of stratification by way of derivation from reality in his epistemology and metaphysics that run from the *Republic* through the *Timaeus-Critias* to the *Laws*, with the dialogues variously focusing on one or another aspect in accordance with the topic of each dialogue (i.e. universal justice, the cosmos, the terrestrial city-state).

Analysis of these issues would be far too large and complicated to undertake in this study and would lead us astray from the main topic at hand, namely, how Plato's treatment of heurematographical myth is reflective of his critical engagement with mathematical Pythagorean philosophical method.[71] Ever since Aristotle notoriously claimed that Plato's philosophy in most aspects accorded with that of the mathematical Pythagoreans, save that he replaced Pythagorean imitation (μίμησις) with participation (μέθεξις)

68. This is not explicitly the case in the treatment of Palamedes and Agamemnon in *Republic* 7, which is far more playful and elusive.

69. Given Socrates's emphasis on attending and investigating the words of Apollo in the *Apology* (21b1–23c1), I take this normative statement seriously in the *Protagoras*.

70. See Morgan 2000: 223–224, who nevertheless does not emphasize the role of τέχνη in mediating between various levels of being.

71. One recent study (Pradeau 2009) on the role of "imitation-participation" in Plato's philosophy discusses this problem in Plato's dialogues without examining in any detail the relationship to Pythagoreanism.

(*Metaphysics* 1.6, 987b10–14), scholars—including Plato's own heirs in the Academy—have found reason to investigate these issues. The discourse of heurematography, which was linked to Pythagoreanism at the latest at the beginning of the fourth century BCE (and possibly earlier in the Aeschylean *Prometheus Bound*), presents a means for understanding the ways Plato criticized and appropriated mathematical Pythagoreanism in the pursuit of a methodology that would facilitate an understanding of the relationship between sensible and intelligible objects, whether we are to call it "imitation" or "participation." As I argued in chapter 5, the status of "number" is a key that makes it possible to account for Aristotle's characterization of the relationship between Platonic and Pythagorean metaphysics. And the discovery of "number" by certain mythological figures could be understood as code, I suggest, for external references to Pythagoreans in Plato's dialogues, especially with regard to the mathematical Pythagoreans who conceptualized or employed "number" in various ways. It will come as no surprise, then, that when Plato has Socrates retell the heurematographical story of the discovery of the divine method in his late dialogue *Philebus* by a "certain Prometheus," he does so, I argue, also with the mathematical Pythagoreans Hippasus of Metapontum, Philolaus of Croton, and Archytas of Tarentum in mind.

After discussing the object of their inquiry, namely "what in the world is good" (*ὅτι ποτ᾽ ἐστὶν ἀγαθόν*)[72] and its relationship to the other goods (especially pleasure and intelligence), Socrates tells his interlocutor Protarchus that the successful investigation into this issue relies on an explication of the "path" or "way" (*ὁδός*) through which each "art" (*τέχνη*) that exists "has been discovered" (*ἀνηυρέθη*) (*Philebus* 16b5–c3).[73] In a reversal of Protagoras's "Great Speech" in Plato's earlier dialogue *Protagoras*, Socrates claims that this "path" to the arts is "not terribly difficult to demonstrate [*δηλῶσαι*], but especially difficult to employ [*χρῆσθαι*]":

> As it appears to me, it was a gift of the gods to humans, cast down from the gods by a certain "Prometheus" along with a most brilliant fire [*διά τινος Προμηθέως ἅμα φανοτάτῳ τινὶ πυρί*]. And our forefathers [*οἱ παλαιοί*], being stronger than we are and living closer to the gods, passed on the tradition that the things that are said to be eternal have come from

72. Pl. *Phlb.* 13e5–7.

73. Don Lavigne points out to me that the name Protarchus (*Πρώταρχος*) is a compound of two words relevant to heurematography and to the investigation into principles, namely "first" (*πρῶτος*) and "origin" (*ἀρχή*). Thus Protarchus's name etymologizes the ontological subplot of the *Philebus*.

> one and many, but they had limit and unlimited innate[74] in themselves [πέρας καὶ ἀπειρίαν ἐν αὑτοῖς σύμφυτον ἐχόντων]. Since things are thus organized [τούτων οὕτω διακεκοσμημένων], we ought to seek out the one Form that we posit in each case for every one of them always, for we will indeed find it there. And once we have got it in our grasp, we must look for two, as the case would have it, or if not, for three or some other number.
>
> (PLATO, *Philebus* 16c5–d4)

As I discussed above, the challenge when reading Plato's heurematographical myths is to identify referents for the various characters, for example the "certain Prometheus" and the anonymous "forefathers" (οἱ παλαιοί) to whom Socrates refers in this passage. This is because the surplus of meaning in Plato's myths—a quality that can be detected in the active double-voiced discourse of Plato's intertextual references—frustrates the desire on the part of the reader to determine a single target. Still, we are invited to consider the identity of these various anonymous figures, especially given the gravity of the passage: Socrates is about to describe what many scholars have taken to be the most mature version of Plato's metaphysics. Who might this "certain Prometheus" have been, and to whom is Socrates referring when he distinguishes him from the "forefathers" who passed down the "gift of the gods"?

It has been extremely tantalizing to speculate about the authorities behind this method of the gods. Since late antiquity, commentators have taken this passage to refer to the Pythagoreans, especially Philolaus of Croton.[75] There is good reason to adopt the consensus view among scholars[76] that the "forefathers"

74. Sayre (2005: 118) translates σύμφυτον as "connaturally" while Gosling (1975: 7) suggests "inherent." Frede (1993: 8) does not translate the συμ-: "having in its nature . . ." It may be said that Plato's usage here recalls the Eleatic Stranger's myth from the *Statesman*, where "destiny and innate [σύμφυτος] desire" take hold of the universe again (*Plt.* 272e6). Also see Pl. *Leg.* 771b7, where σύμφυτον refers to the inborn (or connatural) proclivity for being pious, also known as the "gift of the god."

75. Proclus (*Theol. Plat.* 1.5, p. 26.4–9), Syrianus (*in Metaph.* p. 10.2–4 Kroll), and Damascius (*Pr.* 1.101.1) each asserted the rapport between the πέρας-ἀπειρία pair in the *Philebus* and Philolaus's first principles. For a comprehensive summary of the life of the *Philebus* after Plato, see van Riel 2008: x–lxviii.

76. See Gosling 1975: 83 and 165; Huffman 2001: 70–71; Meinwald 2002: 87–92; Miller 2003: 28; Delcomminette 2006: 93; Thomas 2006: 209, with n. 15, and 220, with n. 39; Kahn 2009: 66. Frede (1993: 8) considers the possibility of Pythagoras, but see below. Barker (1996: 155–156) suggests a developmental interpretation. He argues that, initially, there is no need to associate the passage that refers to the "gift of the gods" to the Pythagoreans, but then he goes on to suggest that later passages (e.g. *Phlb.* 25e) make it clear that the Pythagoreans, especially Philolaus, are intended referents here.

(*οἱ παλαιοί*) in this passage refer to those Pythagoreans who passed on the philosophical method of their ancestors, in particular Philolaus of Croton and his associates, for whom the objects of the universe are constructed out of limiters and unlimiteds, as we see in two fragments from Philolaus's writings:

> Nature in the cosmos was fitted together both out of things which are unlimited and things which are limiting, both the cosmos as a whole and all things in it.
>
> *ἁ φύσις δ᾽ ἐν τῷ κόσμῳ ἁρμόχθη ἐξ ἀπείρων τε καὶ περαινόντων καὶ ὅλος <ὁ> κόσμος καὶ τὰ ἐν αὐτῷ πάντα.*
>
> (Philolaus F 1 Huffman = Diogenes Laertius 8.85; translation after Huffman 1993)

> It is necessary that the things that are be all either limiting, or unlimited, or both limiting and unlimited [*ἀνάγκα τὰ ἐόντα εἶμεν πάντα ἢ περαίνοντα ἢ ἄπειρα ἢ περαίνοντά τε καὶ ἄπειρα*], but not in every case unlimited alone. Well then, since it is manifest [*φαίνεται*] that they are neither from limiting things alone, nor from unlimited things alone, it is clear then that the cosmos and the things in it were fitted together from both limiting and unlimited things [*δῆλον τἆρα ὅτι ἐκ περαινόντων τε καὶ ἀπείρων ὅ τε κόσμος καὶ τὰ ἐν αὐτῷ συναρμόχθη*]. Things in their actions [*ἐν τοῖς ἔργοις*] also make this clear. For, some of them from limiting things limit, others from both limiting and unlimited things both limit and do not limit, others from unlimited things will be manifestly unlimited.
>
> (Philolaus F 2 Huffman = Stobaeus, *Eclogae* 1.21.7a; translation after Huffman 1993)

Some scholars have wished to deny the strong relationships between Socrates's account of the gift of the gods passed down through the "forefathers" (*οἱ παλαιοί*) and the fragments of Philolaus, opting for a rather more general reference to Presocratic philosophers (Melissus, possibly Anaximander?). And it is true that speculation about the nature of the world occurred in Ionia, and that, in spite of Aëtius's testimony (2.1.1 = DK 14 F 21), Pythagoras did not "invent" the term "cosmos."[77] But the fact that nature (*ἁ φύσις*) is assumed to play a central role in the structural organization of limiters and unlimiteds in the cosmos (*διακεκοσμημένων*; i.e. *ἐν τῷ κόσμῳ, ὅ τε κόσμος*) confirms, as Sylvain Delcomminette has argued, that Plato is indeed referring to Philolaus's

77. See Burkert 1972: 77–78.

philosophy in particular when he speaks about the transmission of the gift by the "forefathers" (οἱ παλαιοί).[78]

A question arises, however. if we accept that Plato is indeed referring to the philosophy of Philolaus when he speaks of what the "forefathers" passed down: how would this affect our understanding of the status of the gift of the "certain Prometheus"? Is Socrates referring merely to the philosophical method (ὁδός) of Philolaus, more generally to that of the mathematical Pythagoreans, or even more generally to that of all Pythagoreans? This question is especially difficult to answer because of the surplus of meaning implied by active double-voiced discourse in Plato's "first-discoverer" myths, but I think we can make some advances on it by contextualizing this version of the heurematographical myth with other passages of Plato's writings.

One common interpretation of this myth posits Pythagoras as the "certain Prometheus" who gave humankind the "path," "method," or "solution" (ὁδός) to the problem, which is also known as the gift of the gods.[79] It is worth considering this possibility. In *Republic* 10 (600a9–b4)—the only time Plato actually refers to Pythagoras by name—we hear of a "way of life" (ὁδὸς βίου) as well as "those who come after" Pythagoras (οἱ ὕστεροι), who "even now call their manner of living Pythagorean" (ἔτι καὶ νῦν Πυθαγόρειον τρόπον ἐπονομάζοντες τοῦ βίου) and "in some way seem to be distinctive" (διαφανεῖς πῃ δοκοῦσιν εἶναι) in comparison with other people.[80] These people might be plausibly considered the same as the "forefathers" (οἱ παλαιοί) who pass down the divine method in the *Philebus*. Like his contemporary Isocrates and Aristotle after him,[81] Plato in the *Republic* emphasizes that certain people *speak of themselves* or their lifestyle as "Pythagorean"; and like Heraclitus and Isocrates, Plato highlights the notion that adopting a Pythagorean lifestyle has the potential to render its adherents well-regarded in the public eye.[82] The verb here, ἐπονομάζω, does not in itself indicate skepticism on the part of Plato, although the reference to their "seeming in some way [δοκοῦσιν πῃ] to be distinctive"

78. See Delcomminette 2006: 93, with n. 73, for bibliography.

79. As suggested by Hackforth and Gosling, on which see Huffman 2001: 70–71, with bibliography. It is to Dorothea Frede that we owe the clever translation of ὁδὸς ἐπὶ τὸν λόγον in Protarchus's comments (*Phlb.* 16a8–b1) as "solution to the problem."

80. See chapter 3, section entitled "Pythagoras among the Athenian Philosophers in the Fourth Century BCE."

81. Isocrates (*Busiris*, 28–29) speaks of them as προσποιούμενοι; Aristotle frequently (e.g. *Metaph.* 1.8, 989b29–990a5) calls them καλούμενοι.

82. For Isocrates's treatment of Pythagoras and those who "fashion themselves" Pythagorean, see chapter 3, section entitled "Pythagoras among the Athenian Philosophers in the Fourth Century BCE."

might be thought to contain a hint of criticism (i.e. they aren't *really* worthy of regard, only *esteemed* so by the many, who do not possess knowledge or certain belief).[83] In general, then, Plato's brief description of Pythagoras in *Republic* 10 accords with what other Athenians were saying about Pythagoras and his followers in the critical tradition that stems from Heraclitus.

Now, since both the Prometheus passage in *Philebus* and the description of Pythagoras's teaching in *Republic* 10 refer to a "path" (ὁδός) potentially worth seeking, we might be inclined to think that Pythagoras's art is implied in the description of the "gift of the gods" in the *Philebus*. Carl Huffman, however, has raised some critical objections to this hypothesis that are worth considering.[84] His overall argument consists in the claim that traditions from later antiquity are responsible for elevating Pythagoras to a "semi-divine Promethean figure" and that such an association is unwarranted given the state of evidence from Plato's own writings.[85] Huffman points out that the *Republic* passage does not speak of Pythagoras as divine or semidivine but as an "influential private teacher."[86] He also notes that "in the *Philebus*, the system hurled down from the gods is said to be the basis of all progress in the arts (*technai* 16c2)" and that the emphasis on τέχνη is a holdover from the Prometheus myths that reach back to the *Prometheus Bound*. He also adds: "Burkert's work . . . has shown that the earliest evidence makes clear that Pythagoras was not primarily a mathematician involved in *technai*."[87] For each of these reasons, Huffman argues, we ought to understand that the "certain Prometheus" of the *Philebus* "is not a cover for Pythagoras or for any other philosopher but rather *just Prometheus*."[88]

Someone wishing to object to Huffman might point to the potential significance of the reiteration of the word ὁδός in both passages, which Huffman does not address sufficiently. Recall that in *Republic* 10 (600a9–b4), Plato ascribes to Pythagoras's followers a "manner of living" (Πυθαγόρειον τρόπον τοῦ βίου)

83. Note that there is a consistent theme running through this passage, in which Socrates first speaks of the beloved sage (e.g. Homer, Pythagoras, and the Sophists Protagoras and Prodicus), then describes how their followers hold them in affection, and finally challenges the capacity of those followers to judge correctly whether the wise men are actually wise or not.

84. In two publications: Huffman 1999 and Huffman 2001, which in some ways is a continuation of the earlier piece.

85. Huffman 2001: 71.

86. Ibid.

87. Ibid. This view, of course, has been extensively challenged by Zhmud (2012).

88. Huffman 2001: 71. Italics mine.

that resembles the Homeric "way of life" (ὁδὸς βίου Ὁμηρικῆς).[89] Likewise, in the *Philebus* (16a6–c3), Socrates and his interlocutors are seeking a "way out" (ὁδός)[90] of being at a loss (ἄπορον), or, speaking literally, a "road" that resolves the problem of being "pathless" (ἄ-πορον).[91] There is no reason to assume that when he used the term ὁδός, Plato was not conflating the methodological and ethical implications of the term, since for Plato ethical behavior was coextensive with other aspects of his philosophy and could not be separated from it.[92] Philosophical method and the way of living could not be easily compartmentalized in Plato's philosophy, as the *Philebus* itself demonstrates: it is ostensibly a dialogue that aims to pursue the Good by way of describing a ὁδός to it, but such an inquiry actually results in Plato's longest and most systematic treatment of pleasure. In addition, while Huffman is correct to follow Burkert in seeing Pythagoras as not an early "mathematician" whose philosophical activities can be shown to demonstrate the same sorts of metaphysical speculations involving limiters and unlimited things that Philolaus undertook, we cannot forget that one of the earliest descriptions of Pythagoras's activities, from Heraclitus, explicitly associates him pejoratively with a panoply of τέχναι.[93]

Things are even more complicated if we take into account the surplus of meaning that marks the heurematographical myths of Plato. By juxtaposing the Prometheus myth with what is pretty obviously a reference to Philolaus's natural philosophy, Plato gestures in the direction of comparing the "gift of the gods" with Philolaus and/or the Pythagoreanism that might have been associated with him. But by adding various other details not found in his previous heurematographical myths involving Prometheus, Palamedes, or Theuth, for example the distinguishing of a "certain" Prometheus from a group of followers designated as the "forefathers," Plato complicates any fixed and simple associations.[94] Rather than it being the case that, for Plato, the heurematographical

89. The term ὁδός in Plato's writing is sometimes associated with a μέθοδος, as it is at *R.* 4, 435c9–d3.

90. Protarchus also calls it a "mode and mechanism" (τρόπος καὶ μηχανή).

91. In *Republic* 7 (532b4, e3), Socrates calls the process of seeking out and finally grasping the Good itself, namely "dialectic," a "journey" (πορεία). There is good reason to see all of this as reflective of the activity of philosophical *theoria*, as described by Nightingale (2004: 80–83).

92. See Kamtekar 2007.

93. DK 22 B 129. See above in the section entitled "The Paradigmatic First-Discoverer in the Aeschylean *Prometheus Bound*."

94. See Delcomminette 2006: 93–49.

myth *just* refers to one literary figure (Prometheus), as Huffman holds, I suggest that Plato's appeal to Prometheus in the "gift of the gods" passage here requires the opposite response: it solicits an assortment of possible literary antecedents (including those found in Plato's own *Republic*, *Protagoras*, and *Phaedrus*) and/or historical figures whose methods (1) involve τέχναι that are, broadly speaking, associated with fire and mathematical speculation, and (2) are directly relevant to the task at hand: to discover a philosophical approach or path that can be used to achieve an understanding of the Good by means of both dialectical procedure and empirical inquiry into the universe and its objects. From that point of view, the "forefathers" who passed down the divine gifts could be not only Pythagoreans but also Sophists.[95]

So while we can't count out Pythagoras as a possible object of the reference to the "certain Prometheus" in the heurematographical myth of the *Philebus*, we also cannot firmly fix him as the referent either.[96] If we are to make the interpretive move either to count Pythagoras in or out, we will need to base our decision on consideration of evidence beyond what has been presented up to this point in my argument. One issue I have not yet dealt with in the description of the methodological "gift of the gods" in the *Philebus*, which has troubled commentators including Huffman,[97] is the reference to "fire." Why does Plato refer to the "gift of the gods" as given by a "certain Prometheus along with a most brilliant fire (διά τινος Προμηθέως ἅμα φανοτάτῳ τινὶ πυρί)"? Another way to ask this question is, given the fact that Plato has at his disposal at least three types of culture hero in his own literary repertoire from which to choose (Prometheus, Palamedes, and Theuth), *why does he use Prometheus in this particular philosophical context*? What is it about the Prometheus myth that might be especially fitting for this passage?

It should be noted that fire plays no crucial role in Pythagoras's conceptualization of the cosmic order, at least insofar as we can reconstruct it. But fire retains its importance for Plato's presentation of the Prometheus myth both in the *Philebus* and elsewhere, perhaps as a means to distinguish this

95. Note that in the passage that follows the description of the method of the gods (16e7–17a1), Socrates understates an intellectual lineage—from gods to "forefathers" to "the wise men of today" (οἱ δὲ νῦν τῶν ἀνθρώπων σοφοί), possibly with reference to either contemporary Pythagoreans or even those Sophists who concerned themselves with rhetoric and mathematics, such as Hippias of Elis. Huffman (2002) has effectively argued that some fragments of Archytas should be seen as responding to what he calls the "Sophistic thought pattern." But from the point of view of Plato, Archytas, who apparently wrote treatises (and not dialogues) could easily be seen as engaging in Sophistic discursive practices as well.

96. Also see Meinwald 2002: 88.

97. Huffman 2001: 71 n. 11.

particular culture hero from the others (Palamedes, Theuth). For example, in the "Great Speech" of Protagoras from Plato's *Protagoras*, Prometheus's gift of "wisdom in the practical arts together with fire [ἔντεχνον σοφίαν σὺν πυρί]" is glossed emphatically. There, such "wisdom" is said to be "effectively useless" (ἀμήχανον) without fire.[98] This version of the story is paralleled in the short heurematographical myth in the *Statesman*, which I have reason to treat here as an intermediary between the earlier Prometheus story of the *Protagoras* and the later representation in the *Philebus*. In the *Statesman*'s myth, the Eleatic Stranger describes life in the cosmic cycle that lacks the divine helmsman:

> Since we had been deprived of the god who possessed and pastured us, and since for their part the majority of animals—all those who had an aggressive nature—had gone wild, human beings, by themselves weak and defenseless, were preyed on by them, and in those first times were still without resources and without expertise of any sort [ἀμήχανοι καὶ ἄτεχνοι] . . . they were in great difficulties [ἐν μεγάλαις ἀπορίαις]. This is why the gifts from the gods, of which we have ancient reports, have been given to us, along with the necessity of teaching and education: fire from Prometheus [πῦρ παρὰ Προμηθέως], crafts from Hephaestus and his fellow craftworker, seeds and plants from others. Everything that has helped to establish the human livelihood [πανθ' ὁπόσα τὸν ἀνθρώπινον βίον συγκατεσεύακεν] has come about from these things, once care from the gods, as has just been said, ceased to be available to human beings, and they had to love their lives through their own resources and take care for themselves, just like the cosmos as a whole, which we imitate and follow for all time, now living and growing in this way, now in the way we did then.
>
> (PLATO, *Statesman*, 274b5–d8; translation after Rowe 1995)

This passage aids in determining how modifications to the "first-discoverer" *topos* in Plato's later dialogues are indicative of Plato's changing responses to mathematical Pythagoreanism. Initially, we may note the similarities to the Prometheus myth in the *Protagoras*: the "gifts of the gods" are understood to comprise Promethean "fire," as well as the demiurgic arts of Hephaestus and Athena, and agriculture from unnamed others (Demeter?). The arts come to be out of some sort of natural educational development or "necessity"; human beings are understood to receive these gifts in order to survive without the aid of the gods,

98. See above in the section entitled "Plato's Earlier Treatments of First-Discoverers: Palamedes, Prometheus, and Theuth in *Protagoras*, *Republic*, and *Phaedrus*."

and under the threat of extinction from wild animals.[99] In these ways, then, the heurematographical myth of the *Statesman* does not significantly evolve beyond the earlier version given in the Sophist Protagoras's "Great Speech."

In two ways, however, we see a shift in focus in Plato's treatment of the "first-discoverer" myth in the *Statesman*, both of which suggest a more directly articulated reference to the Pythagoreans.[100] First, there is an explicit association drawn between the human "livelihood" (*βίος*), aided by the divine gifts of the gods (fire, the arts, the causes of agricultural development) and the fruits of proper philosophical investigation. Owing to those gifts, human beings are understood to have a *way out* of their great "confusions" (*ἀπορίαι*), a term that, throughout Plato's middle- and late-period philosophy, is synonymous with difficult and apparently insoluble *philosophical* problems.[101] In the larger context of a dialogue that deals with proper definitional procedure, it should be unsurprising that the myth of the culture hero would feature some application to philosophical method. Themes of the myth itself—chief among them the rotation (*περίοδος*) of the cosmos—concern the issue of the proper way (*ὁδός*) to practice dialectic, a concern that we have seen expressed explicitly in the *Republic*, but without emphasis on the "circularity" that might be a defining factor of dialectical procedure.[102] This

99. Compare Democritus's anthropology (DK 68 B 5.1), in which humans, "taught by necessity" (*ὑπὸ τοῦ συμφέροντος διδασκομένους*), learned to help one another to keep safe from animals and even learned how to communicate with one another through "signs" (*σύμβολα*).

100. For another study that proposes to investigate how the differences between the Prometheus myths in *Protagoras* and *Statesman* indicate developments in Plato's philosophy, see van Riel 2012, especially pp. 157–159.

101. Note that in the Aeschylean *Prometheus Bound*, Prometheus had described his gifts as a *πόρος* that makes accessible the *τέχναι* (*PV* 477). Likewise, in Gorgias's *Defense of Palamedes* (30), Palamedes claims to have "made the human life accessible out of intractable" (*ἐποίησε τὸν ἀνθρώπειον βίον πόριμον ἐξ ἀπόρου*). Politis (2006: 269–272) distinguishes two types of *aporiai* in Plato's earlier writings: (1) those that involve being generally "at a loss" in such a way as to be incapable of knowing that one does not know, which he calls "cathartic," and (2) those that involve a lack of solution to a particular puzzle, which he calls "zetetic." The second "zetetic" type gains prominence as Plato's thought develops, I would argue, and as he pursues more "mathematical" approaches to epistemology. An early occurrence of some significance can be seen in the geometry lesson undertaken by the slave boy in the *Meno* (84c10–11), as Politis convincingly argues.

102. See the dialectical terminology expressed, for example, in the description of the various rotations of the cosmos at *Plt.* 273d4–e9 and the subsequent return to these concerns at 274e4–275a6. The use of the expressions *ὁδός* and *μέθοδος* in order to describe dialectical procedure likely originates in *Republic* 7, where Socrates self-consciously identifies what will become scientific terminology as semantically derived from other, possibly poetic, discourses (see esp. *R.* 7, 532e1–3). It is also the case that Socrates there identifies dialectic as the "only *μέθοδος* that travels (*πορεύεται*) this way, doing away with hypotheses and proceeding to the origin" (*R.* 7, 533c7–d1). Also see *R.* 4, 435c9–d3.

metacritical approach leads us to the second innovation in Plato's treatment of the heurematographical *topos* in the *Statesman*: the concerns with method are made explicitly coordinate with concerns over the structural order and activity of the cosmos—both in terms of space and time and as a consequence of anxieties about what Plato might call the "movement" (φόρα) of the argument, which now comes to be described as a "circle" (περιφορά).[103] Methodology and cosmic physics are thus brought to bear on one another within the larger scope of Plato's philosophy.

As I showed earlier, the association of the overall design of the cosmos with the means to make sense of it is not new, since it appears early in Greek literature, in the Sophistic speech of Prometheus in the Aeschylean *Prometheus Bound* and in the fragments of Heraclitus of Ephesus, both of which emphasize the role that fire plays in making discrimination or judgment of things in the universe. Fire also appears to have a significant role in the philosophy of the mathematical Pythagoreans and those associated with them, starting with their so-called progenitor Hippasus of Metapontum, who, like Heraclitus, was said by Theophrastus to have believed that fire was the first principle and that it was "unified and in movement and limited" (ἓν. . . καὶ κινούμενον καὶ πεπερασμένον).[104] While fire certainly played a central role in Hippasus's philosophy, we cannot be sure whether it is to be identified with any of the other main productive and discriminatory aspects, that is, number and soul, that we hear about in the doxography.[105] If Hippasus approached inquiry into the universe in a way that was similar to that of Heraclitus (with whom he is commonly associated from Aristotle forward), such associations would be unsurprising. Be that as it may, fire is unquestionably important in the cosmology of Philolaus of Croton. For Philolaus, fire plays a central role in the generation and overall physical structure of the universe. It also relates in somewhat obscure ways to the two categories of constituents in Philolaus's philosophy, limiters and unlimiteds. The main evidence comes in the form of these two fragments, one of which I quoted earlier:

103. *Plt.* 274e9–275a6. Also note that the Eleatic Stranger explicitly expands the semantics of φόρα to include "movements" of the soul, body, and "voice" (φωνή). Such concerns were also implicit in Plato's criticism of mathematical Pythagoreanism and Archytas's method in book 7 of the *Republic* (530c8–d9), although Plato's expansive use of the term φόρα in reference to dialectical procedure does not apparently hold there.

104. Arist. *Metaph.* 1.3, 984a6 and Thphr. F 225 FHS&G = Simpl. *in Phys.* p. 23.21–24.12 Diels.

105. E.g. Iambl. *in Nic.* 11, 10.20–24. See Zhmud 2012: 387 n. 2. On Hippasus's doctrines and the problems involved in the doxographical evidence, see chapter 2.

> The first thing fitted together, the one in the center of the sphere, is called the hearth.
>
> *τὸ πρᾶτον ἁρμοσθέν, τὸ ἓν ἐν τῷ μέσῳ τᾶς σφαίρας, ἑστία καλεῖται.*
>
> (Philolaus F 7 Huffman = Stobaeus, *Eclogae* 1.21.8; translation after Huffman 1993)

> Nature in the cosmos was fitted together both out of things that are unlimited and things that are limiting, both the cosmos as a whole and all things in it.
>
> *ἁ φύσις δ' ἐν τῷ κόσμῳ ἁρμόχθη ἐξ ἀπείρων τε καὶ περαινόντων καὶ ὅλος <ὁ> κόσμος καὶ τὰ ἐν αὐτῷ πάντα.*
>
> (Philolaus F 1 Huffman = Diogenes Laertius 8.85; translation after Huffman 1993)

Huffman ingeniously suggests that in Philolaus's metaphysics, relations are being drawn, especially, between position and fire and between "limiters" and "unlimiteds": "the combination of structure with material is precisely what Philolaus means by the combination of limiter and unlimited, so that the 'central' in the central fire refers to its limiting structural element, while the fire refers to its unlimited material element, which has now been limited by being placed in the center."[106] But the question remains: does Philolaus's employment of fire and the first principles limiter/unlimited relate to attainment of knowledge in the universe, as in the "first-discoverer" myths discussed above? If Plato is indeed referring to the philosophical method of Philolaus in the "gift of the gods" passage of the *Philebus*, does this reference fit, given the recurrent emphasis in Plato's treatments of the heurematographical myth on epistemology?

While there is no evidence that Philolaus believed that "knowledge" or "method" was "discovered" or "handed down" by a particular individual (divine or human), as in the case of the heurematographical myths of the Sophists, tragedians, and Plato, he does draw relationships between the harmony of limiters and unlimiteds on the one hand and knowledge, on the other:

> Concerning nature and harmony the situation is this: the being of things, which is eternal, and nature herself admit of divine and not human knowledge—except that it was impossible for any of the things that are and are known by us to have come to be, if the being of the things from

106. Huffman 2007: 89–90. On reference to the hearth in the dialogues as a probable indicator of Plato's evaluation of Philolaus's metaphysics, see chapter 4, section entitled "Plato and Mathematical Pythagorean 'Being' before the *Phaedo*."

> which the cosmos came together, both the limiters and the unlimiteds, did not preexist. But since these beginnings preexisted and were neither alike nor even related, it would have been impossible for them to be ordered, if a harmony had not come upon them, in whatever way it came to be. Well then, like things and related things did not require any harmony additionally, but things that are unlike, being neither related nor of equal speed—it is necessary that such things be bonded together by harmony, if they are going to be held in order.
>
> (PHILOLAUS F 6 HUFFMAN = Stobaeus, *Eclogae* 1.21.7d; translation after Huffman 1993)

A few general remarks on this passage are warranted. First, the closest *comparandum* for Philolaus's epistemology among Presocratics is Heraclitus, who also distinguishes two types of γνῶσις (divine and human) and does not acknowledge the agency of a primordial divinity or human who "discovered" the means to understand the universe.[107] On this point, Philolaus is emphatically agnostic: he claims that the "beginnings" (ταὶ ἀρχαί) that preexisted other things, the limiters and unlimiteds, could not have been organized (since they are unlike one another) without the supervening of harmony (ἁρμονία), "*in whatever way*" (ᾡτινιῶν ἂν τρόπῳ) it supervened. Philolaic harmony is a principle of reconciliation between things that are unlike one another, just as we see in the speculations of Heraclitus, although Heraclitus's epithet for harmony, "back-turner" (παλίντροπος), appears to go further than Philolaus in describing the mechanisms by which harmony reconciles things at variance through agreement.[108]

There are some strong relationships to be drawn between Heraclitus's and Philolaus's philosophical systems, to be sure; and while fire does play a significant role in Philolaus's cosmogony, there is no explicit evidence that associates fire with knowledge, judgment, or the soul, as is attested in the fragments of Heraclitus.[109] It is probable, moreover, that Philolaus would not associate *fire*

107. DK 22 B 41. Also, perhaps, in contradiction to Parmenides (DK 28 B 8.34–36), who claims that one cannot "discover thinking without what is" (οὐ ἄνευ τοῦ ἐόντος εὑρήσεις τὸ νοεῖν). Parmenides thus assumes that thinking can be discovered, but that it cannot occur apart from Being. After Philolaus, Democritus, whose links to Pythagoreanism are well established (see above, chapter 4, note 137) understood two types of γνώμη (see DK 68 B 11).

108. DK 22 B 51. Also see Parmenides's criticism of the path of mortals as a "back-turner" (DK 28 B 6), which might be a reference to Heraclitus's own writings, as argued by Graham (2002: 31). But also see the response of Nehamas (2002: 55–56).

109. The closest Philolaus comes to Heraclitus is his description of the newborn baby (DK 44 A 27), in which the sperm is said to be "what constitutes [κατασκευαστικόν] the body" because it is hot.

itself with harmony in the cosmos, given that it is harmony that apparently gives order to "unlimited" elemental objects such as fire and water, and not the other way around.[110] By introducing the principles of "limiters" and "unlimiteds" as the things ordered by harmony—as well as a somewhat sophisticated system of place and stuff (if Huffman's interpretation is right)—Philolaus makes notable advancements on a system that might have been easy to criticize as absurd by the likes of intellectuals such as Parmenides and Plato. For my purposes, it suffices to say that fire does not *obviously* play a causative or determinative role in the overall metaphysics and epistemology of Philolaus, so we should be hesitant to assume that Philolaus would be the object of Plato's criticism in a strong sense when he refers to the gift of a "certain Prometheus" as being given "along with brilliant fire" (*ἅμα φανοτάτῳ τινὶ πυρί*).

It is obviously more advantageous, then, to associate Philolaus with the "forefathers" who passed down the "gift of the gods" in the *Philebus* rather than with the original culture hero. And there are other reasons to see Philolaus as inheritor of the Promethean gift. By contrast with other myths of the culture hero employed by Plato (such as that of Theuth), stories of Prometheus almost always emphasize the fact that he was punished for divulging the various *τέχναι* associated with fire to human beings. By contrast, there is no historical evidence that Philolaus, in the manner of Prometheus, divulged secrets and was consequently punished for doing so. The only evidence of Philolaus being punished, which indicates that he was put to death for having aimed at tyranny (Diogenes Laertius 8.84), is based on a confusion of the activities of the Sicilian Dion and Philolaus in the doxography.[111] And if Philolaus is the intended object of the reference to the "forefathers," it would be very unlikely that Plato would confuse these people with Prometheus himself.

Perhaps, then, Archytas of Tarentum is intended as a target of Plato's reference in the heurematographical myth of the *Philebus*. As I have already discussed, the joke concerning the invention of number by Palamedes the general in *Republic* 7 (522d1–523a3) occurs in the context of Plato's criticism of Archytas's philosophy in particular. Already by the time of the *Republic*, then, the heurematographical *topos* appears to be associated with Archytas. Second, there is a tradition that traces back at least to Aristotle's student Eudemus of Rhodes, and perhaps earlier, that associates the solution (*εὕρησις*) of the problem of duplicating the cube, which involves the "discovery" (*δύο μέσαι ἀνάλογον*

110. I have speculated that Philolaus thought that harmony and the "being of things" were the same. See chapter 4, section entitled "Growing and Being: Mathematical Pythagoreanism before Plato."

111. See Burkert 1972: 228 n. 48.

ηὕρηνται) of the two mean proportionals, with Archytas.[112] In the science of harmonics, too, Archytas is credited with various discoveries: Theon of Smyrna speaks of Eudoxus, Archytas, and their followers as having participated in the "discovery of the concords" (εὕρησις τῶν συμφωνιῶν).[113] More important, there is a tradition in Iamblichus that likely derives originally from Eudemus's *History of Arithmetic* suggesting that Archytas and Hippasus "initiated the discovery" [ἄρξαντος τῆς εὑρέσεως] of the fourth through sixth mathematical means, which was subsequently completed by Eudoxus.[114] It may be with Eudemus that specific "discoveries" or "solutions" in mathematics and harmonic theory are *systematically* attributed to various Pythagoreans in a catalogue format for the first time.[115] This is because with Aristotle's associates Theophrastus, Eudemus, Dicaearchus, Meno, and Aristoxenus, we get a substantial advancement in the systematization of the historiography of philosophy and science. But the tradition of associating particular discoveries in the methodology of inquiring into the universe to Pythagoras and his followers also derives from Isocrates's and Plato's ruminations on the Pythagoreans, and it filtered through the generation of thinkers who followed them, including those figures who wrote extensively about Pythagoreanism with their own agendas: Speusippus, Xenocrates, Aristotle, and especially Heraclides of Pontus.[116]

112. Archytas T A 14 Huffman and T A 15 Huffman. Generally, on these fragments, see the comprehensive treatment of Huffman (2005: 342–401).

113. Archytas T A 19a Huffman = Theon Sm. *Math.* pp. 60.16–61.23 Hiller.

114. See Huffman 2005: 171–172 and Zhmud 2002: 271–272. The sole remaining fragment of Eudemus's *History of Arithmetic* (F 142 Wehrli) deals with Pythagorean concords, which suggests that it is this text of Eudemus that contained a greater inquiry into the harmonic means. Still, it is difficult to know where exactly to place Hippasus in Eudemus's larger history of mathematics. See Zhmud 2002: 286–288.

115. See, for example, F 136 Wehrli: Εὔδεμος δὲ ὁ Περιπατητικὸς εἰς τοὺς Πυθαγορείους ἀναπέμπει τὴν τοῦδε τοῦ θεωρήματος εὕρεσιν. Note that F 133 Wehrli, from Eudemus's *History of Geometry,* also evinces a teleological history of mathematical discovery starting with Egyptians, passing through various figures such as Pythagoras, Archytas, and Eudoxus, and concluding with Plato's student Philip of Opus. Compare the fascinating, but ultimately ambiguous, statement of Democritus (DK 68 B 144) that "necessity did not call forth music" (μουσικὴν μὴ ἀποκρῖναι τἀναγκαῖον) but that it arose "out of excess" (ἐκ τοῦ περιεῦντος).

116. Heraclides of Pontus, who was associated with the Academy and familiar with Pythagoreanism, may have composed a treatise entitled Συναγωγὴ τῶν <εὑρημάτων> ἐν μουσικῇ. See Zhmud 2006: 50, with n. 24. To this list we might add the mid-fourth-century BCE work *On Discoveries* (FGrHist 70 F 2–5) by the historian Ephorus of Cyme, which deals explicitly with inventions in music. Ephorus (F 104), of course, credited Orpheus with first divulging the mysteries and their rites to the Greeks (πρῶτον εἰς τοὺς Ἕλληνας ἐξενεγκεῖν τελετὰς καὶ μυστήρια).

Another reason why Archytas's philosophical method might be in the scope of Plato's description of the gift of the "certain Prometheus" in the *Philebus* is that, unlike what we see in the cases of Pythagoras and Philolaus, there is a tradition in ancient Italy that makes Archytas into a figure who is punished with an unjust death (apparently by sea) in a way comparable with the deaths of those famous mythological figures Tantalus, Tithonus, Minos, and Euphorbus (Horace, *Odes* 1.28).[117] In what is often referred to by scholars as the "Archytas Ode," punishment by death at sea is coordinate with inability to escape the Erinyes, and especially Persephone.[118] This poem emphasizes the notion that Archytas's attempts to measure the immeasurable natural phenomena and understand the motions of the heavens were undertaken in vain, chiefly because the mind is perishable (*animo morituro*).[119] The proper context for understanding Horace's conceit, I suggest, is not to be found in any other sources regarding Archytas's life and death; instead, what Horace's poem reveals is the complex intertwining of stories in the Italian imagination around the first century BCE about Archytas and Hippasus akin to a passage preserved by Iamblichus that ultimately, I think, owes its provenance to Aristotle.[120] Recall that in his description of the mathematical Pythagorean *pragmateia*, Iamblichus, who is probably summarizing the lost works of Aristotle on the Pythagoreans, claims that Hippasus was "drowned at sea for committing heresy [ἀπόλοιτο κατὰ θάλατταν ὡς ἀσεβήσας], on account of being the first to publish, in written

117. It is not possible that the speaker of Horace's poem could be Archytas, as the scholia to Horace claim. See Huffman 2005: 19–21. Archytas is not *explicitly* said to die by sea, but his tomb is said to be "near the Matine shore" (*prope litus Matinum*), and the reference to the deaths of sailors later on (lines 18–20) may be assumed to include Archytas here. It should also be noted that Iamblichus (*VP* 245, 131.18–25) refers to Tantalus in particular, when he distinguishes between those who are "educated on pure principles" by Pythagoras and "all the rest, as Homer says of Tantalus, [who] might be pained when present in the midst of oral instructions and enjoy nothing" (trans. Dillon and Hershbell 1991).

118. Lines 17–20.

119. Lines 1–6. Lucretius, *DRN* 3.94–135 and 417–424, argues against the (Pythagorean?) claim that the soul is a harmony by attempting to demonstrate that the mind is perishable.

120. As noticed by MacKay (1977), who suggests that the speaker is Hippasus. Huffman (2005: 20 n. 6) objects on the grounds that "none of the details of Horace's presentation of the speaker give any hint that he is Hippasus (e.g. no reference to his divulging of secrets)." While this is not entirely the case (see below on Persephone), Huffman is right to argue that the speaker could not be Hippasus. Of course, seeking a perfect alignment between poetic presentation and historical biography might be a frustrated project from the start. My point is that Horace is confusing the traditions of Hippasus and Archytas and that this has some value for our understanding of how the Hippasus myth developed and became more complex in antiquity.

form, the sphere, which was constructed from twelve pentagons."[121] Hippasus, according to Iamblichus (*On the Pythagorean Way of Life* 246, 132.11–17),[122] was "banished from their common association and way of life" and had "a tomb constructed" to signify the fact that he had been exiled from "life with human beings."[123] Horace's description of the symbolic death of Archytas, with its Epicurean rebuttal of the Platonic/Pythagorean immortality of the mind and soul, correlates with a fragment of unclear provenance that ascribes to Hippasus a theory of the immortality of the soul: "the body is one thing; another—other by a great degree—is the soul, which flourishes in an inactive body, sees in one that is blind, and lives on in one that is dead."[124] Further contextualization comes from the most comprehensive version of this tradition involving punishment for the divulging of the Pythagorean secrets, a widely distributed letter in Doric purported to be from the fifth century BCE Pythagorean Lysis to a certain "Hipparchus"—probably a mistake for "Hippasus"—that seems to have been forged in the second half of the third century BCE.[125] The speaker exhorts "Hipparchus" (Hippasus?) to "remember the divine and human precepts of the famous one [i.e. Pythagoras], [and] not to share the goods of wisdom with those who have their souls in no way purified," on the grounds that "it is not

121. Iambl. *DCM* 25, 77.19–21. Further evidence that some version of this story was in circulation during the first century CE comes from Plutarch's *Life of Numa* (22.4), where Plutarch claims that Numa had all his books buried according to the same reason that the Pythagoreans do not preserve their doctrines in writing: "Some say that the Pythagoreans do not entrust their precepts to writing, but implant the memory and practice of them in living disciples worthy to receive them. And when their treatment of the abstruse and mysterious processes of geometry had been divulged to a certain unworthy person, they said that the gods threatened to punish such lawlessness and impiety with some signal and wide-spread calamity" (trans. Perrin 1914). It is not obvious what Plutarch means by punishment by the vague "some signal and wide-spread calamity" (ἐπισημαίνειν μεγάλῳ τινὶ καὶ κοινῷ κακῷ), but shipwreck could not be dismissed.

122. See Burkert 1972: 457–459. Note that Iamblichus differentiates the authority who speaks about incommensurability from the Aristotelian passage (marked by οἳ δέ φασι; see the comparison with *DCM* and Burkert 1972: 457 with n. 54); and he then returns to the earlier authority (marked by ἔνιοι δέ) by reiterating that they "maintained that the one who broke the news about the irrational and incommensurability suffered this fate" (trans. Dillon and Hershbell 1991).

123. Trans. Dillon and Hershbell 1991. Hippasus is not named in this passage, but the context suggests him.

124. Claudian. Mam. *de Anima* 2.7, in direct quotation from his source: "longe aliud anima, aliud corpus est, quae corpore et torpente viget et caeco videt mortuo vivit." Claudianus refers to "Hippon Metapontinus" here. On Hippo and Pythagoreanism, see Zhmud 2012: 127–128 and 232.

125. See the analysis of Burkert in two publications: 1961 and 1972: 459 n. 63. Zhmud (2012: 189 n. 79) suggests that it should be dated to the first century CE, based on lexical data.

lawful to give any random person things acquired with diligence after so many struggles, or to divulge to the profane the mysteries of the Eleusinian goddesses."[126] The profanation of the mysteries of Demeter and Persephone, the goddesses of Eleusis, is thus closely linked with the distribution of the "goods of wisdom" (τὰ σοφίας ἀγαθά) to the uninitiated. Horace's poem, too, presents a case in which punishment by the Erinyes and Persephone is exacted on Archytas, for reasons unclear.[127] While it is difficult to see with greater precision why, I suggest that themes from Horace's "Archytas Ode" crop up in the context of Hippasus of Metapontum's life and philosophy and are informed by the tradition that makes him a heretic for expressing Pythagorean secrets to the uninitiated.

Even though Horace's presentation of Archytas in the "Archytas Ode" does not perfectly parallel the story of Hippasus's divulging of the Pythagorean secrets, it reflects a larger trend, almost certainly as old as Eudemus of Rhodes and possibly older, of associating the philosophical activities of Archytas and Hippasus. The "Archytas Ode" thus contextualized suggests to us that we might seek to understand more precisely the relationship between Archytas and his philosophical forebear, Hippasus. Conflation of their philosophical activities occurs in two related ways: in the philosophical method that each employed and in the discoveries that occurred as a consequence of their employment of this method, especially in harmonics. It is notable that both Archytas and Hippasus are credited by Aristotle's students with employing empirical observations in order to derive the mathematical properties of the universe, especially in the related fields of geometry and harmonics. As early as the late fourth century BCE, Archytas is credited with having discovered the solution to the problem of finding two mean proportionals in a continuous proportion that exists between two lines, and sometime later on he is credited with various other innovations, including the systematization of mechanics through use of mathematical axioms.[128] In the field of musicology, remarkably, Archytas himself does something very similar to Plato in the *Philebus*: he ascribes the discovery of the relationship between motion and pitch to his predecessors by adapting the heurematographical tradition of the Sophists and tragedians:

> Those who distinguish the sciences [τοὶ περὶ τὰ μαθήματα διαγνώμεν] seem to me to do so well [καλῶς], and there is nothing strange [in suggesting that] they understand individual things correctly, what sort they are. For, after

126. Iambl. *VP* 75, 42.23–43.12; also preserved by Thesleff in his collection of the *Pseudo-Pythagorica* (pp. 111–114). Trans. after Dillon and Hershbell 1991.

127. "Retribution" to be paid (probably to Persephone; see Pindar F 133) is suggested several times in the so-called Orphic-Bacchic Gold Tablets (nos. 6, 7, and 27 Graf and Johnston), which might inform the background of Horace's poem.

128. Archytas T A 1 Huffman = D.L. 8.83. Bowen (1982: 87) mistakenly assigns the authority of this passage to Aristoxenus. See Huffman 2005: 79–83 and 355–357.

> they made good distinctions [καλῶς διαγνὸντες] between the nature of wholes, they were on their way to see well concerning things, what sort they are, part by part [περὶ τῶν κατὰ μέρος, οἷά ἐντι]. In fact, concerning the speed [περὶ ταχυτᾶτος] of the stars and their risings and settings, they handed down to us a clear distinction [παρέδωκαν ἁμῖν σαφῆ διάγνωσιν]; the same goes concerning geometry and numbers and—not least [of all]—music. For these sciences seem to be akin.
>
> So, then, they first undertook to examine [πρᾶτον μὲν οὖν ἐσκέψαντο, ὅτι] the fact that sounds could not exist unless impacts of things against one another were to happen. And they said [ἔφαν], "an impact happens whenever things in motion collide and fall upon one another. Some moving in opposite directions, when they meet, make a sound as each slows the other down [συγχαλᾶντα], but others moving in the same direction but not with equal speed [μὴ ἴσῳ δὲ τάχει], being overtaken by the ones rushing upon them and being struck, make a sound. In fact, many of these sounds cannot be recognized because of our nature [πολλοὺς μὲν δὴ αὐτῶν οὐκ εἶναι ἁμῶν τᾷ φύσει οἵους τε γινώσκεσθαι], some because of the weakness of the blow, others because of the length of the separation from us, and others because of the excess of the magnitude. For the excess of the magnitude of sounds does not steal into our hearing, just as nothing is poured into narrow-mouthed cups, whenever someone pours out too much."
>
> So, then [μὲν οὖν], of the sounds reaching our perception those that arrive quickly [ταχὺ παραγίνεται] and strongly from impacts appear high in pitch [ὀξέα φαίνεται], but those that arise slowly [βραδέως] and weakly seem to be low in pitch [βαρέα δοκοῦντι εἶμεν].
>
> (Archytas F 1 Huffman = Porphyry, *Commentary on Ptolemy's Harmonics* 1.3; translation after Huffman 2005)

Several observations on this fragment present themselves. First of all, as Huffman and Bowen have argued,[129] Archytas's primary concern in the preamble to his discussion of the relationship between speed and pitch is to emphasize the excellence of his (ἁμῖν) predecessors in making distinctions (καλῶς διαγνὸντες), especially those that make it possible to understand how wholes are related to parts.[130] The process of making clear distinctions, an activity that is "handed

129. Huffman 2005: 127–136; Bowen 1982: 85–86.

130. Huffman (2005: 150–151) also acutely notes the strong relationships between descriptions of the goals of διαγιγνώσκειν in Archytas's F 1 and in the Hippocratic treatises *Regimen* (1.2) and *Epidemics* (1.23), which also emphasize the importance of making distinctions between whole and part.

down" (παρέδωκαν) by Archytas's predecessors, is exemplified here by the discernment of the *speed* of the rising and setting of the stars, in the spirit of the heurematographical *topos* found in Prometheus's speech in the Aeschylean *Prometheus Bound* (457–458), where the protagonist claims that he "showed [those who were without γνώμη] the risings and the settings of the stars, hard to discern" (ἔστε δή σφιν ἀντολὰς ἐγὼ / ἄστρων ἔδειξα τάς τε δυσκρίτους δύσεις).[131] Archytas also credits his predecessors with making innovations regarding "numbers" (περὶ ἀριθμῶν), recalling similar attributions of the invention of number to Palamedes (by Aeschylus, Gorgias, and Plato), Prometheus (by the author of *Prometheus Bound*), and Theuth (by Plato).[132] In these ways, Archytas's description of the innovations of his predecessors in making good and clear distinctions in their investigation of the universe does not stray far from the heurematographical paradigm as evinced in the writings of Plato, Gorgias, and the tragedians. Even so, Archytas's heurematographical treatment, as I will show, differs from these in important ways.

The value of drawing such comparisons with the heurematographical tradition, of course, is that we can see in relief what Archytas is doing differently from these accounts. First of all, the primary intellectual context for Archytas's use of the heurematographical *topos* is musicology, with a unique concern over how speed and pitch might be related in acoustical physics.[133] Details of Archytas's theory of musical pitch and its philosophical assumptions have been excellently discussed by Huffman and Bowen and need not concern my argument at this time.[134] Second, Archytas, perhaps surprisingly, does not appeal to the clandestine "wisdom" (σοφία), or for that matter to any ipse dixit model, when describing the innovations of his predecessors. He is not concerned to retain the secrets of the mysteries for a small group of Pythagorean individuals. On the contrary, he appeals to use of the rhombus "in the mysteries" (καὶ τοῖς ῥόμβοις τοῖς ἐν ταῖς τελεταῖς) as an empirical datum that allows one to explain scientifically just how speed correlates with pitch.[135] Moreover, there appears to be no need to refer to those who have made such important discoveries in acoustics by name. Just as in Fragment 3 of Archytas,

131. Griffith (1983: 177) notes the emphasis in *Prometheus Bound* placed on intelligence, in contrast to earlier accounts from Hesiod or the first half of the fifth century BCE.

132. A. F 181a = Stob. *Ecl.* 1 Prooem.; Gorg. *Pal.* 30 = DK 82 B 11a; Pl. *R.* 7, 522c1–d7; Pl. *Phdr.* 274c8; Ps.-A. *PV* 459.

133. In this fragment, Porphyry also quotes Archytas's summary statement, that "high notes move more quickly and low ones more slowly."

134. Huffman 2005: 129–48 and Bowen 1982: 92–98.

135. See lines 35–36 in Huffman's enumeration.

which also featured heurematographical tendencies, there is no explicit authority named for the discovery made, and the emphasis rather is on *discovery through one's self* (τὸ δ' ἐξευρὲν δι' αὔταυτον καὶ ἴδιον), which is described as "difficult and infrequent" (ἄπορον καὶ σπάνιον) but not impossible.[136] In a sense, Archytas has "democratized" discovery by making it available to everyone who is able to gain access to "calculation" (λογισμός) and its art (λογιστική).[137] Because the context of Fragment 3 is the establishment of an art that can be used to promote concord in the city-state for all citizens, regardless of economic status, it is unsurprising that Archytas would reject an ipse dixit model for education in the civic arts. After all, Plato's presentation of the "first-discoverer" Prometheus in Protagoras's "Great Speech" functions primarily as an allegory for the idea that certain private individuals alone, namely Sophists such as Protagoras or Prodicus, have the unique capacity to lead all others to virtue, a position Plato would spend the greater part of his earlier and middle career attacking. Apparently Archytas, too, may have had reasons to criticize the Sophistic paradigm, reasons that both informed his approach to describing a natural history of intellectual discovery and were based, at least in part, on political pretexts. We might recall that the democratization of knowledge went hand in hand with democratic revolutions among the Pythagorean brotherhoods in Magna Graecia, including the so-called Cylonian conspiracy, which had Hippasus of Metapontum as a major actor in the thick of it all.[138]

Where Archytas is most heurematographical, though, is in a sentence that leads up to the discussion of the philosophical activity of "those who distinguish the sciences" (τοὶ περὶ τὰ μαθήματα διαγνώμεν). This sentence, which has not received its due credit in scholarly exegeses of this passage, establishes a new line of thought (μὲν οὖν) following the statement made famous by Plato in the *Republic*—"these sciences seem to be akin" (ταῦτα . . . τὰ μαθήματα δοκοῦντι εἶμεν ἀδελφεά)[139]—and describes the investigative activity of Archytas's predecessors: "So they [i.e. those who handed down the clear distinction

136. See Huffman 2005: 189–90. Note that Archytas continues to use the language of "paths" in the fragment by describing "seeking" as "easy" (εὔπορον), as I analyzed in Plato's discussion of philosophical method in the *Republic*. Compare also Democritus's claim (DK 68 B 5.1) that humans eventually discovered the arts and other things that make the communal life possible "once they had come to know fire" (γνωσθέντος τοῦ πυρός).

137. For λογιστική, which is said to "excel the other arts with regard to wisdom," see Archytas F 4 Huffman.

138. On Timaeus of Tauromenium's association of democratizing and demonstrating the Pythagorean doctrines with the mathematical Pythagoreans, see chapter 3.

139. See Pl. *R.* 7, 530d6–9.

concerning the sciences] first undertook to examine the fact that sounds could not exist unless impacts of things against one another were to happen" (πρᾶτον μὲν οὖν ἐσκέψαντο, ὅτι οὐ δυνατόν ἐστιν εἶμεν ψόφον μὴ γενηθείσας πληγᾶς τινων ποτ' ἄλλαλα).[140] It is particularly difficult, but important for present purposes, to inquire after a precise meaning for the phrase "πρᾶτον … ἐσκέψαντο, ὅτι," since it describes what action Archytas's predecessors did "first" and determines the extent to which Archytas ascribed a heurematographical discovery to his predecessors. Now, the verb σκέπτομαι occurs somewhat infrequently before Plato and Xenophon and never among the Presocratics. It only appears with ὅτι and indicative in a stock phrase that is employed in courtroom contexts: σκέψασθε ὅτι ("see" or "notice that," an injunction to a courtroom audience).[141] In such circumstances, it points to external evidence that a defendant might appeal to in order to substantiate his claims, similar to what we find with the employment of σκέψασθε τόδε in Gorgias's *Defense of Palamedes*.[142] Likewise, the nominal form of this important term is very uncommon in Greek literature before Plato, appearing in Euripides (*Hippolytus* 1323) in a simple grammatical construction, never in Presocratic fragments, and only once in the extant fragments of Antiphon the Sophist—with ὅτι and indicative. In a fragment of his work *On Truth*, Antiphon, after he has just suggested, regarding justice, that it is most advantageous for an individual to obey the laws in the presence of witnesses but to ignore them when one is in a state of nature, reflects on what he has said in this way:

> For the sake of all these [aforementioned] things, there is an examination [σκέψις], that [ὅτι] the majority of what is just according to law

140. I translate ἐσκέψαντο as "undertook to examine" in order to emphasize what I take to be the inceptive quality of the aorist here, also emphasized by the presence of the word πρᾶτον and the shift from the present tense. Huffman has "first they reflected," which does not capture the force of the tense shift or the adverb; Bowen's "first, they observed" is slightly preferable, since it emphasizes the notion that we are dealing with data that are employed in inquiry. A straightforward translation such as "they first noticed that" would be absurd, since we cannot imagine that Archytas's predecessors would have been the "first" to *take notice* of the fact that sounds occur when things are impacted against one another. See below.

141. See something similar in Thrasymachus's fragments without ὅτι (DK 85 B 1) and among the Attic orators with ὅτι (And. *de Pace* 17, Antipho *de Caed. Herodis* 40, D. *contra Calippum* 25). In Thucydides's speeches we see προσκέψασθε ὅτι used in order to present additional evidence for a claim (3.57.1), and σκέψασθε ὅτι used in accordance with conditional logic (3.46.2).

142. Twice (DK 82 B 11a), at secs. 13 and 20. Gorgias also uses σκέψασθαι in the *Encomium of Helen* (11), in a stimulating epistemological passage that associates this activity with investigating what is present (σκέψασθαι τὸ πάρον), in contrast to remembering the past (μνησθῆναι τὸ παροιχόμενον) and prophesying the future (μαντεύεσασθαι τὸ μέλλον).

> and convention is hostile to nature. For laws have been established over the eyes, as to what they must not see; and over the ears, as to what they must and must not hear; and over the tongue, as to what it must and must not say; and over the hands, as to what they must and must not do; and over the feet, as to what they must and must not go after; and over the mind, as to what it must and must not desire.
>
> (ANTIPHON, *On Truth* F 44[a] 2.23–3.11; translation after Pendrick 2002)

Antiphon's employment of σκέψις ὅτι and indicative might be taken to imply that the statement that follows the ὅτι is the conclusion of the premises that precede this passage, as Pendrick has suggested.[143] If so, it only appears in order to initiate a new discussion that focuses on the consequences of this conclusion, in which Antiphon provocatively states that the various functions of the human body are ruled by law, rather than nature. Similarly, in Archytas's prelude to the quotation from his anonymous predecessors, the examination that they undertake is meant to investigate more fully, by appeal to evidence of some sort, the observed fact that sounds cannot exist unless things impact one another. This is clear from the passage that follows in direct quotation, attributed to Archytas's predecessors, which emphasizes the notion that human "nature" plays a role in the "recognition" of (τᾳ̃ φύσει οἵους τε γινώσκεσθαι), and failure to recognize, sounds. As I argued with regard to Aristotle's descriptions of the mathematical Pythagoreans, the focus here is on how Archytas's predecessors take the observed fact (i.e. that sounds occur when things impact one another) and provide an explanation for that particular phenomenon.[144] The explanation provided by the predecessors refers to the speeds of the objects with regard to one another, that is, whether one object is faster or slower than the other or whether the two objects are "equal" (i.e. ἴσῳ) in speed. This explanation also accounts for why some sounds aren't heard in various ways, submitting that one reason for this could be human "nature" (ἁμῶν τᾳ̃ φύσει). Quantitative mathematics plays a strong role in this explanation, as the predecessors claim that it is on account of the "weakness of blow" (διὰ τὰν ἀσθένειαν τᾶς πλαγᾶς), "length of separation" (διὰ τὸ μᾶκος τᾶς . . . ἀποστάσιος), and "excess of magnitude" (διὰ τὰν ὑπερβολὰν τοῦ μεγέθεος) that we do not hear every sound. One might speculate that Archytas's predecessors were referring to deficiency in perception, even though there is no clear reference to it; and this is precisely how Archytas interprets their words

143. Pendrick 2002: 327.

144. See chapter 1.

when he switches *back* from indirect discourse (marked by a return to the indicative and the transitional formula μὲν οὖν) and speaks in technical language of "perception" (αἴσθασις), "appearance" (φαίνεται), and "seeming" (δοκοῦντι), all in the same sentence. In that sentence, which marks the return to his own voice, Archytas also shifts the theme of the argument, from the speed of objects and its relationship to the magnitude of sound, to the speed of objects and its relationship to the height or depth of the pitch note. Magnitude is no longer the criterion for receptability of the note, on Archytas's interpretation.

When we look at the words Archytas attributes to his predecessors, those figures who "concern themselves with the sciences" (τοὶ περὶ τὰ μαθήματα), we might be surprised to note that there is no concern with the epistemic status of sensation or perception, which is somewhat obscurely articulated in Archytas's other fragments[145] and had been an important issue for various Presocratic philosophers, including Heraclitus, Parmenides, Zeno, Anaxagoras, and Democritus, as well as for the Sophists Gorgias and Protagoras. Nor do Archytas's predecessors appear to posit any *determinate* relationship between speed of colliding objects and height or depth of pitch. It is also notable that Archytas's predecessors also appear to have no knowledge of the Sophistic debate between law and nature, in which a variety of contemporary interlocutors of Archytas had participated, including Antiphon and the Sophist known as Anonymous Iamblichi.[146] By all counts, the predecessors' interrogation into the fact that sound cannot exist without impact of objects is rudimentary, developed out of the application of simple arithmetical relationships (including distance between objects and the "magnitude" of sound) to observational data, and supported by analogies from daily life (e.g. the comparison between sounds of large "magnitude" being unable to enter our ears and what can be poured into a narrow-mouthed cup). It is true that Archytas goes on to employ techniques of scientific inquiry and argument similar to those practiced by his predecessors in the remainder of Fragment 1, but in a way that demonstrates a greater level of sophistication and philosophical nuance, as he adapts the advances made in scientific method by his predecessors by shaping them into a philosophical system that is informed by contemporary intellectual debates in the early fourth century BCE.[147]

145. Archytas F 1 Huffman: καλῶς ὀψεῖσθαι, etc.; F 4 Huffman: δοκεῖ ἁ λογιστικά, ἐναργεστέρω. Huffman has argued (2005: 236–237 and 246–247), by reference to F 4, that Archytas values the visible or sensible over the intelligible.

146. See Huffman 2002.

147. For an excellent analysis of how Archytas continued to employ empirical observations in order to justify mathematical ideas, see Barker 2007: 292–299.

Just as is the case in the various modalities of the heurematographical myth in Plato's writings, the predecessors of Archytas remain anonymous. As Huffman has argued, because each of them pursued knowledge in the sciences to which Archytas refers in Fragment 1 (geometry, astronomy, numbers, music), these predecessors could include the mathematicians Hippocrates of Chios and Theodorus of Cyrene, as well as the astronomers Oenopides of Chios and Meton and Euctemon of Athens. They might, moreover, include the Pythagorean Philolaus of Croton, who made advances both in astronomy and harmonics, and the polymath Democritus of Abdera, who wrote books on geometry, arithmetic, and acoustics (DK 68 A 1), was probably affiliated with Philolaus and/or other Pythagoreans (Diogenes Laertius 9.38), and wrote speculatively about the nature of hearing (DK 68 A 126).[148] To this list we might add Empedocles (DK 31 A 86 = Theophrastus *On Sense Perception* 7.9.12–13) and Anaxagoras (DK 59 A 106 = Aëtius 4.19.6), who apparently espoused theories of sound that had to do with impacts in which there is a movement of a sound into a pore, as well as the Hippocratic *Fleshes* (15 and 18), where sound is understood to be something "directed" (*ἀπερείδονται*) toward the ear.[149]

In the argument being made in Archytas's Fragment 1, however, the emphasis lies particularly in the innovations made in scientific *inquiry* as a consequence of examination of the fact that high or low sounds occur when things are impacted.[150] While each of these figures could make a general claim to be influential over Archytas's philosophy, none of them was known for expressly theorizing about the precise quantitative value of a sound as obtained through observation of the ways things can impact one another. The figures who are indeed associated with such theorizing are the fifth-century BCE music theorist Lasus of Hermione[151] and Hippasus of Metapontum.[152] One lacunose and problematic passage, from Theon of Smyrna (fl. ca. 115–140 CE), testifies explicitly to this:

> Some people thought it best to derive these concords from weights, others from magnitudes, others from movements <and numbers>, and others from vessels. Lasus of Hermione, so they say, and the followers of Hippasus of Metapontum, the Pythagorean man, pursued the speeds and slownesses of movements, through which concords . . . [lacuna] . . . thinking

148. Democritus claims that hearing is a "receptacle of words" (*ἐνδοχεῖον μύθων*), which "penetrates and flows in" (*εἰσκρίνεται καὶ ἐνρεῖ*), by contrast with sight.

149. As suggested to me by Andrew Barker.

150. Pace Huffman 2005: 132–140.

151. For Lasus and his role in early musicology, see Barker 2007: 19–20 and 79–80.

152. Huffman 2005: 135. See Barker 2007: 305–306 and Burkert 1972: 441, with n. 84.

> that these sorts of ratios come from numbers, he [i.e. Hippasus? Lasus?] derived them from vessels. For, using vases all equal and of like figure, he left one empty and filled another half-way full of water; he struck them together and produced concord of an octave. And again, leaving one of the vases empty, he filled up another one-fourth of the way, and striking them together he produced concord of a fourth. And he produced the concord of the fifth when he filled up one-third of another. Thus the emptiness of the first vase was in a relation to the second of 2:1 in the concord of an octave, and 3:2 in the concord of a fifth, and 4:3 in the concord of a fourth.
> (DK 18 F 13 = THEON OF SMYRNA, *Mathematics Useful for Reading Plato* p. 59.4–15 Hiller)

Several uncertainties present themselves. We cannot be sure where Theon is deriving his information, although the Aristotelian *Problems* (19.50) describes a very similar type of experimentation, suggesting that the problem was at least known to the early Peripatetics.[153] Moreover, when Theon refers to the percussive activity of filling up vessels and striking them together in order to discover concords, it is difficult to know whether he is referring to activities of Hippasus of Metapontum or of Lasus of Hermione.[154] A very unfortunate lacuna prevents us from being absolutely certain. Still, while there is no external evidence that supports the claim that Lasus is to be credited with this activity, we do hear, on good authority, of Hippasus performing similar types of percussive experiment involving the collision of objects in order to understand the phenomena of resonance.[155] The deduction of mathematical ratios (which correspond with basic concords) made possible by objects colliding together was thus one of the discoveries in musical theory attributed to Hippasus of Metapontum by the musicologist Aristoxenus of Tarentum, who was, somewhat paradoxically, both a Pythagorean (at least early in life)[156] and an associate of Aristotle:[157]

153. It should be noted that the experiment described here will not produce the results that are listed, as Barker notes (1989: 31–32, with n. 11). Zhmud (2012: 276) suggests that the source comes from the fourth century BCE, but he cannot prove it.

154. On the problems involved in interpreting this passage, see especially Burkert 1972: 377–378, with n. 36.

155. Of course, the discovery of the relationship between numerical intervals and concords was attributed by Xenocrates to Pythagoras (F 87 IP = Porph. *in Harm.* 30.1–10), but he does not explain how this would have been done.

156. On Aristoxenus's Pythagorean heritage, see Huffman 2008: 106.

157. See Barker 1989: 30–32 and Burkert 1972: 206–207.

> For a certain Hippasus made four bronze discs in such a way that while their diameters were equal, the thickness of the first disc was epitritic in relation to that of the second, hemiolic in relation to that of the third, and double that of the fourth, and when they were struck they produced a concord.
>
> (DK 18 F 12 = ARISTOXENUS F 90 WEHRLI = Scholium to Plato's *Phaedo* 180d4; translation by Barker 1989)

Aristoxenus's fragment testifies that Hippasus investigated the sounds that could result as a consequence of striking bronze disks whose dimensions were proportional in terms of thickness; the results were empirically successful, in that if the bronze disks featured equal diameters, the thicknesses of each would be proportional to their unique pitches.[158] Aristoxenus, who in particular was concerned with establishing a historiography of competing musical theorists (both mathematical theorists of a Pythagorean sort and stricter empiricists),[159] may have seen Hippasus as one of those figures who, in the words of Barker, "are said to judge what is true and false in harmonics by the criterion of reason (which here usually means reasoning of a mathematical sort), and to rely as little as possible on perception, since it is fallible and can easily mislead us."[160] But Archytas's description of his predecessors' theory of inquiry into the universe, if indeed it refers to Hippasus, betrays no knowledge of a separation between reality and perception, which suggests the possibility that Hippasus might have actually toed the line between what would, in post-Platonic harmonics, have been a fundamental distinction alien to an early musical theorist.

If my argument holds water, then we can conclude that Hippasus of Metapontum is the most likely candidate for being the first person to engage in empirical experimentation that led to the association of mathematical ratios and musical intervals.[161] His experiments and discoveries sought to demonstrate that speed of blow (i.e. quickness or slowness) combined with the mass of the colliding objects elicits a particular pitch, whether it can be heard by a listener or not.[162]

158. For a useful recent analysis of this passage and of Hippasus's activities more generally, see Creese 2010: 93–97.

159. In his *On Arithmetic* (F 23 Wehrli = Stob. *Ecl.* 1 Prooem. 6), Aristoxenus claims that Pythagoras advanced the *pragmateia* concerning numbers, but he does not strictly attribute *the discovery* of number to Pythagoras. Instead, the discovery of numbers is associated with Hermes-Thoth, at least according to "the Egyptians."

160. Barker 2009: 165–166.

161. See Zhmud 2012: 276.

162. See Huffman 2005: 139–140.

The relationship of the numerical values of these pitches is expressed as a mathematical ratio. Moreover, this project of "those who concern themselves with the sciences," in the eyes of Archytas, was significant for the innovation it presented as a *mode of inquiry* into the universe.[163] While Huffman is probably right in seeing a panorama of predecessors, both Pythagorean and non-Pythagorean, who influenced Archytas in the development of his philosophical method, it would be infelicitous to underestimate the likelihood that Archytas, when he is at his most heurematographical (in Fragment 1), refers to the innovations of Hippasus of Metapontum in phonological inquiry. For reasons that have to do with Archytas's philosophy of knowledge, ethics, and political theory, he probably did not refer to Hippasus by name; and Archytas's apparent rejection of the ipse dixit model might be considered symbolic of what had been for the mathematical Pythagoreans a great departure, on philosophical and political lines, from the acousmatic Pythagoreans, as I argued in chapter 3. In both cases, it is the name of Hippasus of Metapontum that is assigned to the heretical "first-discoverer" among the Pythagoreans of Southern Italy, who was legendary for having betrayed the Pythagorean brotherhood by developing new approaches to investigating the universe through empirical investigation and by revolting against their political caste. The idea that Hippasus might be chiefly intended when Archytas speaks of his predecessors "who concern themselves with the sciences" thus prompts consideration of whether Hippasus of Metapontum might be the target of Plato's reference to a "certain Promètheus with most brilliant fire" who handed down the gift of the gods to the "forefathers" in the *Philebus*.

THE MATHEMATICAL PYTHAGOREANS AND MUSICAL DIALECTICS IN THE *TIMAEUS* AND *PHILEBUS*

If my arguments have persuasively shown that the chief figure in the mind of the mathematical Pythagorean Archytas when he refers in Fragment 1 to his predecessors, who "concern themselves with the sciences" (*τοὶ περὶ τὰ μαθήματα*), is Hippasus of Metapontum, then we are compelled to allow for the possibility that Iamblichus, apparently looking at a passage of Aristotle, was historically correct (or at least had it on an authority of some Pythagoreans) in asserting that the unique *pragmateia* of the mathematical Pythagoreans originated with Hippasus.[164] As I have argued, moreover, Plato's reference to the

163. Pace Creese 2010: 96–97.

164. Zhmud (2012: 256) accepts the tradition, ultimately from Eudemus of Rhodes forward, that associated Hippasus's and Archytas's innovations in mathematics, but he does not think that the tradition involving the division into mathematical and acousmatic Pythagorean is original with Aristotle or is reliable.

philosophical method of the "forefathers" (οἱ παλαιοί) in the *Philebus* constitutes a characterization of a type of inquiry into the universe that can be associated especially with the mathematical Pythagorean Philolaus of Croton. Several comprehensive studies of the Philolaic Pythagoreanism implied in the "gift of the gods" in Plato's *Philebus*, especially those written by Thomas, Miller, Meinwald, and Barker, emphasize two significant aspects concerning the "certain Prometheus": that the "certain Prometheus" in question might be Pythagoras (or at least Pythagorean) and that the method described could not properly function without data derived from empirical examination.[165] When Socrates is faced with confusion on the part of Protarchus concerning the relationship between the limiter/unlimited pair and philosophical method, he attempts to describe the philosophical method using the familiar paradigms of letters and music. In particular, Socrates is interested in the use of sensory data, that is, knowledge of musical intervals, inasmuch as it leads up to a more advanced form of inquiry into the unity and plurality of objects in the universe:

> But, my friend, whenever you grasp the intervals in the quickness or slowness [ὀξύτητός τε πέρι καὶ βαρύτητος] of sound, both their numerical *quantity* [ὁπόσα τὸν ἀριθμόν] and *quality* [ὁποῖα], in the limits[166] of their intervals [τοὺς ὅρους τῶν διαστημάτων], and in however many arrangements [ὅσα συστήματα] come about from these—observing [all] these things, our forefathers passed down to us, their followers [οἱ πρόσθεν παρέδοσαν ἡμῖν τοῖς ἑπομένοις ἐκείνοις], the name "harmonies";[167] and, in turn, they declare [αὖ φασι] that other sorts of features that also come to be in the motions of a body [ἔν τε ταῖς κινήσεσιν αὖ τοῦ σώματος] ought to be measured in numbers and called "rhythms" and "measures," and, likewise, that we ought to realize that this is the proper way to make an inquiry into everything, both unified and many [ἐννοεῖν ὡς οὕτω δεῖ περὶ παντὸς ἑνὸς καὶ πολλῶν σκοπεῖν]. For

165. Thomas 2006: 224–225, with nn. 49 and 52; Miller 2003: 27–30; Meinwald 2002: 93, with n. 15, is suggestive, but not explicit on this count; Barker 1996: 143–164. Huffman (2001) does not explicitly discuss empirical knowledge, but he does not count it out in relation to this passage either.

166. Barker (1996: 147–148) points out that a great deal rests on how we interpret and translate the term ὅροι: "As to the ὅροι of the διαστήματα, ὅρος is another technical term in both musicology and mathematics, but an ambiguous one. Are these ὅροι the points of pitch forming the boundaries of a quasi-spatial interval, or are they to be understood in their alternative mathematical sense as the 'terms' of a relation such as a numerical ratio? The choice turns out to matter quite a lot."

167. By "harmonies" (ἁρμονίας), Plato apparently means something like "arrangement of intervals within an octave," as Aristoxenus (*El. Harm.* 36.31) defines this term.

> when you have grasped these things in this way, then you become wise [in them]; and whenever you have grasped the unity of any of the other things by examining them [σκοπούμενος] in this way, you become intelligent concerning it in this way.
>
> (PLATO, *Philebus* 17c11–e3)

It is important to make sense of this passage, since it points us in the general direction of understanding how quantity and quality inform our ability to "grasp the unity" of any class of objects, a necessary condition for obtaining knowledge.[168] I have attempted to preserve in my rendering what I take to be the basic binary opposition that appears to underlie much of this passage: the difference between *quantity* and *quality* in the intervals with regard to the "number" (τὰ διαστήματα τὸν ἀριθμόν).[169] It is difficult to know for sure whether Plato intended pitch height or depth to be numerically quantifiable, if indeed this is the right way to read this passage.[170] It was the suggestion of Gosling that Plato was referring to the enumeration of musical notes, but as Mary Louise Gill notes, the number of notes could be infinite on a continuum.[171] As I will argue with regard to the *Timaeus*, it is possible to see a third way between the two, in which the "number" of notes can be limited quantitatively *because it is shown to repeat*. A straightforward reading of the Greek text of the *Philebus* would acknowledge that a quantifiable number of *intervals* of phonic sounds is possible to enumerate, but this forces us to understand the "number" to which Socrates refers not simply as a positive integer,[172] but as a *numerical ratio*, which is apparently how the mathematical Pythagorean Archytas would have described a "high" or "low" pitch.[173]

168. Plato had, of course, already described a similar process in the *Phaedo* (100a3–7 and 101d3–5), on which see the compelling arguments of Bailey 2005.

169. Contrast Xenocrates's description of Pythagoras's discovery of the intervals (F 87 IP): "Pythagoras . . . discovered that the intervals in music, too, do not arise in separation from number: for they are a blending of quantity with quantity [σύγκρισις ποσοῦ πρὸς ποσόν]." Translation by Barker.

170. Barker's excellent article (1996) is aporetic about what the quantity in "number" is. My interpretation follows that of Gill (2009: 43), which is exemplified in her translation: "Well, my friend, [you will become expert] when you grasp the intervals of high sound and low, how many they are in number and of what sorts they are."

171. Gill 2009: 46, with n. 36, contra Gosling 1975: 172–173.

172. Of course, each integer can be expressed as a ratio of two other numbers, with the exception of the number 1, which is obviously a more complicated case (i.e. it is the ratio of any same two numbers, which is perhaps why it is often elicited as the paradigm for commensurability).

173. See Creese 2010: 117–128 and 150, who also quotes the testimony of Aristoxenus (*El. Harm.* 32.24–26).

Such a reading is strengthened by a later passage that more explicitly describes what Socrates has in mind when he describes the "class" (γέννα)[174] that limits, which, when combined with the "class" that is unlimited, produces a quantifiably high or low "number":

> SOCRATES: I am referring to the class of the equal and the double, that is, however many things stop [παύει] [opposite] things from being in a state of difference from one another; and the class that, by imposing a number on them [ἐνθεῖσα ἀριθμόν], makes things completely commensurate and concordant [σύμμετρα καὶ σύμφωνα ἀπεργάζεται].
>
> PROTARCHUS: I understand. For you seem to me to be saying that certain generations [γενέσεις τινάς] occur from the mixture of these things in each circumstance.
>
> SOCRATES: Your impression is correct.
>
> PROTARCHUS: Go on, then.
>
> SOCRATES: Is it not the case that the correct combination [ὀρθὴ κοινωνία] of these [opposites] gives rise to [ἐγγένησεν] the nature of health in [ἐν] sick people?
>
> PROTARCHUS: Certainly.
>
> SOCRATES: Is it not also the case that the same things are generated[175] in the same way in the high and the low, in the fast and the slow [ἐν ὀξεῖ καὶ βαρεῖ καὶ ταχεῖ καὶ βραδεῖ], which are unlimited? For at the same time they completely establish a limit [πέρας ἀπηργάσατο] and thereby constitute a music that is entirely perfect.
>
> PROTARCHUS: Beautifully spoken.
>
> (PLATO, *Philebus* 25d11–26a5)

This passage helps us to secure a better sense of what Plato means by grasping the intervals "in number" (τὸν ἀριθμόν) from the Promethean passage. In particular, we see that intervals are made up of two constituents: what is unlimited and what limits the unlimited. The "class" of limiters, which apparently includes numerical relations (πρὸς ἀριθμὸν ἀριθμός) based on equivalence and doubleness, is said to "pause" or "stop" (παύει) things from being in a state of opposition by putting them into a state of commensurability.[176] Unlimited things, by contrast, are associated with the "greater and the lesser" as well as, importantly,

174. This is the only occurrence of this unusual word in Plato's oeuvre. It should be considered in relation to the more common γένος.

175. Retaining the manuscript reading of ἐγγιγνόμενα, contra Burnet.

176. See Pl. *Phlb.* 25a6–b4.

the "faster and the slower."[177] This activity of "pausing" what is unlimited is understood to be the imposition of a limiting "number" (ἐνθεῖσα ἀριθμόν) on it, which results in the "completion" (ἀπεργάζεται; ἀπηργάσατο) of what Socrates later calls the "correct combination" (ὀρθὴ κοινωνία), a term that elicits comparisons with Plato's metaphysics, which focuses on the communion of various Forms in opposition.[178] What is "correct" (ὀρθή) in the combination is indicated by two Pythagorean watchwords, "commensurate" and "concordant" (σύμμετρα καὶ σύμφωνα), as argued by Andrew Barker.[179] The newly formed unity, which is constituted of an unlimited that has been determined by a limiter, has the capacity to "generate" or "give rise to" things "in" other things. Two examples are given. It gives rise to "health" in sick people, and it generates a perfected "music," as he argues, "in the high and the low, in the fast and the slow" (ἐν ὀξεῖ καὶ βαρεῖ καὶ ταχεῖ καὶ βραδεῖ), that is, in the unlimited entities such as sound that, without something to limit them, remain subject to the greater and the lesser. It is notable that when Socrates refers to the pitch values of "the high and the low" here, he speaks of them as being coordinate with the modalities of "the fast and the slow," just as argued above with regard to Archytas's harmonic theory.[180]

It is pretty clear that Plato's description of the generation of a complex entity such as "health" or "music" that is made up of a factor that limits the unlimited in the *Philebus* is coordinate with other late presentations of the cosmic generation of entities marked by the qualities of being concordant and symmetrical, especially what is found in Plato's *Timaeus*. Timaeus's description of συμφωνίαι there obtains a dialectical tenor, when Timaeus claims that their formation occurs when slower sounds

> catch up [to swifter sounds] they do not disturb their motion by imparting a different one but impart the beginning of a lower motion in conformity with that of the swifter sound, when the latter is fading. By attaching [to one another] in a similarity[181] [ὁμοιότητα προσάψαντες], they are

177. See Pl. *Phlb.* 25c8–d3.

178. See my brief discussion of Plato's *Sophist* below.

179. See Barker 1996: 155.

180. Pace Barker 1996: 153.

181. This is my best attempt to render this difficult phrase, which Barker (1989: 62 n. 31) glosses by saying "the main difficulty is in the interpretation of 'similar.'" His "attaching a similarity" does not make good sense of προσάψαντες. Zeyl, who renders this phrase "they graft onto the quicker movement" doesn't translate ὁμοιότητα at all. Fronterotta's (2006: 387) "portandoli all'uniformità" is probably the best of recent translations, in that his rendering, like mine, reads ὁμοιότητα as an accusative of respect with προσάψαντες.

> blended together into a single effect, derived from the high and the low [μίαν ἐξ ὀξείας καὶ βαρείας συνεκεράσαντο πάθην]. Hence they provide pleasure to people of poor understanding, and delight to those of good understanding, because of the imitation of the divine *harmonia* that comes into being in mortal movements.
>
> (PLATO, *Timaeus* 80b2–8; translation after Barker 1989)

So, in this presentation, concordance is understood to be a "single effect" (μία πάθη) that occurs when swift and slow sounds impact one another. The physics that underlies the production of a concordant sound is based on Archytan assumptions about the identification of pitch and velocity of movement of sound, as discussed above.[182] So far, Plato is not deviating much from Archytas; but the concluding statement, which emphasizes how people (both intelligent and nonintelligent) derive pleasure from this experience of concordance "because of the imitation of the divine *harmonia*" (διὰ τὴν τῆς θείας ἁρμονίας μίμησιν), exhibits the unique stamp of Platonic metaphysics.

Similarly, in the famous generation of the world-soul in Plato's *Timaeus*, we see that mathematical Pythagorean approaches to musical theory inform the design of the universe, in accordance with the principles of symmetry and concordance.[183] The Demiurge, who shapes the world-soul, is described as being a "discoverer" (ηὕρισκεν) of the fact that things that lack reason cannot be better than things that have reason, after he has "calculated" (λογισάμενος) that it is not permitted (οὐ θέμις) for anything that is supremely good to do anything other than what is best.[184] Now, as I discussed in chapter 5, the invocation of teleological causation is probably original to Plato, but there are still several reasons to take seriously the traditions that link the Demiurge's activities with the mathematical Pythagoreanism of Archytas.[185] As in Archytas's Fragment 3, calculation (λογισμός) is understood to be a prerequisite for discovery (τὸ ἐξευρέν) through one's self.[186] Once he

182. See Barker 1989: 62 n. 31.

183. See Pl. *Tim.* 31b8–c4, where emphasis is placed on the idea that the generated cosmos is organized according to principles of proportioning (ἀναλογία).

184. Pl. *Tim.* 29e1–30c1. Remarkably, this story is also reported "by the wise men" (παρ' ἀνδρῶν φρονίμων). Timaeus (*Tim.* 28c3–5) also associates the telling of the story of the making of the universe to his own "discovery" (εὑρεῖν; εὑρόντα).

185. As suggested by Geoffrey Lloyd (1990: 169 n. 18).

186. See above in the section entitled "The Mathematical Pythagoreans and the Heurematographical Tradition." It should be noted that the Demiurge emphasizes the fact that things come to be "through himself" in a memorable etymologization of his own name, when he addresses the gods: δι' ἐμοῦ γενόμενα. See Regali 2010: 260.

takes the decision to generate the world-soul, the Demiurge achieves this task by dividing up the stuff of the universe, compounded into a strip, according to various intervals (διαστήματα), including the hemiolic, epitritic, and epogdoic intervals, which are in mathematical ratios of 3:2, 4:3, and 9:8, respectively.[187] These are, of course, the same intervals said to have been discovered by Hippasus of Metapontum in his percussive experimentation with bronze discs (on the reliable account of the late fifth-century BCE music historian Glaucus of Rhegium) and assumed in Philolaus's and Archytas's studies of the relationship between these intervals and the octave.[188] Indeed, as Barker has argued, the Demiurge's activity of division is based on the classification of means and proportions advanced by Archytas in Fragment 2.[189] It remains only a speculation, but we can nevertheless see Hippasus of Metapontum hiding in the background of Archytas's classification, informing both Archytas's approaches to music theory and Plato's approaches to generation of the world-soul.

Throughout these passages of the *Timaeus* and the *Philebus*, then, there is an emphasis on how giving numerical order (or order in ratios) to what is otherwise without order, boundary, or shape constitutes a spatiotemporal "pause." This pause is understood, in its most basic sense, to refer to the assimilation of one thing to another that had previously been different, or alternatively to the placing of things in opposition in a relationship of concordance. It occurs in dialectical method as well as in physics and metaphysics, which are coextensive in Plato's later dialogues.[190] We might speculate that the basic principle of the spatiotemporal "pause" that creates commensurability and harmony in the universe could be derived from Plato's theorizing about the monochord.[191] Now David Creese has recently argued that it is likely that Philolaus and Archytas, and perhaps Plato, did not employ the monochord; it is certainly the case that

187. Pl. *Tim.* 35b4–36b6. On the musical background of this passage, see Barker 2007: 318–321.

188. Aristoxenus F 90 Wehrli.

189. See Barker 2007: 320.

190. See Miller 2003: 29–30. The emphasis on attaining a "pause" (παύεσθαι) also appears in Plato's *Statesman* in two related contexts: first (285a4–c2), in the context of the Eleatic Stranger's attempt to describe two types of measurement that can be employed in division, namely that which cuts according to those things subject to the greater and the lesser, and that which cuts according to due measure, and second (273e3–274b3), in the context of the turning of the cosmos, when, after the tremors and confusion consequent of the divine helmsman letting go of the rudder of the universe, it stops and calms itself in such a way as to become ordered again.

191. Here recall the strip of undifferentiated stuffs in the *Timaeus* that the Demiurge measures out according to the intervals that bound the octave.

Philolaus and Archytas *did not need* the monochord, as it is described in the *Sectio Canonis* (roughly 300 BCE), in their investigations of musical intervals.[192] Their discoveries in music theory could have been achieved either through calculation (λογισμός, as Archytas calls it) or by employing experiments of the bronze disc sort that Hippasus likely undertook.[193] But the evidence in Plato's writings is rather more suggestive, and certain aspects of Plato's later philosophy make best sense when we imagine that he had in mind something like the monochord when he refers to measuring according to "due measure" (τὸ μέτριον). Such a claim is necessarily speculative, but it helps us to make better sense of a passage in the *Statesman* (284e6–8), in which "due measure" (τὸ μέτριον) is said to coordinate with "all that has been withdrawn from the extremes to the middle" (πανθ' ὁπόσα εἰς τὸ μέσον ἀπῳκίσθη τῶν ἐσχάτων). According to Kenneth Sayre,

> Generally speaking, measures of this sort mark off a middle ground between their relevant extremes. In the language of [*Statesman*] 284e7–8, this is the middle to which they will have been withdrawn as part of the process of bringing Limit to the extremes concerned. The middle in question is the locus of moderation. Any artful endeavor must be pursued within the middle ground established by its appropriate measures; and any departure would be a return to either excess or deficiency.[194]

With regard to the monochord, at any rate, the "pause" that occurs is the placement of a movable bridge at a specific location on the continuum, which, in Plato's terminology, is representative of the unlimited that is also characterized by excess and deficiency. That "pause," if properly located, establishes an interval as a mathematical ratio based on units of string lengths, and such ratios could be used to build up a tetrachord.[195] For example, if one places the movable bridge at precisely the middle of the string, the result will be a note that is exactly one octave above the original note. In this way, musical relations are quantifiable, and a pitch-value can be expressed as a ratio of two numbers. If Mitchell Miller is right, and if Plato is thinking of the so-called Dorian mode when Timaeus describes (*Timaeus* 35b4–36b6) the Demiurge dividing the span

192. As argued convincingly by Creese (2010: 104–130).

193. To be sure, it is surprising that Creese does not address the explicit reference to a κανών in Archytas's fragments (F 3 Huffman), which is said to be one of the functions of calculation (λογισμός) in mediating between the rich and the poor.

194. Sayre 2006: 234.

195. Generally, my description of the monochord is owed to Creese 2010: 1–21.

of the universe according to the Pythagorean ratios that make up the octave, then the matrix of notes that results from a two-octave stretch of string—appended by a single note that is at an interval of one tone lower than the last note in the lowest mode—creates a set of fifteen notes, with a "middle" note precisely one octave from the first and the fifteenth notes.[196] On this paradigm, the other seven notes would recur in each octave, establishing a *repeating order* in the continuum. Plato might describe this activity as bringing a limit based in "due measure" to bear on what is otherwise unlimited, the continuum that lacks proper measurement and is thereby neither "commensurate" nor "concordant" without it. Dialectic, cosmology, and metaphysics are thus understood in Plato's *Timaeus* and *Philebus* to conform to the rules of mathematics, both harmonic and calculative, *and* are understood to be informed by empirical observation.

CONCLUSIONS

This chapter began by hypothesizing that a standard locus for Plato's engagement with the critical approaches of his intellectual contemporaries is the so-called first-discoverer myths of Prometheus, Palamedes, and Theuth. By employing these "heurematographical" myths in various dialogues throughout his life (*Protagoras*, *Republic*, *Phaedrus*, *Statesman*, and *Philebus*), Plato recurrently treats several themes central to his unique investigation into what philosophy and its practice ought to be: truth versus imitation, reality versus mere image, knowledge versus belief, and so on. At stake in these heurematographical stories is the status of the *τέχναι* invented by the *πρῶτοι εὑρεταί*—which are often broadly construed as the arts of number—as well as the value that they offer for human society both at the community and the individual levels. The heurematographical myth presents Plato with recurring opportunities to evaluate the intellectual discoveries made by various competitors in the world of fifth- and fourth-century BCE philosophy, especially Sophists such as Protagoras of Abdera and Hippias of Elis and the mathematical Pythagoreans Philolaus of Croton and Archytas of Tarentum, who, in the mind of Plato, were probably closer in kind than scholars might initially assume. This is the case because, like the Sophists, the mathematical Pythagoreans practiced various modes of demonstration and did so both (1) by appeal to sensible objects, and (2) in written treatise formats. Concerning these formats (2), the fragments of Philolaus of Croton and Archytas of Tarentum exhibit qualities that align them with the Sophists, chiefly the fact that they did not (in the tradition of the Presocratic cosmologists) write in meter and that they did not employ the dialogue format.

196. Miller 2003: 28–30 and 55 n. 31.

Concerning the appeal to sensible objects (1), Plato's use of the heuremato-graphical myth allows us to see how he appears to have modified his thoughts throughout his career about the value of sensible objects in the pursuit of the Good, as demonstrated especially in two dialogues that explicitly deal with that problem: the *Republic* and the *Philebus*. In *Republic* 7, Plato appears to be poking fun at Archytas of Tarentum in particular when he makes reference to the discovery of number by Palamedes, and the implicit criticism here of Archytas occurs within the larger claim that the mathematical Pythagoreans fail to employ the objects of the senses in order to pursue what is best: the Good. By contrast, the *Philebus*, a dialogue that aims to present Plato's most nuanced articulation of the intellective method that leads one up to the Good, associates that method with the famous "gift of the gods" as passed down by a "certain Prometheus with most brilliant fire" to the philosophical "forefathers." The method of using a limit to determine what would otherwise be infinite (in metaphysics, in harmonics, and, perhaps by extension, in dialectic) associated with the philosophical "forefathers" refers to the extant fragments of the mathematical Pythagoreans Philolaus of Croton and Archytas of Tarentum. But the determination of the identity or identities of the "certain Prometheus with most brilliant fire" encounters serious problems, in part due to the phenomenon of surplus meaning in Plato's heurematographical myths, which are punctuated by the active double-voiced discourse that characterizes Plato's parody. While Pythagoras must remain a possible referent here, the emphasis on quantification of intervals by inquiry into the speeds of objects in the pursuit of a method that leads one to become "wise" suggests Hippasus of Metapontum as a possible source for the divine method ascribed to the "certain Prometheus," who employed fire in order to discover the science of number. Such a science of number leads to an understanding of the universe in accordance with due measure and produces the qualities of symmetry and concordance, which mark the Good and everything that has been crafted in its image.

AFTERWORD

> The primary philosophy of the Pythagoreans finally died out, initially because it was enigmatic, and then because their writings were in Doric—a dialect that itself is somewhat obscure—so that the recorded doctrines in Doric were not fully comprehended; and they were counterfeited and finally rendered spurious by those who published them, who were not true Pythagoreans. After this, the Pythagoreans say, Plato, Aristotle, Speusippus, Aristoxenus, and Xenocrates appropriated what was fruitful, with slight modifications, but they collected some superficial or inconsequential things and recorded as the particular doctrines of the sect whatever was brought forward by those later malicious slanderers, in their desire to refute and mock the school.
>
> PORPHYRY, *Life of Pythagoras*

Porphyry, in a passage whose provenance cannot be securely identified,[1] tells a fine tale that builds on the evidence I have surveyed throughout this book. While figures such as Timaeus of Tauromenium, Neanthes of Cyzicus, and Aristotle testified to the publication of the secrets of the Pythagoreans, Porphyry's source here expands by adding two important ideas: (1) that Plato, Aristotle, Speusippus, Aristoxenus, and Xenocrates had access to the "recorded doctrines" (*ἀνιστορούμενα δόγματα*) of the Pythagoreans, and, importantly, (2) that the "recorded doctrines" they received had been written in Doric by figures who were mendicant Pythagoreans. The unnamed source states that Plato, Aristotle, Speusippus, Aristoxenus, and Xenocrates all had access to Pythagorean doctrines that had been recorded in Doric and that, according to

1. The most likely candidates would be Moderatus or Nicomachus, but we cannot be sure. See Zhmud 2012: 75 n. 59, with bibliography, and Dillon 1996: 346. The epigraph to this chapter is from Porphyry, *Life of Pythagoras* 53; translation after Burkert/Minar.

this later tradition at least, had been written in order to slander the true Pythagoreans.[2] It is curious to find several of the leading figures of fourth-century BCE Athenian philosophy set side by side in this account. The identity of these "Pythagoreans," who claim that the Academics and Peripatetics adapted spurious Pythagorean doctrines, is difficult to infer: they locate themselves in a broader historical tradition that imagined certain figures who wrote in Doric as pretender Pythagoreans, counterfeit actors who forever changed the history of Pythagoreanism in their malicious desire to subvert the true doctrines of Pythagoras. The tone sought after is tragic, and we are expected to lament with the author the capitulation of Pythagoreanism, which failed to survive intact, first because of its obscurity, and second because of the adulteration of the original principles. Pythagoreanism seems to suffer a folkloric death, at the hands of the malicious frauds who made it the property of the many. And the herald who proclaims its death remains an anonymous self-proclaimed Pythagorean himself.

But did Pythagoreanism really die? It would have to have lived a life first. I return to the statement with which this entire study began, the words of Epicharmus's comic actor, who probably stood in the shining, newly built theater in the Temenites section of the northwest part of Syracuse,[3] pointing absurdly at his own masked face, and clamored out in Doric Greek: "whatever naturally is in a process of exchange and never remains the same should be always different from what it had been changed from" (DK 23 B 2 = D.L. 3.9). If indeed this gesture is to be considered the first step toward the death of true Pythagoreanism, it is a fundamental step forward not just for Pythagorean philosophy but for philosophy more universally. The question of the stability of personal identity has continued to fascinate people who take up the Delphic mandate to "know yourself," as it did in the ancient world with Plato, Aristotle, Chrysippus, Plutarch, and Seneca.[4] Today, the problem of the continuous self retains its relevance for philosophers and filmmakers alike.[5] Am I one, or am I many? Was I someone else yesterday, or am I the same one today? It is a fascinating problem, because it brings mathematics to bear on

2. I have already argued (Horky: forthcoming) that Theophrastus's view of Pythagoreanism has been channeled through Xenocrates's works on the Pythagoreans, and that he cannot be relied on simply as a witness to either mathematical or acousmatic Pythagoreanism.

3. On the identity of this theater, see Robinson 2011: 88–89.

4. See Sorabji 2006: 38–42.

5. The topic has been one of central significance to Derek Parfit (1984), as well as the subject of numerous treatments in popular media, including Christopher Nolan's *Memento* (2000) and *The Prestige* (2006).

metaphysical and psychological questions of stability and continuity, precisely the sorts of questions that tend to be prompted by the assumption, for example, that a soul can outlast a body. From a certain point of view, metempsychosis demands a response to the question of the discontinuity of the numerical identity of the person, and the principles of symmetry, if they are to be considered axiomatic for our understanding of the nature of the universe and ourselves, must be able to respond to the apparent facts that over time we grow, then diminish; that our heath flourishes, then fades; that our memories quicken to the light, then stumble in the fog. These fundamental problems of stability harrowed Plato, who, in a crooked twist of fortune, was said to have been losing his memory when he was called out and challenged by Aristotle at the advanced age of eighty.[6] Aristotle and his gang were said to have defeated and displaced the sage in the ensuing contest, probably because, like Parmenides in the eponymous dialogue (*Parm.* 137a4–5), he could not recall the arguments of old. The author of one of the most important and influential theories of memory and knowledge was crippled intellectually by a disease that attacked his memory.[7] Within a year or so, Plato was dead, and the history of the reinterpretation of his doctrines, which continues to this day, was born.

All of this leads me to wonder a bit about the continuity of philosophical ideas and the challenges we run into when we attempt to grasp them and preserve their stability—to discern within the broader milieu of historical evidence a consistent and continuous set of concepts to which we can assign a name. Can we really say, as Porphyry's anonymous source claims, that "Pythagoreanism" died? That would require the assumption that "Pythagoreanism" ever existed as a continuous and numerically integrated set of beliefs, ideals, doctrines, actions, rituals, and life practices. It would require, I suspect, the proposition that what people call "Pythagoreanism" derives, at some fundamental and essential level, from Pythagoreanism *as such*. What the "Growing Argument" of the mathematical Pythagorean Epicharmus does—when applied beyond the confines of personal identity—is to challenge the very assumption that there could be a hidden, original, causative, and essential Pythagoreanism underneath all the aggregation of historical evidence. It forces us to consider the ways we could possibly obtain knowledge of such a thing, were it to exist. Mathematical Pythagoreanism, from its apparent historical inception in the first quarter of the fifth century BCE, was a multivocal force that organized itself around the

6. Ael. *VH* 3.19. For a serious consideration of this anecdote, see Dillon 2003: 3–4.

7. As I look around myself now, in 2012, at the master historians of ancient philosophy who have been or are now suffering from this affliction, I am struck by its unfortunate recurrence.

critical analysis and explanation of a set of enigmatic beliefs. Because it was not dogmatic, it sought to multiply and diversify the various ways we could respond to, and develop a rational discourse concerning, the role of mathematics (broadly construed) in our own life experiences. And if Plato is to be considered a Pythagorean, as he was throughout much of antiquity, he should be considered a Pythagorean chiefly insofar as he took up the project of developing new ways of explaining how the universe could be, in one way or another, unified and subsistent, but still subject to growth and diminishing. Epicharmus's mask was yet another aggregate through which Plato's soul looked out, and by which it might, through careful examination, discussion, and reflection, finally come to discover itself.

BIBLIOGRAPHY

Ademollo, F. *The Cratylus of Plato: A Commentary*. Cambridge: Cambridge University Press, 2011.

Álvarez Salas, O. "I frammenti 'filosofici' di Epicarmo: Una rivisitazione critica." In *Studi Italiani di Filologia Classica* 5.1 (2007), 23–72.

Afonasin, E. "The Pythagorean Way of Life in Clement of Alexandria and Iamblichus." In E. Afonasin, J. Dillon, and J. F. Finamore (eds.), *Iamblichus and the Foundations of Late Platonism* (Leiden: Brill, 2012), 13–36.

Afonasin, E., Dillon, J., and Finamore, J. F. (eds.). *Iamblichus and the Foundations of Late Platonism*. Leiden: Brill, 2012.

Alesse, F. (ed.). *Philo of Alexandria and Post-Aristotelian Philosophy*. Leiden: Brill, 2008.

Algra, K., van der Horst, P., and Runia, D. (eds.). *Polyhistor*. Leiden: Brill, 1996.

Allen, R. E. *Plato's "Euthyphro" and the Earlier Theory of Forms*. London: Routledge and Kegan Paul, 1970.

Anceschi, B. *Die Götternamen in Platons Kratylos: Ein Vergleich mit dem Papyrus von Derveni*. Studien zur klassischen Philologie 158. Frankfurt am Main: Peter Lang, 2007.

Annas, J. *Aristotle's Metaphysics Books M and N*. Oxford: Oxford University Press, 1976.

Bailey, D. T. J. "Logic and Music in Plato's *Phaedo*." *Phronesis* 50.2 (2005), 95–115.

Baltussen, H. "Playing the Pythagorean: Ion's *Triagmos*." In V. Jennings and A. Katsaros (eds.), *The World of Ion of Chios* (Leiden: Brill, 2007), 295–318.

Baltussen, H. *Theophrastus against the Peripatetics and Plato: Peripatetic Dialectic in the De Sensibus*. Leiden: Brill, 2000.

Bárány, I. "From Protagoras to Parmenides: A Platonic History of Philosophy." In M. M. Sassi (ed.), *La costruzione del discorso filosofico nell'eta dei Presocratici* (Pisa: Scuola Normale Superiore, 2006), 305–328.

Barker, A. "Aristoxenus and the Early Academy." In C. A. Huffman (ed.), *Aristoxenus of Tarentum: Discussion* (New Brunswick, N.J.: Transaction, 2012), 298–324.

Barker, A. "Shifting Conceptions of <<schools>> of Harmonic Theory, 400 BC–200 AD." In M. M. Martelli (ed.), *La Musa dimenticata: Aspetti dell'esperienza musicale greca in età ellenistica* (Pisa: Scuola Normale Superiore, 2009), 165–190.

Barker, A. *The Science of Harmonics in Classical Greece*. Cambridge: Cambridge University Press, 2007.

Barker, A. "Plato's *Philebus*: The Numbering of a Unity." In E. Benitez (ed.), *Dialogues with Plato: Apeiron*, supp. 39.4 (Edmonton, 1996), 143–164.

Barker, A. *Greek Musical Writings II: Harmonic and Acoustic Theory*. Cambridge: Cambridge University Press, 1989.

Barnes, J. *The Presocratic Philosophers*. London: Routledge, 1982.

Barney, R. "History and Dialectic (*Metaphysics* A 3, 983a24–984b8)." In C. Steel (ed.), *Aristotle's Metaphysics Alpha: Symposium Aristotelicum* (Oxford: Oxford University Press, 2012), 69–104.

Barney, R. *Names and Nature in Plato's Cratylus*. London: Routledge, 2001.

Baron, C. A. *Timaeus of Tauromenium and Hellenistic Historiography*. Cambridge: Cambridge University Press, 2012.

Bastianini, G., and D. N. Sedley (eds.). *Commentarium in Platonis Theaetetum* (P.Berol. inv. 9782). In *Corpus dei papiri filosofici* III (Florence: Olschki, 1995), 227–562.

Battezzato, L. "Pythagorean Comedies from Epicharmus to Alexis." In *Aevum Antiquum*, new ser., 8 (2008), 139–164.

Baxter, T. M. S. *The Cratylus: Plato's Critique of Naming*. Leiden: Brill, 1992.

Berger, S. "Democracy in the Greek West and the Athenian Example." *Hermes* 117 (1989), 303–314.

Berti, E. (ed.). *Aristotle on Science: The Posterior Analytics, Proceedings of the Eighth Symposium Aristotelicum*. Padua: Editrice Antinore, 1981.

Betegh, G. "What Makes a Myth Eikos?" In R. D. Mohr, K. Sanders, and B. Sattler (eds.), *One Book, the Whole Universe: Plato's Timaeus Today* (Las Vegas: Parmenides, 2010), 213–226.

Betegh, G. *The Derveni Papyrus: Cosmology, Theology, and Interpretation*. Cambridge: Cambridge University Press, 2004.

Bluck, R. S. *Plato's Phaedo*. London: Routledge and Kegan Paul, 1955.

Bobonich, C. *Plato's Utopia Recast*. Oxford: Oxford University Press, 2002.

Bodnár, I., and W. W. Fortenbaugh (eds.). *Eudemus of Rhodes*. New Brunswick, N.J.: Transaction, 2002.

Bonazzi, M. "Towards Transcendence: Philo and the Renewal of Platonism in the Early Imperial Age." In F. Alesse (ed.), *Philo of Alexandria and Post-Aristotelian Philosophy* (Leiden: Brill, 2008), 233–252.

Bonazzi, M., and V. Celluprica (eds.). *L'eredita platonica: Studi sul platonismo da Arcesilao a Proclo*. Naples: Bibliopolis, 2005.

Bonitz, H. *Index Aristotelicus*. 2nd ed. Berlin: G. Reimer, 1970.

Bostock, D. *Plato's Phaedo*. Oxford: Oxford University Press, 1986.

Bowen, A. C. "The Foundations of Early Pythagorean Harmonic Science: Archytas, Fragment 1." *Ancient Philosophy* 2 (1982), 79–104.

Bowie, E. L. "Le portrait de Socrates dans les <<Nuées>> d'Aristophane." In M. Trédé and P. Hoffmann (eds.). *Le rire des anciens* (Paris: Presses de l'École normale supérieure, 1995), 53–66.

Boyancé, P. "La doctrine d'Euthyphron dans le *Cratyle*." *Revue des études grecques* 54 (1941), 141–175.
Boys-Stones, G. R. "Alcinous, *Didaskalikos* 4: In Defense of Dogmatism." In M. Bonazzi and V. Celluprica (eds.). *L'eredita platonica: Studi sul platonismo da Arcesilao a Proclo* (Naples: Bibliopolis, 2005), 203–234.
Boys-Stones, G. R. *Post-Hellenistic Philosophy: A Study of Its Development from the Stoics to Origen*. Oxford: Oxford University Press, 2001.
Boys-Stones, G. R., and J. H. Haubold (eds.). *Plato and Hesiod*. Oxford: Oxford University Press, 2010.
Brisson, L. "Chapter 18 of the *De communi mathematica scientia*: Translation and Commentary." In E. Afonasin, J. Dillon, and J. F. Finamore (eds.), *Iamblichus and the Foundations of Late Platonism* (Leiden: Brill, 2012), 51–62.
Brisson, L. "*Epinomis*: Authenticity and Authorship." In K. Döring, M. Erler, and S. Schorn (eds.), *Pseudoplatonica* (Stuttgart: Franz Steiner Verlag, 2005), 9–24.
Brisson, L. *Lectures de Platon*. Paris: Vrin, 2000.
Brisson, L., and A. P. Segonds. *Jamblique: Vie de Pythagore*. 2nd ed. Paris: Les Belles Lettres, 2011.
Brown, T. S. *Timaeus of Tauromenium*. Berkeley: University of California Press, 1958.
Burkert, W. "Prehistory of Presocratic Philosophy in an Orientalizing Context." In P. Curd and D. W. Graham (eds.), *The Oxford Handbook of Presocratic Philosophy* (Oxford: Oxford University Press, 2008), 55–85.
Burkert, W. *Babylon, Memphis, Persepolis*. Cambridge, Mass.: Harvard University Press, 2004.
Burkert, W. "Craft versus Sect: The Problem of the Orphics and the Pythagoreans." In B. Meyer and E. P. Sanders (eds.), *Jewish and Christian Self-Definition*, vol. 3, *Self-Definition in the Graeco-Roman World* (London: SCM Press, 1982), 1–22.
Burkert, W. *Lore and Science in Ancient Pythagoreanism*. Trans. E. Minar. Cambridge, Mass.: Harvard University Press, 1972.
Burkert, W. "Review of M. von Albrecht, *Iamblichi De vita Pythagorica liber*, Zürich 1963." *Gnomon* 37 (1965), 24–26.
Burkert, W. "Hellenistische Pseudopythagorica." *Philologus* 105 (1961), 16–43, 224–246.
Burnet, J. *Early Greek Philosophy*. London: Routledge and Kegan Paul, 1945.
Burnet, J. *Plato: Euthyphro, Apology of Socrates, Crito*. Oxford: Oxford University Press, 1924.
Burnyeat, M. F. "Archytas and Optics." *Science in Context* 18 (2005), 35–53. Cited in the text as Burnyeat 2005a.
Burnyeat, M. F. "Eikôs Mythos." *Rhizai* 2 (2005), 143–165. Cited in the text as Burnyeat 2005b.
Burnyeat, M. F. "Platonism and Mathematics: A Prelude to Discussion." In A. Graeser (ed.), *Mathematics and Metaphysics in Aristotle* (Bern: Paul Haupt, 1987), 213–240.
Burnyeat, M. F. "Aristotle on Understanding Knowledge." In E. Berti (ed.), *Aristotle on Science: The Posterior Analytics, Proceedings of the Eighth Symposium Aristotelicum* (Padua: Editrice Antinore 1981), 97–139.

Cairns, D., F.-G. Herrmann, and T. Penner (eds.). *Pursuing the Good*. Edinburgh: Edinburgh University Press, 2007.

Cassio, A. C. "The Language of Doric Comedy." In A. Willi (ed.), *The Language of Greek Comedy* (Oxford: Oxford University Press, 2002), 51–84.

Cassio, A. C. "Two Studies on Epicharmus and His Influence." *Harvard Studies in Classical Philology* 89 (1985), 38–51.

Caston, V., and D. Graham (eds.). *Presocratic Philosophy: Essays in Honour of Alexander Mourelatos*. Aldershot, England: Ashgate, 2002.

Cavalieri, M. C. "La *Rassegna dei Filosofi* di Filodemo: Scuola eleatica ed abderita (PHerc. 327) e scuola pitagorica (PHerc. 1508)?" In M. Capasso (ed.), *Dal restauro dei materiali allo studio dei testi: Aspetti della ricerca papirologica, Papyrologica Lupiensa* 11 (Congedo, 2002), 17–53.

Cavini, W. "Aporia 11." In M. Crubellier and A. Laks (eds.), *Aristotle's Metaphysics Beta* (Oxford: Oxford University Press, 2009), 175–188.

Cherniss, H. *Aristotle's Criticism of Plato and the Academy*. Baltimore: Johns Hopkins Press, 1944.

Cherniss, H. *Aristotle's Criticism of Presocratic Philosophy*. Baltimore: Johns Hopkins Press, 1935.

Collobert, C., P. Destrée, and F. J. Gonzalez (eds.). *Plato and Myth: Studies on the Use and Status of Platonic Myths*. Leiden: Brill, 2012.

Coope, U. *Time for Aristotle*. Oxford: Oxford University Press, 2005.

Cooper, J. M., and D. S. Hutchinson (eds.). *Plato: Complete Works*. Indianapolis: Hackett, 1997.

Creese, D. *The Monochord in Ancient Greek Harmonic Science*. Cambridge: Cambridge University Press, 2010.

Crubellier, M., and A. Laks (eds.). *Aristotle's Metaphysics Beta*. Oxford: Oxford University Press, 2009.

Curd, P., and D. W. Graham (eds.). *The Oxford Handbook of Presocratic Philosophy*. Oxford: Oxford University Press, 2008.

Dancy, R. "Xenocrates." In E. N. Zalta (ed.), *The Stanford Encyclopedia of Philosophy*. http://plato.stanford.edu/archives/fall2011/entries/xenocrates/. Fall 2011.

Darbo-Peschanski, C. "The Origin of Greek Historiography." In J. Marincola (ed.), *A Companion to Greek and Roman Historiography*, vol. 1 (Oxford: Blackwell, 2007), 27–38.

Delatte, A. *Essai sur la politique pythagoricienne*. Liège: Imp. H. Vaillant-Carmanne, 1922.

Delatte, A. *Études sur la littérature pythagoricienne*. Paris: Champion, 1915.

Delcomminette, S. *La Philèbe de Platon*. Leiden: Brill, 2006.

Demand, N. "Plato, Aristophanes, and the Speeches of Pythagoras. *Greek, Roman, and Byzantine Studies* 23 (1982), 179–184.

Denyer, N. *Plato: Protagoras*. Cambridge: Cambridge University Press, 2008.

Denyer, N. "Sun and Line: The Role of the Good." In G. R. F. Ferrari (ed.), *The Cambridge Companion to Plato's Republic* (Cambridge: Cambridge University Press, 2007), 284–309.

Depew, D., and T. Poulakos (eds.). *Isocrates and Civic Education*. Austin: University of Texas Press, 2004.

Deubner, L. "Bemerkungen zum Text der *Vita Pythagorae* des Iamblichos." *Sitzungberichte der Preussischen Akademie der Wissenschaften zu Berlin* (1935), 612–690, 824–827.

Dillon, J. "Pedantry and Pedestrianism? Some Reflections on the Middle Platonic Commentary Tradition." In H. Tarrant and D. Baltzly (eds.), *Reading Plato in Antiquity* (London: Duckworth, 2006), 19–32.

Dillon, J. *The Heirs of Plato: A Study of the Old Academy (347–274 BC)*. Oxford: Oxford University Press, 2003.

Dillon, J. "Theophrastus' Critique of the Old Academy in the *Metaphysics*." In W. W. Fortenbaugh and G. Wöhrle (eds.), *On the Opuscula of Theophrastus* (Stuttgart: Franz Steiner Verlag, 2002), 175–187.

Dillon, J. *The Middle Platonists: 80 BC–AD 220*. Ithaca, N.Y.: Cornell University Press, 1996. (Originally published London: Duckworth, 1977.)

Dillon, J. *Alcinous: The Handbook of Platonism*. Oxford: Oxford University Press, 1993.

Dillon, J., and L. Brisson (eds.). *Plato's Philebus: Selected Papers from the Eighth Symposium Platonicum*. St. Augustin: Academia, 2009.

Dillon, J., and J. Hershbell. *Iamblichus: On the Pythagorean Way of Life*. Atlanta: Scholars Press, 1991.

Dixsaut, M. (ed.). *La Felure du Plaisir: Études sur le Philèbe de Platon*. Vol. 2. *Contextes*. Paris: Vrin, 1999.

D'Ooge, M. L. (trans.). *Nicomachus of Gerasa: Introduction to Arithmetic*. New York: Macmillan, 1926.

Dorandi, T. *Filodemo, Storia dei filosofi: Platone e l'Academia*. Naples: Bibliopolis, 1991.

Döring, K., M. Erler, and S. Schorn (eds.). *Pseudoplatonica*. Stuttgart: Franz Steiner Verlag, 2005.

Eucken, C. *Isokrates: Seine Positionen in der Auseinandersetzung mit den zeitgenössischen Philosophen*. Berlin, 1983.

Ferrari, G. R. F. (ed.). *The Cambridge Companion to Plato's Republic*. Cambridge: Cambridge University Press, 2007.

Ferrari, G. R. F. (ed.). *Listening to the Cicadas: A Study in Plato's Phaedrus*. Cambridge: Cambridge University Press, 1987.

Finamore, J. F., and J. M. Dillon. *Iamblichus: De Anima*. Philosophia Antiqua 42. Atlanta: Scholars Press, 2002.

Flach, P. A., and A. C. Kakas (eds.). *Abduction and Induction: Essays on Their Relation and Integration*. Dordrecht: Kluwer Academic, 2000.

Fortenbaugh, W. W., and E. Schütrumpf (eds.). *Dicaearchus of Messana*. New Brunswick, N.J.: Transaction, 2001.

Fortenbaugh, W. W., and G. Wöhrle (eds.). *On the Opuscula of Theophrastus*. Stuttgart: Franz Steiner Verlag, 2002.

Frank, E. *Plato und die sogenannaten Pythagoreer: Ein Kapitel aus der Geschichte des griechischen Geistes*. Halle: Verlag von Max Niemeyer, 1923.

Frede, D. *Plato: Philebus*. Indianapolis: Hackett, 1993.

Frede, D., and B. Reis (eds.). *Body and Soul in Ancient Philosophy*. Berlin: de Gruyter, 2009.

Fronterotta, F. *Platone: Timeo*. 2nd ed. Milan: BUR, 2006.
Gallop, D. *Plato: Phaedo*. Oxford: Oxford University Press, 1975.
Gerson, L. *Aristotle and Other Platonists*. Ithaca: Cornell University Press, 2005.
Giannantoni, G. *Socratis et socraticorum reliquiae*. Vols. 1–4. Naples: Bibliopolis, 1990.
Gill, M. L. "The Divine Method in Plato's *Philebus*." In J. Dillon, and L. Brisson, (eds.), *Plato's Philebus: Selected Papers from the Eighth Symposium Platonicum* (St. Augustin: Academia, 2009), 36–46.
Gill, M. L., and P. Pellegrin (eds.). *A Companion to Ancient Philosophy*. Oxford: Blackwell, 2006.
Gorman, P. "The 'Apollonios' of the Neoplatonic Biographies of Pythagoras." *Mnemosyne*, 4th ser., 38.1/2 (1985), 130–144.
Gosling, J. C. B. *Plato: Philebus*. Oxford: Oxford University Press, 1975.
Gotthelf, A. "First Principles in Aristotle's Parts of Animals." In A. Gotthelf and J. G. Lennox (eds.), *Philosophical Issues in Aristotle's Biology* (Cambridge: Cambridge University Press, 1987), 167–198.
Gotthelf, A., and J. G. Lennox (eds.). *Philosophical Issues in Aristotle's Biology*. Cambridge: Cambridge University Press, 1987.
Gottschalk, H. B. *Heraclides of Pontus*. Oxford: Oxford University Press, 1980.
Graeser, A. (ed.). *Mathematics and Metaphysics in Aristotle*. Bern: Paul Haupt, 1987.
Graham, D. (ed.). *The Texts of Early Greek Philosophy*. Vols. 1–2. Cambridge: Cambridge University Press, 2010.
Graham, D. "Heraclitus and Parmenides." In V. Caston and D. Graham (eds.), *Presocratic Philosophy: Essays in Honour of Alexander Mourelatos* (Aldershot, England: Ashgate, 2002), 27–44.
Graham, D. "Symmetry in the Empedoclean Cycle." *Classical Quarterly*, new ser., 38 (1988), 297–312.
Griffith, M. *Aeschylus: Prometheus Bound*. Cambridge: Cambridge University Press, 1983.
Griswold, C. L. *Self-Knowledge in Plato's Phaedrus*. New Haven, Conn.: Yale University Press, 1986.
Guthrie, W. K. C. (ed.). *Aristotle: On the Heavens*. Loeb Classical Library. Cambridge, Mass.: Harvard University Press, 1939.
Hackforth, R. *Plato: Phaedo*. Cambridge: Cambridge University Press, 1955.
Herrmann, F.-G. *Words and Ideas: The Roots of Plato's Philosophy*. Swansea, England: Classical Press of Wales, 2007.
Hintikka, J., D. Gruender, and E. Agazzi (eds.). *Theory Change, Ancient Axiomatics, and Galileo's Methodology*. Dordrecht: Reidel, 1981.
Holmes, B. *The Symptom and the Subject: The Emergence of the Physical Body in Ancient Greece*. Princeton: Princeton University Press, 2010.
Horky, P. S. "Theophrastus on Platonic and 'Pythagorean' Imitation." *Classical Quarterly*, forthcoming.
Horky, P. S. "On the Phylogenetics of Wisdom: A Response to Alexis Pinchard, *Les langues de sagesse dans la Grèce et l'Inde anciennes*." *Antiquorum Philosophia* 5 (2011), 149–163.

Horky, P. S. "Persian Cosmos and Greek Philosophy: Plato's Associates and the Zoroastrian *Magoi*." *Oxford Studies in Ancient Philosophy* 37 (winter 2009), 47–103.

Horky, P. S. "The Imprint of the Soul: Psychosomatic Affection in Plato, Gorgias, and the 'Orphic' Gold Tablets." *Mouseion* 3.6 (2006), 383–398.

Huffman, C. A. (ed.). *Aristoxenus of Tarentum: Discussion*. New Brunswick, N.J.: Transaction, 2012.

Huffman, C. A. "A New Mode of Being for Parmenides: A Discussion of John Palmer, *Parmenides and Presocratic Philosophy*." *Oxford Studies in Ancient Philosophy* 41 (2011), 289–305.

Huffman, C. A. "Pythagoreanism." In E. N. Zalta (ed.), *The Stanford Encyclopedia of Philosophy*. http://plato.stanford.edu/entries/Pythagoreanism/. Summer 2010.

Huffman, C. A. "The Pythagorean Precepts of Aristoxenus: Crucial Evidence for Pythagorean Moral Philosophy." *Classical Quarterly*, new ser., 58.1 (2008), 104–120.

Huffman, C. A. "Philolaus and the Central Fire." In. S. Stern-Gillet and K. Corrigan (eds.), *Reading Ancient Texts I: Presocratics and Plato, Essays in Honour of Denis O'Brien* (Leiden: Brill, 2007), 57–94.

Huffman, C. A. "Aristoxenus' Pythagorean Precepts: A Rational Pythagorean Ethics." In M. M. Sassi (ed.), *La costruzione del discorso filosofico nell'eta dei Presocratici* (Pisa: Scuola Normale Superiore, 2006), 103–121.

Huffman, C. A. *Archytas of Tarentum: Pythagorean, Philosopher, and Mathematician King*. Cambridge: Cambridge University Press, 2005.

Huffman, C. A. "Archytas and the Sophists." In V. Caston and D. Graham (eds.), *Presocratic Philosophy: Essays in Honour of Alexander Mourelatos* (Aldershot, England: Ashgate, 2002), 251–270.

Huffman, C. A. "The Philolaic Method: The Pythagoreanism behind the *Philebus*." In A. Preus (ed.), *Essays in Ancient Greek Philosophy VI: Before Plato* (Albany, N.Y.: State University of New York Press, 2001), 67–85.

Huffman, C. A. "Limite et Illimité chez les premiers philosophes grecs." In M. Dixsaut (ed.), *La Felure du Plaisir: Études sur le Philèbe de Platon*, vol. 2, *Contextes* (Paris: Vrin, 1999), 11–31.

Huffman, C. A. *Philolaus of Croton: Pythagorean and Presocratic*. Cambridge: Cambridge University Press, 1993.

Humm, M. *Appius Claudius Caecus: La République Accomplie*. Rome: Ecole Francaise de Rome, 2005.

Hutchinson, D. S., and M. R. Johnson. "Authenticating Aristotle's *Protrepticus*." *Oxford Studies in Ancient Philosophy* 29 (2005), 193–294.

Inwood, B. *The Poem of Empedocles*. 2nd ed. Toronto: University of Toronto Press, 2002.

Irby-Massie, G. L. "Prometheus Bound and Contemporary Trends in Greek Natural Philosophy." *Greek, Roman, and Byzantine Studies* 48 (2008), 133–157.

Isnardi Parente, M. *Senocrate-Ermodoro: Frammenti*. Naples: Bibliopolis, 1982.

Jaeger, W. *Aristotle: Fundamentals of the History of His Development*. Trans. R. Robinson. 2nd ed. Oxford: Oxford University Press, 1948.

Jennings, V., and A. Katsaros (eds.). *The World of Ion of Chios*. Leiden: Brill, 2007.

Johnson, M. "The Aristotelian Explanation of the Halo." *Apeiron* 42 (2009), 325–357.

Johnson, M. *Aristotle on Teleology*. Oxford: Oxford University Press, 2005.

Josephson, J. R. "Smart Inductive Generalizations Are Abductions." In P. A. Flach and A. C. Kakas (eds.), *Abduction and Induction: Essays on Their Relation and Integration* (Dordrecht: Kluwer Academic, 2000), 31–44.

Judson, L., and V. Karasmanis (eds.). *Remembering Socrates: Philosophical Essays*. Oxford: Oxford University Press, 2006.

Kahn, C. H. "Dialectic, Cosmology, and Ontology in the *Philebus*." In J. Dillon and L. Brisson (eds.), *Plato's Philebus: Selected Papers from the Eighth Symposium Platonicum* (St. Augustin: Academia, 2009), 56–67.

Kahn, C. H. *Pythagoras and the Pythagoreans: A Brief History*. Indianapolis: Hackett, 2001.

Kahn, C. H. *Plato and the Socratic Dialogue: The Philosophical Use of a Literary Form*. Cambridge: Cambridge University Press, 1996.

Kahn, C. H. *The Art and Thought of Heraclitus*. Cambridge: Cambridge University Press, 1979.

Kahn, C. H. "Language and Ontology in the *Cratylus*." In E. N. Lee, A. P. D. Mourelatos, and R. M. Rorty (eds.), *Exegesis and Argument: Studies in Greek Philosophy Presented to Gregory Vlastos* (Assen: Van Gorcum, 1973), 152–176.

Kamtekar, R. "The Good and Order: Does the *Republic* Display an Analogy between a Science of Ethics and Mathematics? Reply to Christopher Gill, 'The Good and Mathematics.'" In D. Cairns, F.-G. Herrmann, and T. Penner (eds.), *Pursuing the Good* (Edinburgh: University of Edinburgh Press, 2007), 275–278.

Kerferd, G. B. *The Sophistic Movement*. Cambridge: Cambridge University Press, 1981.

Kingsley, P. *Ancient Philosophy, Mystery, and Magic: Empedocles and the Pythagorean Tradition*. Oxford: Oxford University Press, 1995.

Klein, J. *Greek Mathematical Thought and the Origin of Algebra*. Cambridge, Mass.: MIT Press, 1968.

Knorr, W. R. "On the Early History of Axiomatics, the Interaction of Mathematics and Philosophy in Greek Antiquity." In J. Hintikka, D. Gruender, and E. Agazzi (eds.), *Theory Change, Ancient Axiomatics, and Galileo's Methodology* (Dordrecht: Reidel, 1981), 145–186.

Knorr, W. R. *The Evolution of the Euclidean Elements*. Dordrecht: Reidel. 1975.

Laks, A., and G. Most (eds.). *Studies in the Derveni Papyrus*. Oxford: Oxford University Press, 1997.

Lateiner, D. *The Historical Method of Herodotus*. Toronto: University of Toronto Press, 1989.

Lee, E. N., A. P. D. Mourelatos, and R. M. Rorty (eds.). *Exegesis and Argument: Studies in Greek Philosophy Presented to Gregory Vlastos*. Assen: Van Gorcum, 1973.

Lennox, J. G. *Aristotle on the Parts of the Animals* I–IV. Oxford: Oxford University Press, 2001. Cited in the text as Lennox 2001a.

Lennox, J. G. *Aristotle's Philosophy of Biology: Studies in the Origins of Life Science*. Cambridge: Cambridge University Press, 2001. Cited in the text as Lennox 2001b.

Lesher, J. "Early Interest in Knowledge." In A. A. Long (ed.), *The Cambridge Companion to Early Greek Philosophy* (Cambridge: Cambridge University Press, 1999), 225–249.

Lesher, J. *Xenophanes of Colophon*. Toronto: University of Toronto Press, 1992.

Leunissen, M. *Explanation and Teleology in Aristotle's Science of Nature*. Cambridge: Cambridge University Press, 2010.

Lévêque, P., and P. Vidal-Naquet. *Cleisthenes the Athenian*. Trans. D. A. Curtis. Atlantic Highlands, N.J.: Humanities, 1996.

Lévy, I. *Recherches sur les sources de la légende de Pythagore*. Paris: Ernest Leroux, 1926.

Livingstone, N. *A Commentary on Isocrates' Busiris*. Leiden: Brill, 2001.

Lloyd, A. B. *Herodotus, Book II*. Vols. 1–3. Leiden: Brill, 1976–88.

Lloyd, G. E. R. *Methods and Problems in Greek Science*. Cambridge: Cambridge University Press, 1991.

Lloyd, G. E. R. "Plato and Archytas in the Seventh Letter." *Phronesis* 35.2 (1990), 159–174.

Lloyd, G. E. R. *Magic, Reason, and Experience*. Cambridge: Cambridge University Press, 1979.

Lloyd, G. E. R. *Polarity and Analogy*. Cambridge: Cambridge University Press, 1966.

Long, A. A. "Heraclitus on Measure and the Explicit Emergence of Rationality." In D. Frede and B. Reis (eds.), *Body and Soul in Ancient Philosophy* (Berlin: de Gruyter, 2009), 87–109.

Long, A. A. "Parmenides on Thinking Being." In G. Rechenauer (ed.), *Frühgriechisches Denken* (Göttingen: Vandenhoeck & Ruprecht, 2005), 227–251.

Long, A. A. (ed.). *The Cambridge Companion to Early Greek Philosophy*. Cambridge: Cambridge University Press, 1999.

Long, A. A. "The Principles of Parmenides' Cosmology." *Phronesis* 8 (1963), 90–107.

Long, A. A., and D. Sedley. *The Hellenistic Philosophers*. Vols. 1–2. Cambridge: Cambridge University Press, 1987.

Luzzatto, M. T. "Dialettica o retorica? La *polytropia* di Odisseo da Antistene a Porfirio." *Elenchos* 17 (1996), 275–357.

MacKay, L. A. "*Horatiana*: Odes 1.9 and 1.28." *Classical Philology* 72.4 (1977), 316–318.

Macris, C. *Le Pythagore des néoplatoniciens: Recerches et commentaires sur Le mode de vie pythagoricien de Jamblique*. Doctoral thesis, la Section des Sciences Religieuses de l'Ecole Pratique des Hautes Études. Paris, 2004.

Mansfeld, J. "Doxography of Ancient Philosophy." In E. N. Zalta (ed.), *The Stanford Encyclopedia of Philosophy*. http://plato.stanford.edu/archives/sum2012/entries/doxography-ancient/. Summer 2012.

Mansfeld, J. *Heresiography in Context: Hippolytus' Elenchos as a Source for Greek Philosophy*. Philosophia Antiqua 56. Leiden: Brill, 1992.

Mansfeld, J. *Studies in the Historiography of Greek Philosophy*. Assen/Maastricht: Van Gorcum, 1990.

Mansfeld, J., and D. T. Runia. *Aëtiana: The Method and Intellectual Context of a Doxographer*. Vol. 1. *The Sources*. Philosophia Antiqua 73. Leiden: Brill, 1997.

Marcovich, M. *Clementis Alexandrini Protrepticus*. Leiden: Brill, 1995.

Marcovich, M. *Heraclitus: Greek Text with a Short Commentary*. Ed. Maior. Merida: Los Andes University Press, 1967.

Marincola, J. (ed.). *A Companion to Greek and Roman Historiography*. Vol. 1. Oxford: Blackwell, 2007.

Martelli, M. M. (ed.). *La Musa dimenticata: Aspetti dell'esperienza musicale greca in età ellenistica*. Pisa: Scuola Normale Superiore, 2009.

McDiarmid, J. B. "Theophrastus on the Presocratic Causes." *Harvard Studies in Classical Philology* 61 (1953), 85–156.

McKirahan, R. D. "Signs and Arguments in Parmenides B 8." In P. Curd and D. W. Graham (eds.), *The Oxford Handbook of Presocratic Philosophy* (Oxford: Oxford University Press, 2008), 189–229.

McKirahan, R. D. *Philosophy before Socrates*. Indianapolis: Hackett, 1994.

McKirahan, R. D. *Principles and Proofs: Aristotle's Theory of Demonstrative Science*. Princeton: Princeton University Press, 1992.

Meinwald, C. "Plato's Pythagoreanism." *Ancient Philosophy* 22.1 (2002), 87–101.

Menn, S. "On Socrates' First Objections to the Physicists (*Phaedo* 95e8–97b7)." *Oxford Studies in Ancient Philosophy* 38 (2010), 37–68.

Menn, S. "Collecting the Letters." *Phronesis* 43.4 (1998), 291–305.

Merlan, P. *From Platonism to Neoplatonism*. 3rd ed., rev. The Hague: Martinus Nijhoff, 1968.

Meyer, B., and E. P. Sanders (eds.). *Jewish and Christian Self-Definition*. Vol. 3. *Self-Definition in the Graeco-Roman World*. London: SCM Press, 1982.

Miller, M. "The *Timaeus* and the 'Longer Way.'" In G. Reydams-Schils (ed.), *Timaeus as Cultural Icon* (Notre Dame, Ind.: University of Notre Dame Press, 2003), 17–59.

Minar, E. L. *Early Pythagorean Politics in Practice and Theory*. Baltimore: Waverly Press, 1942.

Mohr, R. D., K. Sanders, and B. Sattler (eds.). *One Book, the Whole Universe: Plato's Timaeus Today*. Las Vegas: Parmenides, 2010.

Moravcsik, J. *Plato and Platonism*. Oxford: Blackwell, 1992.

Morgan, K. "Inspiration, Recollection, and Mimēsis in Plato's *Phaedrus*." In A. W. Nightingale and D. Sedley (eds.), *Ancient Models of Mind: Studies in Human and Divine Rationality* (Cambridge: Cambridge University Press, 2010), 45–63. Cited in the text as Morgan 2010a.

Morgan, K. "The Voice of Authority: Divination and Plato's *Phaedo*." *Classical Quarterly* 60.1 (2010), 63–81. Cited in the text as Morgan 2010b.

Morgan, K. "The Education of Athens: Politics and Rhetoric in Isocrates (and Plato)." In D. Depew and T. Poulakos (eds.), *Isocrates and Civic Education* (Austin: University of Texas Press, 2004), 125–154.

Morgan, K. *Myth and Philosophy from the Pre-Socratics to Plato*. Cambridge: Cambridge University Press, 2000.

Morrison, J. S. "Pythagoras of Samos." *Classical Quarterly* 50 (1956), 135–156.

Morrow, G. R., and J. M. Dillon (eds. and trans.). *Proclus' Commentary on Plato's Parmenides*. Princeton: Princeton University Press, 1987.

Mourelatos, A. P. D. "The Concept of the Universal in Some Later Pre-Platonic Cosmologists." In M. L. Gill and P. Pellegrin (eds.), *A Companion to Ancient Philosophy* (Oxford: Blackwell, 2006), 56–76.

Musti, D. *Magna Grecia: Il quadro storico*. Rome: Editori Laterza, 2005.
Musti, D. *Strabone e la Magna Grecia*. Padua: Edizioni Scientifiche Italiane, 1988.
Mueller, I. "Greek Arithmetic, Geometry, and Harmonics: Thales to Plato." In C. C. W. Taylor (ed.), *From the Beginning to Plato*, Routledge History of Philosophy, vol. 1 (London: Routledge: 1997), 271–322.
Mueller, I. "Aristotle's Approach to the Problem of Principles in *Metaphysics* M and N." In A. Graeser (ed.), *Mathematics and Metaphysics in Aristotle* (Bern: Paul Haupt, 1987), 241–259.
Nehamas, A. "Parmenidean Being/Heraclitean Fire." In V. Caston and D. Graham (eds.), *Presocratic Philosophy: Essays in Honour of Alexander Mourelatos* (Aldershot, England: Ashgate, 2002), 45–64.
Nehamas, A. "Predication and the Forms of Opposites in Plato's *Phaedo*." *Review of Metaphysics* 26.3 (1973), 461–491.
Netz, R. *The Shaping of Deduction in Greek Mathematics: A Study in Cognitive History*. Cambridge: Cambridge University Press, 1999.
Nightingale, A. W. *Spectacles of Truth in Classical Greek Philosophy: Theoria in Its Cultural Context*. Cambridge: Cambridge University Press, 2004.
Nightingale, A. W. *Genres in Dialogue: Plato and the Construct of Philosophy*. Cambridge: Cambridge University Press, 1995.
Nightingale, A. W., and D. Sedley (eds.). *Ancient Models of Mind: Studies in Human and Divine Rationality*. Cambridge: Cambridge University Press, 2010.
Nussbaum, M. C. (ed.). *Logic, Science, and Dialectic: Collected Papers in Greek Philosophy*. Ithaca: Cornell University Press, 1986.
Nussbaum, M. C. "Eleatic Conventionalism and Philolaus on the Conditions of Thought." *Harvard Studies in Classical Philology* 83 (1979), 63–108.
Ober, J. *Political Dissent in Democratic Athens: Intellectual Critics of Popular Rule*. Princeton: Princeton University Press, 1998.
O'Brien, D. *Empedocles' Cosmic Cycle*. Cambridge: Cambridge University Press, 1969.
Olson, S. D. *Broken Laughter: Selected Fragments of Greek Comedy*. Oxford: Oxford University Press, 2007.
O'Meara, D. *Pythagoras Revived: Mathematics and Philosophy in Late Antiquity*. Oxford: Oxford University Press, 1989.
Owen, G. E. L. "*Tithenai ta phainomena*." In M. C. Nussbaum (ed.), *Logic, Science, and Dialectic*. Ithaca: Cornell University Press, 1986), 240–263.
Owens, J. *The Doctrine of Being in the Aristotelian Metaphysics*. Toronto: Pontifical Institute of Mediaeval Studies, 1951.
Palmer, J. *Parmenides and Presocratic Philosophy*. Oxford: Oxford University Press, 2009.
Palmer, J. *Plato's Reception of Parmenides*. Oxford: Oxford University Press, 1999.
Parfit, D. *Reasons and Persons*. Oxford: Oxford University Press, 1984.
Pearson, L. *The Greek Historians of the West: Timaeus and His Predecessors*. Atlanta: Scholars Press, 1987.
Pendrick, G. *Antiphon the Sophist: The Fragments*. Cambridge: Cambridge University Press, 2002.
Perrin, B. (ed. and trans.). *Plutarch Lives I*. Loeb Classical Library. Cambridge, Mass.: Harvard University Press, 1914.

Peterson, S. *Socrates and Philosophy in the Dialogues of Plato*. Cambridge: Cambridge University Press, 2011.
Philip, J. A. *Pythagoras and Early Pythagoreanism*. Toronto: University of Toronto Press, 1966.
Pinchard, A. *Les langues de sagesse dans la Grèce et l'Inde anciennes*. Geneva: Droz, 2009.
Politis, V. "*Aporia* and Searching in Early Plato." In L. Judson and V. Karasmanis (eds.), *Remembering Socrates: Philosophical Essays* (Oxford: Oxford University Press, 2006), 88–109.
Pradeau, J.-F. *Platon, l'imitation de la philosophie*. Paris: Aubier, 2009.
Preus, A. (ed.). *Essays in Ancient Greek Philosophy VI: Before Plato*. Albany: State University of New York Press, 2001.
Primavesi, O. "Second Thoughts on Some Presocratics (*Metaphysics* A 8, 989a18–990a32)." In C. Steel (ed.), *Aristotle's Metaphysics Alpha: Symposium Aristotelicum* (Oxford: Oxford University Press, 2012), 225–260.
Radicke, J. *Die Fragments der Griechischen Historiker (FGrHist) Continued*. Vol. 4A. *Biography*. Collection 7. *Imperial and Undated Authors*. Leiden: Brill, 1999.
Rapp, C. "Eleatischer Monismus." In G. Rechenauer (ed.), *Frühgriechisches Denken* (Göttingen: Vandenhoeck & Ruprecht, 2005), 290–315.
Rashed, M. "Aristophanes and the Socrates of the *Phaedo*." *Oxford Studies in Ancient Philosophy* 36 (2009), 107–136.
Raven, J. *Pythagoreans and Eleatics*. Cambridge: Cambridge University Press, 1948.
Rechenauer, G. (ed.). *Frühgriechisches Denken*. Göttingen: Vandenhoeck & Ruprecht, 2005.
Regali, M. "Hesiod in the *Timaeus*: The Demiurge Addresses the Gods." In G. Boys-Stones and J. H. Haubold (eds.), *Plato and Hesiod* (Oxford: Oxford University Press, 2010), 259–275.
Reydams-Schils, G. (ed.). *Timaeus as Cultural Icon*. Notre Dame, Ind.: University of Notre Dame Press, 2003.
Rhodes, P. J. *A Commentary on the Aristotelian Athenaion Politeia*. Oxford: Oxford University Press, 1991.
Riedweg, C. *Pythagoras: His Life, Teaching, and Influence*. Trans. S. Rendall. Ithaca: Cornell University Press, 2005.
Riginos, A. S. *Platonica: The Anecdotes Concerning the Life and Writings of Plato*. Leiden: Brill, 1976.
Robinson, E. *Democracy beyond Athens: Popular Government in the Greek Classical Age*. Cambridge: Cambridge University Press, 2011.
Robinson, R. *Plato's Earlier Dialectic*. 2nd ed. Oxford: Clarendon Press, 1953.
Rodríguez-Noriega Guillén, L. *Epicarmo de Siracusa: Testimonios y Fragmentos*. Oviedo: Universidad de Oviedo Servicio de Publicaciones, 1995.
Ross, W. D. *Aristotle: Metaphysics*. Vols. 1–2 Oxford: Clarendon Press, 1924.
Rostagni, A. *Il Verbo di Pitagora*. Turin: Il Basilisco, 1924.
Rostagni, A. "Pitagora e i Pitagorici in Timeo." *Atti della R. Academia delle Scienze in Torino* 49 (1913–14), 373–395.
Rowe, C. *Plato: Symposium*. Warminster: Aris & Phillips, 1998.
Rowe, C. *Plato: Statesman*. Warminster: Aris & Phillips, 1995.

Rowe, C. *Plato: Phaedo*. Cambridge: Cambridge University Press, 1993.
Sartori, F. *Le eterie nella vita politica ateniese del VI e V sec. A. C.* Rome, 1967.
Sassi, M. M. (ed.). *La costruzione del discorso filosofico nell'eta dei Presocratici*. Pisa: Scuola Normale Superiore, 2006.
Sayre, K. *Metaphysics and Method in Plato's Statesman*. Cambridge, 2006.
Sayre, K. *Plato's Late Ontology: A Riddle Resolved*. 2nd ed. Las Vegas: Parmenides Press, 2005.
Schäfer, C. "Das Pythagorasfragment des Xenophanes und die Frage nach der Kritik der Metempsychosenlehre." In D. Frede and B. Reis (eds.), *Body and Soul in Ancient Philosophy* (Berlin: de Gruyter, 2009), 45–70.
Schibli, H. S. "On 'The One' in Philolaus, Fragment 7." *Classical Quarterly*, new ser., 46.1 (1996), 114–130.
Schiefsky, M. *Hippocrates: On Ancient Medicine*. Leiden: Brill, 2005.
Schofield, M. "Pythagoreanism: Emerging from the Presocratic Fog (*Metaphysics* A 5)." In C. Steel (ed.), *Aristotle's Metaphysics Alpha: Symposium Aristotelicum* (Oxford: Oxford University Press, 2012), 141–160.
Schofield, M. "ARXH." *Hyperboreus* 3.2 (1997), 218–236.
Schofield, M. "The Dénouement of the *Cratylus*." In M. Schofield and M. C. Nussbaum (eds.), *Language and Logos* (Cambridge: Cambridge University Press, 1982), 61–81.
Schofield, M., and M. C. Nussbaum (eds.). *Language and Logos*. Cambridge: Cambridge University Press, 1982.
Scott, D. (ed.). *Maieusis: Essays in Ancient Philosophy in Honour of Myles Burnyeat*. Oxford: Oxford University Press, 2007.
Sedley, D. "Equal Sticks and Stones." In D. Scott (ed.), *Maieusis: Essays in Ancient Philosophy in Honour of Myles Burnyeat* (Oxford: Oxford University Press, 2007), 68–86.
Sedley, D. *Plato's Cratylus*. Cambridge: Cambridge University Press, 2003.
Sedley, D. "Alcinous' Epistemology." In K. Algra, P. van der Horst, and D. Runia (eds.), *Polyhistor* (Leiden: Brill, 1996), 300–312.
Sedley, D. "The Dramatis Personae of Plato's *Phaedo*." *Proceedings of the British Academy* 85 (1995), 3–26.
Sedley, D. "The Stoic Criterion of Identity." *Phronesis* 27 (1982), 255–275.
Sharples, R. W. *Theophrastus of Eresus: Sources for His Life, Writings, Thought, and Influence*. Vol. 3.1. *Sources on Physics (Texts 137–223)*. Leiden: Brill, 1998.
Smith, R. "Aristotle's Theory of Demonstration." In G. Anagnostopoulos (ed.), *A Companion to Aristotle* (Oxford: Blackwell, 2009), 51–65.
Smith, R. *Aristotle: Topics I, VIII, and Selections*. Oxford: Oxford University Press, 1997.
Sommerstein, A. "The Prologue of Aeschylus' *Palamedes*." *Rheinisches Museum für Philologie* 143 (2000), 118–127.
Sorabji, R. *Self: Ancient and Modern Insights about Individuality, Life, and Death*. Chicago: University of Chicago Press, 2006.
Steel, C. (ed.). *Aristotle's Metaphysics Alpha: Symposium Aristotelicum* Oxford: Oxford University Press, 2012.
Steel, C. "Plato as Seen by Aristotle (*Metaphysics* A 6)." In C. Steel (ed.), *Aristotle's Metaphysics Alpha: Symposium Aristotelicum* (Oxford, 2012), 167–200.

Stern-Gillet, S., and K. Corrigan (eds.). *Reading Ancient Texts I: Presocratics and Plato, Essays in Honour of Denis O'Brien*. Leiden: Brill, 2007.

Struck, P. *Birth of the Symbol: Ancient Readers and the Limits of Their Texts*. Princeton: Princeton University Press, 2004.

Tarán, L. *Collected Papers 1962–1999*. Leiden: Brill, 2001.

Tarán, L. *Speusippus of Athens*. Leiden: Brill, 1981.

Tarrant, H., and Baltzly, D. (eds.). *Reading Plato in Antiquity*. London: Duckworth, 2006.

Taylor, A. E. *Plato: The Man and His Work*. London: Methuen, 1937.

Taylor, C. C. W. *The Atomists: Leucippus and Democritus*. Toronto: University of Toronto Press, 1999.

Taylor, C. C. W. (ed.). *From the Beginning to Plato*. Routledge History of Philosophy. Vol. 1. London: Routledge, 1997.

Taylor, C. C. W. "Forms as Causes in the *Phaedo*." *Mind* 78 (1969), 45–59.

Thanassas, P. "Doxa revisitata." In G. Rechenauer (ed.), *Frühgriechisches Denken* (Göttingen: Vandenhoeck & Ruprecht, 2005), 270–289.

Thesleff, H. (ed.). *The Pythagorean Texts of the Hellenistic Period*. Åbo: Åbo Akademi, 1965.

Thiel, D. *Die Philosophie des Xenokrates im Kontext der Alten Akademie*. Munich: K. G. Saur, 2006.

Thomas, C. J. "Plato's Prometheanism." *Oxford Studies in Ancient Philosophy* 31 (2006), 203–232.

Thomas, R. *Herodotus in Context: Ethnography, Science, and the Art of Persuasion*. Cambridge: Cambridge University Press, 2000.

Timpanaro Cardini, M. *Pitagorici: Testimonianze e Frammenti*. Vol. 3. Florence: La Nuova Italia Editrice, 1964.

Trédé, M., and P. Hoffmann (eds.). *Le rire des anciens*. Paris: Presses de l'École normale supérieure, 1995.

Trépanier, S. *Empedocles: An Interpretation*. London: Routledge, 2004.

Tsantsanoglou, K. "The First Columns of the Derveni Papyrus and Their Religious Significance." In A. Laks and G. Most (eds.), *Studies in the Derveni Papyrus* (Oxford: Oxford University Press 1997), 93–128.

van der Waerden, B. L. *Die Pythagoreer: Religiöse Bruderschaft und Schule der Wissenschaft*. Zurich: Artemis, 1979.

van Raalte, M. *Theophrastus: Metaphysics*. Leiden: Brill, 1993.

van Riel, G. "Religion and Morality: Elements of Plato's Anthropology in the Myth of Prometheus (*PROTAGORAS*, 320d–322d)." In C. Collobert, P. Destrée, and F. J. Gonzalez (eds.), *Plato and Myth: Studies on the Use and Status of Platonic Myths* (Leiden: Brill, 2012), 145–164.

van Riel, G. *Damascius: Commentaire sur le Philèbe de Platon*. Paris: Les Belles Lettres, 2008.

Vasunia, P. *The Gift of the Nile: Hellenizing Egypt from Aeschylus to Alexander*. Berkeley: University of California Press, 2001.

Vattuone, R. "Western Greek Historiography." In J. Marincola (ed.), *A Companion to Greek and Roman Historiography*, vol. 1 (Oxford: Blackwell, 2007), 189–199.

Vattuone, R. *Sapienza d'Occidente: Il pensiero storico di Timeo di Tauromenio.* Bologna: Patron Editore, 1991.
Visconti, A. *Aristosseno di Taranto: Biografia e formazione spirituale.* Naples: Centre Jean Bérard, 1999.
Vlastos, G. *Platonic Studies.* 2nd ed. Princeton: Princeton University Press, 1981.
von Fritz, K. "Mathematiker und Akusmatiker bei den alten Pythagoreern." In *Sitzungsberichte der bayerischen Akademie der Wissenschaften.* Munich, 1960.
von Fritz, K. "The Discovery of Incommensurability by Hippasus of Metapontum." *Annals of Mathematics*, 2nd ser., 46 (1945), 242–264.
von Fritz, K. *Pythagorean Politics in Southern Italy.* New York: Columbia University Press, 1940.
Walbank, F. *An Historical Commentary on Polybius.* Vols. 1–3. Oxford: Oxford University Press, 1957–59.
White, D. A. *Myth and Metaphysics in Plato's* Phaedo. London: Associated University Presses, 1989.
White, S. "Milesian Measures." In P. Curd and D. W. Graham (eds.), *The Oxford Handbook of Presocratic Philosophy* (Oxford: Oxford University Press, 2008), 89–133.
White, S. "*Principes Sapientiae*: Dicaearchus' Biography of Philosophy." In W. W. Fortenbaugh and E. Schütrumpf (eds.), *Dicaearchus of Messana* (New Brunswick, N.J.: Transaction, 2001), 195–236.
Willi, A. *Sikelismos: Sprache, Literatur und Gesellschaft in Griechischen Sizilien (8.–5. Jh. v. Chr.).* Basel: Schwabe Verlag, 2008.
Willi, A. (ed.). *The Language of Greek Comedy.* Oxford: Oxford University Press, 2002.
Wuilleumier, P. *Taranto dalle origini alla conquista romana.* Trans. G. Ettorre. Taranto, Italy: Mandese Editore, 1987.
Zanatta, M. *Aristotele: I Dialoghi.* Milan: BUR, 2008.
Zhmud, L. *Pythagoras and the Early Pythagoreans.* Trans. K. Windle and R. Ireland. Oxford: Oxford University Press, 2012.
Zhmud, L. "Mathematics vs. Philosophy: An Alleged Fragment of Aristotle in Iamblichus." *Hyperboreus* 13 (2007), 77–88.
Zhmud, L. *The Origin of the History of Science in Classical Antiquity.* Trans. A. Chernoglazov. Berlin: de Gruyter, 2006.
Zhmud, L. "Eudemus' History of Mathematics." In I. Bodnár and W. W. Fortenbaugh (eds.), *Eudemus of Rhodes* (New Brunswick, N.J.: Transaction, 2002), 263–306.
Zhmud, L. "Some Notes on Philolaus and the Pythagoreans." *Hyperboreus* 4 (1998), 243–270.
Zhmud, L. "All Is Number?" *Phronesis* 39 (1989), 270–292.

INDEX LOCORUM

GENERAL INDEX

Printed in the USA/Agawam, MA
September 30, 2014

598164.027